Growing explanations

Science and Cultural Theory

A series edited by Barbara Herrnstein Smith

and E. Roy Weintraub

Growing explanations

Historical perspectives on recent science

M. Norton Wise, Editor

Duke University Press   Durham and London 2004

© 2004 Duke University Press   All rights reserved
Printed in the United States of America on acid-free paper ∞
Typeset in Quadraat by Keystone Typesetting, Inc.
Library of Congress Cataloging-in-Publication Data
appear on the last printed page of this book.

Contents

Growing explanations

Introduction: dynamics all the way up

M. Norton Wise

Put in the broadest terms, "growing explanations" refers to what may be a sea change in the character of much scientific explanation. Over the past forty years, the hierarchy of the natural sciences has been inverted, putting biology rather than physics at the top, and with this inversion emphasis has shifted from analysis to synthesis. In place of the drive to reduce phenomena from higher-order organization to lower-lying elements as the highest goal of explanation, we see a new focus on understanding how elementary objects get built up—or better, are "grown up"—into complex ones and even a reconsideration of the nature of things regarded as elementary, like particles and genes. The essays in this volume reflect on the history of this development.

Because the essays deal with quite diverse subjects, readers looking for a point of entry suitable to their own knowledge and taste may do well to begin with the following survey of all of the essays and their relation to the theme of growing explanations. Its further exploration can then proceed in whatever order best satisfies individual interests. A brief afterword will collect several common threads.

Some markers of the change we are pursuing are the newfound prestige of biology; new institutes being established at major research universities to study building-up problems in both biological and physical sciences; and claims in fields from condensed matter physics to ecology that the problems of synthesis are every bit as fundamental as those of elementary particle physics, which has for some time been able to reserve the term *fundamental* for itself. In this scenario, the project to map the human genome, once touted as the answer to understanding human biology, has become a mere starting point, a catalogue compiled by mechanical means from which the real problems and answers will arise. Perhaps even more telling of how prominent this shift in perspective has become is the recent call from educators and scientists, reported on the front page of the *New*

York Times and widely elsewhere, for a reordering of the high school science curriculum. Students have long proceeded from biology to chemistry and then to physics as the culmination of their science education, reflecting the view that explanation ought to proceed from description and classification at the macro level to fundamental law and causation at the micro level. Inversion of this standard order would instead arrange the high school science curriculum according to the building-up principle—physics, then chemistry, then biology—beginning with unorganized matter and universal principles and culminating with higher states of organization and differentiation. At first sight, therefore, it seems surprising that the leading advocate of "physics first" has been Leon Lederman, Nobel laureate and prime organizer of that ill-fated emblem of elementary particle physics, the Super-Conducting Supercollider, and the author of a book promoting its virtues, appropriately titled The God Particle.[1]

But inverting the teaching order says nothing in itself about what is most fundamental. Lederman, for example, continues to hold reduction as the goal of science. Nevertheless, other advocates of "physics first" believe it reflects the importance of building up to complexity. From both perspectives, elementary units remain the building blocks of higher order structures: protons and neutrons for nuclei, atoms for molecules, amino acids for proteins, and so on. In this sense elementary things and the rules that govern them remain fundamental. But for those concerned with complexity, the elementary units are not the gods that will explain more highly organized structures. Instead, they put strong constraints on what structures are possible at the higher level. They establish necessary conditions but typically not sufficient conditions for determining what the higher structures will be or how they will behave.[2] "Complexity" names this problem. It is usually taken to go beyond the merely complicated to the irreducibly complex. That is the sense in which complexity is used throughout this volume.

It has often been said that the only way to understand an organism is to grow it. In an expanded sense of "growing," the same can be said of complex systems generally. Since the behavior of a complex system cannot adequately be explained in terms of its constituents, little option remains but to put it in action and observe the results. A full spectrum of results obtained from often-repeated trials may then provide considerable understanding of how the system works, in terms of shapes and periodicities, critical points for phase transitions or fractures, the function of particular parts in their interactions with other parts, and the form of those interactions, especially their nonlinearity.

Such features have always been difficult to observe, analyze, or model in a controlled way. So it is not surprising that the fascination with complexity in the post–World War II era has been closely linked with new technologies: microwaves, X rays, and electron beams to produce visual images of physical and biological structures; powerful means for sequencing and amplifying DNA; and a variety of mathematical techniques. Most important of all, of course, has been the computer, without which it would be impractical either to handle the large amounts of information required to map most complex behavior or to circumvent physical and mathematical complexity with computational schemes.

This power of the computer gives it a special place for the theme of growing explanations. When it is too difficult or expensive to grow up real systems, it is often possible to simulate their growth. Many of the sciences of complexity depend essentially on high-speed computers to iterate algorithms, thereby to produce finely controlled simulations of processes developing in time. Such virtual systems have the great advantage of transparency, at least in principle. Knowing the model on which a program has been built, one can hope to understand how it works, in the sense of seeing how it produces the range of particular outcomes that it does, even when predicting them in detail is not possible. That is, one grows their explanation. In a variation on the age-old theme that we know what we can make, contemporary investigators sometimes prefer the simulation to the real thing. A striking example reached the general public recently from Clyde A. Hutchison, a microbiologist involved, along with a number of others including J. Craig Venter, the controversial motor of the Celera Corporation, in decoding the genomes of very small bacteria. Hutchison offered his views on understanding life: "The ultimate test of understanding a simple cell, more than being able to build one, would be to build a computer model of the cell, because that really requires understanding at a deeper level."

Interestingly, although Hutchison emphasizes that the relation of genotype to phenotype is complicated, he himself believes that "life can be explained from a reductionist point of view" and he and his coworkers suggest that the question "What is Life?" should be rephrased in genomic terms: "What is the minimal set of essential cellular genes?"[3] Others disagree, believing that the nature of life lies in the higher orders of organization that would emerge in building up any organism, whether real or virtual. This belief in the fundamental nature of complexity for living systems often appears in public forums (as it did in this case)[4] as though it were an inherently religious matter. Most scientists, however, take a thoroughly naturalistic view, even when drawing on religious metaphors. A rich perspective on formulations of this kind can be obtained by considering the nature of life as represented in the purely computational form of "artificial life," or Alife. This volume will culminate in three such reflections.

Part I. Mathematics, physics, and engineering

Before considering the biological sciences, where the theme of growing seems most obvious but where issues of complexity involve the imposing subject of life itself, we begin with a series of essays on the recent history of the mathematical sciences. Although written from diverse perspectives about diverse subject matter, they will draw out several recurrent themes. One of these is the quite prominent role of topology as the mathematical route to understanding morphology, or how shapes develop in space and time. Closely related is a focus on the dynamic nature of the things we observe as more or less stable. And in this regard, phenomena and materials of the everyday world, in all of their diversity and contingency, have gained newfound respect and attention. Finally, related in different ways with each of these themes is the role of hybrids of all kinds: hybrid objects, hybrid research, and hybrid disciplines. Since morphology, dynamical objects, mundane phenomena, and hybridity are often linked almost exclusively with the sciences of complexity, it will be instructive to start out by looking at how some of those topics have appeared in the field that most often stands as the antithesis of complexity: elementary particle physics.

Elementary particles?

If the traditional ideal of explanation in the sciences of the twentieth century has been reduction and unification—reduction to elementary entities and unification under general laws governing those entities—the paradigmatic reference for that ideal has been elementary particle physics. And yet, recent developments suggest that even this field is undergoing a rather dramatic transformation. This is particularly true with respect to the nature of the objects regarded as "elementary." Only a decade ago, the elementary units were point particles without extension or shape. Now, if string theory may be taken as the reigning ideal for the future, they are entirely different sorts of things. As Peter Galison describes the recent history of string theory in "Mirror Symmetry: Persons, Values, Objects," the most elementary objects have become strings in ten dimensions whose "compactified" shape and size are critical parameters for the kinds of particles that exist in the familiar world of three space dimensions and one of time.

Strings, if they exist at all, are topological objects, or better, adopting the language of biology, morphological, characterized by their form and its development. A good historical analogue for the conceptual shift from point particles to strings is the attempt in the second half of the nineteenth century, especially by British theorists of electromagnetic fields (William Thomson [Lord Kelvin], J. C. Maxwell, J. J. Thomson, and Joseph Larmor), to represent the point masses and electric charges of action-at-a-distance theories as manifestations of vortices in an underlying, continuous substratum. Arrangements of these vortices would constitute not only the luminiferous ether but also chemical atoms, with the table of the elements being built up from increasingly complicated structures of interlinked and knotted vortices. And although the vortices themselves would not be accessible to empirical observation, their vibrations and oscillations would be detectable in the characteristic spectra of electromagnetic waves associated with the different elements. In other words, the vortices were characterized by their morphological properties in space and time: their shapes and periodicities. Like strings, but easier to visualize, they were essentially dynamical processes, in sharp contrast to the static point particles of previous theory.

In addition to this dynamical character, and again like strings, if the full mathematical articulation of vortices had met the requirements of established physical theory, they would have unified physical nature, reducing all three kinds of matter then thought to exist—ether, electricity, and ponderable matter—to a common foundation, including the forces between them, perhaps extending even to gravity. But there the problem stuck. On the one hand, no adequate mathematics existed for the full articulation of vortex motion, and on the other, existing physical theory and experimentation were not adequate to specify its properties.

This is the sort of situation that Galison finds particularly intriguing for strings, although at a much more esoteric level of both mathematics and physics. At the moment, the status of strings is uncertain, though they are full of promise. Neither fully mathematical objects nor fully physical ones, they live in a state of possibility. Most interesting of all, the people who investigate them find themselves in a similarly indefinite state. Having approached string theory originally from either the mathematical or physical side, they are uncertain of whether they are doing mathematics or physics and they find it difficult to communicate with their opposite members. Thus they have set out quite consciously to develop modes of communication that will make it possible to exploit the capacities of both mathematics and physics. Galison places their activities in the "trading zone" that he has analyzed in depth elsewhere. Meanwhile, their peers in both disciplines are sometimes highly critical of their enterprise, whether because it fails to meet the standards of mathematical rigor or because it is not sufficiently constrained by empirical evidence. Thus, as Galison interprets it, the status of the objects, the identity of the investigators, and the values espoused by their respective disciplines are all in play at once and in intimately connected ways.

This shifting complex of objects, identities, and values appears to be characteristic as well for many other areas of contemporary science, some of which are discussed in this volume. They operate in the borderlands between one established discipline and another, using techniques that have yet to become fully naturalized. As a result, the reality of the objects they describe remains as uncertain as their own identities, whether the objects are fuzzy logic, the biological self, or artificial life.

Nonlinear dynamics and chaos

In a wide-ranging discussion of "Chaos, Disorder, and Mixing: A New Fin-de-Siècle Image of Science?" Amy Dahan captures some of the main historical developments in the mathematical and physical sciences that have upset traditional ideals and that many, both inside and outside the sciences, believe have given science a new face. Much of the action, once again, has occurred at the crossroads of the disciplines, joined there by computational exploration and the all-important computer as the engine of discovery for the modern world. Dahan is interested in how technical developments in many different areas have converged to produce the new air du temps of science. She is equally interested in how, during the course of this convergence, meanings have shifted. Chaos, for example, has for many centuries implied the antithesis of order, but through its mathematical and computational treatment in nonlinear dynamics and its experimental investigation in fields from cardiology to meteorology, order and disorder have come to be seen as intimately related. The same holds for randomness and nonrandomness, determinacy and nondeterminacy. Much current research focuses on this interdependence and on the subtle crossover between previously antithetical states, as epitomized in the now standard expression, "deterministic chaos." We may be authorized to speak once again of a dialectics of nature, but in a sense entirely different from that of the polar forces and higher syntheses of Schelling, Hegel, and other Naturphilosophen of the early nineteenth century. Not polarities but contingencies, not syntheses but hybrids populate the dialectics of complexity in the sciences of the late twentieth century.

The most obvious victim of complexity has been the reductionist ideal of explaining the properties of complex entities in terms of the properties of simple elementary entities. Not that anyone doubts that reductions can be carried out to yield successively molecules, atoms, nucleons, and quarks, but the reverse processes seem to defy any simple summation of parts. In the process of synthesis, symmetry breaking is often invoked, connecting the hierarchy of elementary particles with the phase transitions of condensed matter physics. More generally, one speaks of emergent properties, typically resulting from nonlinearity in the equations governing interactions. But most profoundly, as Dahan stresses, simplicity and complexity are no longer antithetical. Very simple systems, in mathematical terms, exhibit complexity, while complex ones exhibit simplicity.

Coupled with the surprising properties of simple systems has come a renewed interest in phenomena of everyday life. Not only clouds and ocean currents but dripping faucets and the forms of snowflakes have acquired an interest of their own, almost independent of the water molecules to which they once were supposedly reducible. It is the subtle dynamical processes responsible for these mundane phenomena in all of their irregular regularity that draws attention. And with that focus on dynamical processes comes also a renewed interest in the historical contingency of things. In slightly different circumstances they might look radically different. The task of scientists, Dahan observes, then becomes similar to that of every historian, to understand the processes of temporal development as they occur in the real world of variation and perturbation, where remarkable stability and order coexist with instability and disorder.

The roots of this shift in meanings and goals have been many and diverse. Nevertheless, to get at some of its characteristic mix of new mathematical practices with changing views of how the world is, David Aubin takes us back to "Forms of Explanation in the Catastrophe Theory of René Thom: Topology, Morphogenesis, and Structuralism." Catastrophe theory generated widespread enthusiasm in the late 1960s and early 1970s and has left an indelible mark on the vocabulary of complexity and nonlinear dynamics, despite its collapse in the late 1970s as a program for research.

As for so many later mathematical investigators (including string theorists) topology provided Thom's technical resource base. Although thoroughly trained in the formalist mathematical structuralism of the French group that named itself Bourbaki, Thom favored intuitive geometrical conceptions over rigorous proofs. Even as one of its most successful practitioners, he subverted the ethos of Bourbaki with his emphasis on shapes and their transformations. Morphological development of dynamical systems constituted the subject of catastrophe theory. The "catastrophes"—fold, cusp, swallowtail, and so forth—named the singularities that could occur in the transformations of a particular kind of topological objects, which Thom thought characteristic of natural systems. He sought the mathematical correlates, perhaps the mathematical reality, of the morphologies of the everyday world, like "the cracks in an old wall," arguing that naturally occurring shapes had primacy in themselves and should not be reduced to any more primitive elements. These shapes, he supposed, exhibited both a dynamical stability of structure and a set of possibilities for development that were governed by the topology they instantiated.

One of the most important sources for Thom's morphogenetic interpretation of nature, Aubin argues, derived from biology, particularly from embryological processes like cell division. A promising project was to interpret, in the intuitive topological terms of catastrophe theory, C. H. Waddington's earlier notion of an "epigenetic landscape" describing the relation between the genotype of an organism and the possible developmental pathways (valleys in the landscape) of its phenotype. Rejecting any reduction to molecular biology and focusing on the stability of the organism's final forms, Thom interpreted the branching points of valleys in the landscape as elementary catastrophes and he coined the term *attractor* to describe the state of dynamical stability of an organism in a valley or "basin of attraction." The terms have become ubiquitous in chaos theory even though Thom's belief in the generality of the elementary catastrophes and of dynamical stability has not held up.

As intriguing as the embryology-mathematics relation in catastrophe theory, Aubin explains, was Thom's attempt to reformulate the structuralist linguistics and social theory of the 1960s, if only heuristically, in the topological terms that he had extracted from Bourbakist structuralism. With this foray into semiotics Thom hoped to attain a theory of meaning itself. If his efforts ultimately undermined linguistic structuralism, as they had formalist mathematical structuralism, he nevertheless joined both enterprises in articulating an antireductionist view of knowledge that continues to resonate in the view that "reality presents itself to us as phenomena and shapes." The task of science, in this view, is to understand the dynamics of shapes, their formation and destruction.

Many investigators of nonlinear dynamics and chaos continue this morphological emphasis. But free of the faith in structural stability, which turned out not to be generalizable, and lacking Thom's commitment to qualitative analysis, they pursue quantitative mathematical and experimental investigations aimed at practical understanding and control of complexity. As Dahan has emphasized, this work occurs at the boundaries of order and disorder, determinacy and indeterminacy.

Coping with complexity in technology

The attention of nonlinear dynamics to complexity in the everyday world carries with it a close interaction between the mathematical sciences and engineering. But nonlinearity is not the only locus of complexity nor the only one in which engineers have played a leading role. This volume offers examples in two other widely divergent areas: finite element analysis and fuzzy logic.

Finite element analysis (FEA) is a technique of structural analysis that many engineering students today, using prepackaged computer programs, have encountered by their second year. The programs take a proposed structure as input and produce a stress-strain analysis as output, with little independent action on the part of the user, much as calculators have replaced long division. But it was not always so. Ann Johnson investigates the circumstances of the development of FEA in "From Boeing to Berkeley: Civil Engineers, the Cold War, and the Origins of Finite Element Analysis." She draws our attention to the fact that, as in so many other areas of complexity, it has been a combination of military requirements, industrial opportunity, and academic interest that has provided the motivation and funding for this computationally sophisticated but user-friendly vehicle of analysis.

In the Cold War environment of the 1950s, FEA addressed the problems presented by developing airframes for jet aircraft. They demanded maximum strength with minimum weight. Their costly construction made destructive testing of alternate designs unattractive, however, while the perceived need for Cold War advantage made increased speed in moving from design to prototype a critical consideration. For all of these reasons, Johnson explains, companies like Boeing sought better ways to model the performance of their designs prior to any material embodiment of them and prior to experimental testing in a wind tunnel. Accurate calculations of the strains that the airframe would experience under the loading conditions of jet engines and high speed would solve the problem, but existing means of structural analysis lacked the required accuracy.

The situation presented an opportunity for academic civil engineers, a group that had largely been excluded from the rewards in funding and prestige of what Eisenhower dubbed the "military-industrial complex." Johnson narrates how Ray Clough, a young Berkeley engineer with a summer position at Boeing, took up the challenge in 1952. In cooperation with his supervisor at Boeing, Clough identified the skin of an airplane as a critical structural element and found ways to model it through what he would soon name "finite element analysis." The procedure begins by dividing up a structure, such as the skin of a wing, into small elements forming a mesh, finer in critical areas and coarser in others. Applying linear equations of elasticity to each element generates a large set of simultaneous equations. The strain in the whole wing can then be calculated using matrix methods on a computer, effectively matching elements across their boundaries to build up an accurate simulation of the wing's response.

Johnson discusses three features of this method that are of interest here. First, the complexity that engineers refer to in the finite element calculation enters in the first instance through the practical requirement of accuracy in modeling and only secondarily through the physical nonlinearity of the problem, although underlying nonlinearity strongly affects the form and density of an adequate mesh. To put it differently, the method leads to an understanding of physical complexity through the complexity of the mesh and its associated calculation. Second, the technique would have been intractable for all but very simple structures without the digital computers that IBM leased to Boeing and a small number of other aerospace companies but that were not originally available in academia. These two features involve a third: the priority placed on accuracy and computing power correlates with a notable shift in engineering values, from mathematical elegance to utility in modeling.

Dahan also points out this instrumentalist orientation, which seems to arise quite generally in the shift of values from simplicity to complexity and unity to diversity, since it involves a widespread turn to understanding and controlling the strange properties of mundane things through computational means. In common parlance, the computations proceed "from the bottom up," beginning from a specification of rather simple actions or interactions of some operative units. The units may be natural (electrons) or artificial (finite elements), with the latter often obtained by conceiving the problem as a grid, patchwork, or network. Then the computer program simulates the behavior of the entire system by performing the actions and interactions of the units, typically in a stepwise fashion through the continued repetition of a set of algorithms. The procedure thus grows the behavior of the system as a whole, often revealing emergent properties that could not be predicted from the behavior of the parts, especially when the interactions are nonlinear. Ironically, although this emergent holism is often regarded by its critics as romantic, it is equally open to their charge of mere instrumentalism because its primary criterion of validity is accuracy of simulation. As a historical matter, we should no doubt attribute the conflict to the shifting sands of "simplicity," "elegance," and "fundamentality."

That those sands are shifting can be seen also in a means of coping with complexity very different from FEA, namely, fuzzy logic. It too mixes its validity as a description of how the world is with its utility in controlling technological devices, now made more user friendly by fuzzy logic. In "Fuzzyfying the World: Social Practices of Showing the Properties of Fuzzy Logic," Claude Rosental examines how its proponents have spread their message through various modes of "showing." These heterogeneous modes might normally be called "proofs," "histories," and "machines," but Rosental is concerned with how they are used in concert to make the virtues and properties of fuzzy logic seem manifest. He therefore describes their function as that of mediators, mediators which themselves disappear in the course of "showing," or making immediately visible, the power of fuzzy logic.

Fuzzy logic operates on the premise that statements about the world are not simply true or false but have a continuous range of truth values, so that a statement and its contradiction may both be partly true. Showing how a famous contradiction or paradox in binary logic gets resolved in a natural way by fuzzy logic therefore provides an effective strategy for proponents. Rosental observes that its effectiveness depends not only on abstract reasoning but also on its staging through a series of familiarizing manipulations and translations that are displayed materially as inscriptions on a page. The reader is supposed to experience the proof visually through its performance. Its manifest credibility can then be appropriated into other means for establishing the universal status of fuzzy logic, such as a cultural history that shows the demise of Western binary logic in self-contradictions and the rise of Eastern thinking in continuous-valued logic. Detractors, of course, mount their own performances to depict the proofs as deceptive and fuzzy logic as nothing fundamental, offering no properties that traditional logics cannot provide. The stakes in the contest are high; besides intellectual rewards, they include funding from industrial and military sources for academics and sales of control mechanisms for design companies and manufacturers.

Among the most interesting mediators of showing that Rosental discusses in this context are "demos" of consumer products like cameras and vacuum cleaners. Such demos have been the most important means of promoting fuzzy logic in both academic and business settings. For that purpose, a demo depends on a skillful demonstrator who acts as the representative of a device in operation and exhibits its properties as embodied properties of fuzzy logic. Success, Rosental argues, depends first on making the properties of fuzzy logic emanate from the device as its epitome (while effacing many other components and properties) and then on detaching them from the device as transcendent properties available for reasoning in other contexts, as well as for other control mechanisms of many kinds. Not only must the particular device disappear; the demonstrator too must become immaterial to the demonstration.

Rosental's analysis thus delves into the mechanisms through which unfamiliar aspects of complexity may become naturalized in the world by interrelating truth and utility, or specifically here, truth value and use value. More globally, it makes explicit how such a project of naturalization may incorporate a thorough rewriting of intellectual and social history.

Part II. The organism, the self, and (artificial) life

Just as the physical sciences have been reevaluating in the last several decades the ways in which they formerly ascribed priority to particle physics as "fundamental physics," and even within that domain the new units called strings are neither so intuitively elementary nor particle-like, so in the biological sciences, the former citadel of reduction to elementary genes, molecular biology, has been finding that "the gene" is not so easily localizable and that its relation to phenotypic traits is complex indeed. In the broad fields of biology and medicine beyond the molecular domain, complexity is even more apparent. It is with those broader fields in view that the essays in Part II reflect on questions so basic to biology and medicine: What is an organism? What, speaking immunologically, is the self? And what is life? These are all questions with long histories that have found new formulations in contemporary research, particularly with re-spect to both actual and metaphorical uses of the computer and with respect to the problem of self-organization.

Self-organization

To introduce the contemporary scene in historical perspective, Evelyn Fox Keller discusses "Marrying the Premodern to the Postmodern: Computers and Organisms after World War II." She succinctly recovers some of the main episodes in the rising and falling fortunes of self-organization over the last two hundred years. As in much else, Immanuel Kant set out with particular clarity the concept of an autonomous entity in which every part is both end and means in relation to the other parts. Highly developed as a nonvitalist research tradition in the first half of the nineteenth century (dubbed "teleomechanism" by Timothy Lenoir),[5] the Kantian agenda de-clined in the second half of the century with the rise of more strictly mechanist physiology and then with genetics, although variations on the theme survived in areas like embryology, where questions of the internal dynamics of development remained crucial.

In the United States especially, the Morgan school of genetics reigned su-preme by the 1930s. They and their successor molecular geneticists even claimed to rule embryology itself. The discovery of the double-helical struc-ture of DNA in the early 1950s seemed to confirm that self-replication and development would soon be reduced to the mechanics of particle-like genes. Meanwhile, something quite different was going on. The problem of aiming the antiaircraft guns of World War II gave rise to "cybernetics." Innocent of molecular biology but schooled in the mathematics and engi-neering of feedback mechanisms, the promoters of this new science en-visaged self-controlled, goal-seeking machines with digital computers at their heart. To articulate this vision, Keller has shown, they adopted the older biological language of self-organization and organicism that the molecular biologists were expunging from biology itself.

It remains one of those fascinating "accidents" of history that cybernetics, though it prospered for twenty years after the war, enrolling biologists with a bent for embryogenesis and pioneering connection machines and neural nets, had virtually died out by 1970, the apparent victim, on the one hand, of its own lack of tangible success, and on the other, of the remarkable successes of sometimes stridently antiorganicist research strategies in both molecular biology and computer science. Equally fascinating is its rebirth in new forms in the last ten to fifteen years. The history of this period is only now being written—indeed, being made—but two features seem apparent. First, molecular biology itself has become much more complex. Despite the continued promotion, especially in popular writing, of the myth of a simple one-to-one relation between genes and phenes—one gene, one protein—detailed research has made it increasingly difficult to maintain this central tenet of the older molecular faith. At every level, from DNA sequences to transcription into RNAs to translation into proteins, the notion of the singular gene with its unique product has become untenable. Even which sections of DNA are to count as genes is made problematic by the fact that the vast majority of sections do not code for proteins but serve a myriad of regulatory and other functions, mostly unknown, so that any given product results from a complex web of interactions at various stages involving many different sections of DNA. Similarly, a great deal of cutting and pasting of pieces has been shown to occur at the level of transcription, such that one sequence can ultimately give rise to many different products. And translation may be no simpler, with a variety of modifications being required to convert an original translation product into its functional counterpart. Summarizing this situation, the historian Hans-Jörg Rheinberger has remarked: "It has become evident that the genome is a dynamic body of ancestrally tinkered pieces and forms of iteration."[6] As he interprets it, this new organicism results quite naturally from the increasingly complex experimental systems that are producing the future of molecular biology. Nevertheless, it has brought with it attempts to reincorporate the older cybernetic vision.

That attempt, as Keller observes, involves a second and most obvious feature of the changed conceptual landscape, the computer. The personal computer has brought complexity down to earth, within the purview of all those whose work it performs and whose lives it interconnects with such immediacy. And atop this everyday functionality is built the broad spectrum of computer-based fields encompassed under names like nonlinear dynamics, computational biology, and cognitive science. Everyone now wants to know how to think about complex systems and how to work with them. Biological systems are the Holy Grail.

Among various proposals for reconceptualizing biological complexity, or the organism, one of very general scope has been articulated by Humberto Maturana and Francisco Varela under the name of *autopoiesis*. Keller observes, in fact, that they drew inspiration for their theory directly from participants in the old cybernetics. The rebirth, however, has a much expanded domain for action. The other authors in part II have taken up two such areas, which are exemplary for both the promise and the problems of making sense of the biological "self" and of "life." They are immunology and artificial life.

Immunology

The problem of the self, in philosophical terms, has been the problem of identity: In what sense, and how, does an individual remain the same individual over a lifetime of constantly changing material and mental existence? In the biological domain, the self lies in the territory of immunologists, but in a very special sense. Throughout the twentieth century they have been attempting to understand how the immune system acquires the capacity to distinguish the agents that perform the normal repertoire of bodily functions from the agents that disrupt those functions, or self from nonself. The dominant metaphors for describing this relation between self and nonself have been invasion and war, invasion of antigenic foreign agents and the war against them carried on by a domestic defense system that somehow recognizes otherness and deploys such weapons as a specialized army of white blood cells, the T4 lymphocytes. But the immune system is notoriously complicated and alongside the images of war have always stood alternative pictures in which the self represents the relatively constant properties of a shifting network of interactions. Self and nonself, in such views, are part of the same process of recognition, so that what counts as the one or the other depends on context. The proving ground for these alternatives has been autoimmunity, in which the system seems to turn on itself and attack its own agents. Is autoimmunity simply an extreme of the normal process of recognition, a symptom of the dynamic system having lost its self-generating stability? Or is it the pathological behavior of a system gone completely awry?

In "Immunology and the Enigma of Selfhood," Alfred I. Tauber advocates a complex-systems view. He believes immunology has outlived the reified self, with its fixed boundary between self and nonself. In its place he would put a dynamic, process-oriented description of a system whose continuing growth, in constant interaction with both its inner and outer environment, yields the emergent properties commonly associated with the self. Tauber writes in a highly personal vein, exceeding the usual conventions of either history or science. But he believes that the reified self is a holdover from a "modernist" ideology and that his own trajectory in coming to a "postmodern" view, while surely atypical, may nevertheless represent trends that have been developing since the 1970s, when Niels Jerne presented his "idiotypic network theory" of the immune system as a contrast to the "clonal selection theory" of Macfarlane Burnet dominant in the 1950s. Other important resources for Tauber have been the perspectives of Francisco Varela (although he is critical of the autopoiesis interpretation per se), Antonio Coutinho, and others analyzing the immune system as analogous to the nervous system and to language. This work has sometimes been done at the Santa Fe Institute, where so much of complex-systems analysis has germinated. Tauber began, however, from the early history of immunology and from a historical reinterpretation of the works of Nobelist Elie Metchnikoff, whose import, he argues, has been obscured within the modernist tradition.

An empirically minded historian who doubts immunologists have much enthusiasm at present for a complex-systems approach is Ilana Löwy. In "Immunology and AIDS: Growing Explanations and Developing Instruments," she takes a practical approach to evaluating competing views of autoimmunity by looking at the most famous case of it. She shows that in the early history of the disease clinicians observing the response of patients to infection with HIV (human immunodeficiency virus) and their subsequent development of full-blown AIDS (acquired immunodeficiency syndrome) leaned toward the complex-network approach, especially because the viral load in the bloodstream of infected patients seemed to be very low, far too low to support the image of warfare between virus and T4 cells. And yet, as patients developed AIDS, their T4 cell counts plummeted, suggesting that the immune system had lost control of its stabilizing dynamics and was destroying its own defenses. The T4 cell count thus seemed to be the best marker of the disease's progression, as a disease of autoimmunity. From about 1990, however, new technologies for detecting the virus became available. They revealed not only much higher levels of HIV in the blood of AIDS patients than had previously been thought but also that during the prior period of supposed dormancy the virus was actually multiplying rapidly in the lymph glands, out of sight of the earlier tests. As the new tests gained precision and reliability, the warfare image quickly resurfaced as the dominant explanation of infection. The war going on in the lymph glands was depleting the T4 cells in the blood. With that, the disease reentered more conventional virology as an infectious disease, with viral load as its measure, and the previous emphasis on the mechanisms of disruption of the immune system disappeared.

This is a cautionary tale. It may very well be, Löwy agrees, that the immune system—and with it the biological self—will ultimately require for their adequate understanding the concepts and tools of complexity theory, but in the particular case of AIDS, the disease can best be diagnosed and represented as an all-out viral attack that ultimately overwhelms the immune system's capacity for defense. This conception, moreover, has major implications for treatment. Drugs that kill the virus directly are, at present, the only effective agents. This practical lesson carried by the instruments of diagnosis and treatment stands in uneasy relation with the complexities of discriminating between the self and nonself and the ephemeral status of the postmodern self.

Artificial life

At the very edge of the real in the age of complex systems stands artificial life. How can this life be narrated? What are the rhetorical resources that artificial life researchers join with the digital computer in projecting life into the little "animals" that populate the screen? And how might life be defined? These are the questions taken up by literary critic Richard Doyle, anthropologist Stefan Helmreich, and biologist Claus Emmeche.

Alife makes Richard Doyle nervous, as he says in "Artificial Life Support: Some Nodes in the Alife Ribotype." He would like to understand how the little creatures of Alife come to seem lively—at least to some observers— and narratable as organisms moving through space and time. Why would anyone imagine that a pattern of flashing lights on a screen, representing visually the states of a Boolean network as widely distributed as one cares to imagine, have anything to do with life? Adopting an analogy from theoretical biology, Doyle argues that just as the translation of genotypes into phenotypes has been said to depend on an intermediary set of "ribotypes" that translate the information encoded in DNA (genotype) into the functional proteins and higher-order structures of a living organism (phenotype), so the translation of digital states into lifelike virtual creatures in Alife depends on a complex of translational mechanisms or rhetorical ribotypes. The rhetoric of localization, for example, produces the intuition that something inside the computer is alive, while that of ubiquity makes it seem that almost anything could be alive.

To explore these modes of actualization more deeply, Doyle employs C. S. Peirce's notion of "abduction." Abduction is the method of projecting into the future the realization of a present expectation or hypothesis. It allows a novel concept or subject, a new unity, to stand in the place of an otherwise disconnected set of predicates. Doyle shows how the rhetoric of Alife uses a form of abduction to project "life-as-it-could-be" into the open future of life as simulacrum, not simply a simulation of biological life but a creation by simulation of life in silicon. But producing this "life effect" depends also on the sheer contingency and unexpectedness of what befalls the creatures of Alife because their generation from nonlinear and evolving algorithms precludes the possibility of knowing their fate in advance. They must be run or grown. Furthermore, their survival depends on their interaction with the humans who help to narrate their life and thereby continue the process of its abduction.

These issues of narrative effects and their problems are familiar territory for literary critics. Doyle finds fruitful analogies in the writings of Philip K. Dick. He also finds useful, as do so many others attempting to come to grips with complexity, the concept of autopoiesis as developed by Maturana and Varela. Autopoiesis might provide the autonomy and interiority of the living organism that initially seems so lacking in the distributed networks of Alife.

Stefan Helmreich is also concerned with how Alife converts the virtual into the actual, but in "The Word for World Is Computer: Simulating Second Natures in Artificial Life" he investigates the transformation through an ethnography of Alife researchers, basing his analysis on fieldwork at the Santa Fe Institute in New Mexico. He has explored the cultural resources they invoke to render computer simulations as "worlds" or "second natures," maintaining that the transubstantiation depends on conjoining familiar mythologies of European and American culture with computer science and complex-systems theory. Such a world, however, is not merely a dream world. It depends on the reasoned view that the sets of machine instructions making up the self-replicating creatures in a simulation are the same sort of thing as the amino acids that make up a biological genome. More generally, it depends on the controversial view that our familiar world is just a mammoth computer that continuously generates and carries out its own complex instructions. This view is grounded formally in the physical Church-Turing thesis, claiming that any physical process can be thought of as a computation. So for those who accept the strong claim, the artificial world inside the computer is indeed a real world, however different from the usual one.

Still, the artificial world gains much of its intuitive reality from its assimilation to everyday cultural myths and metaphors by Alife researchers. The "green" and natural organic world of ordered processes developing in evolutionary progression is one such metaphorical resource. Another calls on the creation stories of Judeo-Christian theology to put the Alife programmer in the role of a deistic god who creates the world and its laws, which then run on their own until the god destroys them. Similarly prominent in informal discussions among Alifers, Helmreich shows, are science fiction stories of travels in cyberspace, frontier imagery, and colonization. Many of these stories involve a thoroughly masculine discourse of creation, exploration, and conquest. The effectiveness of the imagery, however, depends always on the very lifelike simulations made visually present on the screen, as though seen through the glass of an aquarium.

If Doyle and Helmreich make us aware of the narrative and cultural conventions on which Alife relies and of how important those resources are to its liveliness, Claus Emmeche attempts to unpack more formally the implicit conceptions of life that may lend theoretical coherence to its claims. In "Constructing and Explaining Emergence in Artificial Life: On Paradigms, Ontodefinitions, and General Knowledge in Biology," he draws from biology two possible definitions of life, which he calls "onto-definitions" to emphasize their status as root metaphors for the sorts of things that can count as living. These he calls "life as the natural selection of replicators" and "life as an autopoietic system." They both treat life as an emergent phenomenon but in different senses.

The first view—that life can be regarded as a property of entities that self-replicate by a process of information transfer and that evolve by random variation and environmental selection—seems to be widespread among biologists, although largely as an implicit assumption rather than a definition. Emmeche makes it explicit, emphasizing an important distinction between the mere self-replication of information by a computer program and the full self-reproduction going on in cells, which is carried out by self-maintaining material and causal processes in the cell. Whether the silicon-based replication in computer simulations can be regarded as "real" self-reproduction may be the central debating point for Alife.

Focusing on this autonomy of the cell, the same interiority to which Doyle refers, leads Emmeche also to the theory of autopoiesis as presented by Maturana and Varela and to his second ontodefinition: life as an auto-poietic system. In this scheme, which is as yet entertained by only a small minority of biologists, the metabolic processes in a cell constitute an organizationally closed network of self-reference which produces itself and its own components, as well as its boundary, while drawing energy and materials from its surroundings. In this self-referencing character of the biological network lies the emergence of biological life from physical mechanism. The theory, however, does not provide a physicochemical analysis of the structure of the network. Rather, it offers an account of the characteristics of a network that could be called autopoietic and therefore autonomous. Likewise, questions of reproduction and evolution are secondary. But the theory is general enough to include autopoiesis in the virtual space of a computer as well as in physical space and therefore may be of increasing interest as the simulations of Alife become ever more realistic.

First, however, Emmeche believes we need a better understanding of key notions, beginning with emergence. To this end he presents an analysis by Nils Baas. Baas's view is of interest because it incorporates explicitly the role of an observer as a necessary ingredient for a particularly strong form of emergence, that in which the properties of a complex system are not simply computable from known interactions. The "observer" need not be a living agent but may be simply the environment with which the system interacts (reminiscent of the role of observation in quantum mechanics in producing distinct states of momentum or position). When humans are involved, however, readers will want to recall Doyle's concern with the role of observers in the continuing narration of Alife creatures, their survival and evolution.

Conclusion

It would be impossible to capture in a few sentences the changing character of scientific explanation in the last several decades. The essays in this volume seek rather to explore some of the diverse strands of that change and to present them in a readable form. Although they cross the broad spectrum from strings to Alife, however, they do exhibit some representative features. Repeatedly we find objects understood as dynamical processes rather than static units, objects defined by their topological or morphological properties. They seem to be best understood in terms of self-stabilizing systems whose properties can sometimes be captured by simulating them on high-speed computers. In this generative computational effort, the bottom-up approach to explanation has come into its own. It gives concrete substance to the notion of growing explanations, perhaps heralding a sweeping transformation of the reductionist program that has dominated scientific explanation in the modern period. Instead of particles all the way down, it would be dynamics all the way up. That scenario, however, remains very much in flux. It continues to arouse the uncertainties, and often the passions, of people both inside and outside the specialized sciences. And it involves a great deal of boundary crossing between those specialities. These features make the history of contemporary science an ideal location for viewing the intimate relationship between the content and goals of the sciences as well as the identities and values of their practitioners.

Notes

1 Details available through the websites of "Physics First" and "ARISE" (American Renaissance in Science Education); Tamar Lewin, "A Push to Reorder Sciences Puts Physics First," *New York Times*, January 24, 1999, 1; representative response letters, January 27, A24; Marjorie G. Bardeen and Leon M. Lederman, "Coherence in Science Education," *Science*, 281 (1998), 178–79; Leon Lederman with Dick Teresi, *The God Particle: If the Universe Is the Answer, What Is the Question?* (Boston: Houghton Mifflin, 1993).

2   A more radical philosophical suggestion, called "metaphysical pluralism," would be that the higher levels constrain the lower ones as much as the reverse and should be regarded as independent; to make the building-up problem fundamental is simply to maintain the old metaphysics of reduction in a new form, called "supervenience." See Nancy Cartwright, *The Dappled World: A Study of the Boundaries of Science* (Cambridge: Cambridge University Press, 1999), especially "Fundamentalism versus the Patchwork of Laws," 23–34.

3   Nicholas Wade, "Life is Pared to Basics; Complex Issues Arise," *New York Times*, 14 December 1999, F3; Clyde A. Hutchison III et al., "Global Transposon Mutagenesis and a Minimal Mycoplasma Genome," *Science* 286 (1999), 2165.

4   Mildred K. Cho, David Magnus, Arthur L. Caplan, Daniel McGee, and Ethics of Genomics Group, "Ethical Considerations in Synthesizing a Minimal Genome," *Science* 286 (1999), 2087–2090.

5   Timothy Lenoir, *The Strategy of Life: Teleology and Mechanics in Nineteenth Century German Biology* (Dordrecht: Reidel, 1982).

6   Hans-Jörg Rheinberger, "Gene Concepts: Fragments from the Perspective of Molecular Biology," in P. J. Buerton, R. Falk, and H.-J. Rheinberger, *The Concept of the Gene in Development and Evolution: Historical and Epistemological Perspectives* (Cambridge: Cambridge University Press, 2000), 219–239.

PART I   Mathematics, physics, and engineering

Elementary particles?

1 Mirror symmetry: persons, values, and objects

Peter Galison

At the millennium, there was much talk about "the end of physics." Many physicists believed that their enterprise was coming to a final phase of its history, but they interpreted "the end" in numerous ways. At the end of the cold war some made dire predictions about the discipline because military research and development were downsized, the $15 billion superconducting supercollider was canceled in 1993, the National Science Foundation offered less to physics, and the Department of Energy budget was significantly reallocated. These same words—"the end of physics"—took on a second meaning in the 1980s and 1990s, when some high-energy particle physicists turned against a small but growing minority of theorists who embraced string theory. For these critics string theory appeared a false salvation, a mathematical chimera that abandoned experiment, tempted the young, distorted pedagogy, and ultimately threatened the existence of physics as science. A third meaning of the "end of physics" emerged within string theorists' own ambitions: many argued that a remarkable series of discoveries within the mathematical physics of strings provided grounds, the best ever, for an account of all the known forces including gravity, completing the historical mission of fundamental physics. It would be, its most enthusiastic backers argued, a "theory of everything."

Physics at the millennium

Hopes and fears for finality in physics are not new. Distinctive is the string-theoretical vision in which mathematics itself came to stand where experiment once was: the view that the powerful constraints of mathematical self-consistency would hem theory in so tightly that, at the end, only one theory would stand, and that an elegant, compact theory would cover the world by predicting all the basic forces and masses that constitute and bind matter. Theoretical physics would end, in this third sense, because the ancient search for physical explanation in terms of basic building blocks had reached its last station.

Rejecting the doomsayers, the new alliance between physics and mathematics saw the opportunity to restructure the bounds of both fields, opening a window on twenty-first century mathematics that might, imminently, produce a unified account of gravity and particle physics. But the optimism came with warnings from some quarters of both math and physics—what would be the proper standards of demonstration in this new interzone between disciplines? Would this new "speculative mathematics" sacrifice experiment on the physics side and intrude physical argumentation into the heartland of mathematics on the other? Would it compromise both the physicists' demand for laboratory confirmation and the mathematicians' historical insistence on rigor?

These discussions over the place of string theory are enormously instructive. They offer a glimpse into theoretical physics during a remarkable period of transition, for in the realignment of theory toward mathematics the meaning of both *theory* and *theorist* were in flux: First, in the 1990s a new category of theorist was coming into being, part mathematician and part physicist. Second, theorists ushered a new set of conceptual objects onto the stage, not exactly physical entities and yet not quite (or not yet) fully mathematical objects, either. Finally, alongside the shift in *theorist* and *theory*, there arose, in the trading zone between physics and mathematics, a style of demonstration that did not conform either to older forms of physical argumentation familiar to particle physicists or to canonical proofs recognizable to "pure" mathematicians.[1]

By introducing new categories of persons, objects, and demonstrations, string theory made manifest conflicts over the values that propel and restrict the conduct of research. These debates, joined explicitly by both mathematicians and physicists, focused on the *values* that ought to guide research, and the disagreements were consequential. At stake were the principles according to which students should be trained, how credit for demonstrations should be partitioned, what research programs ought be funded, and what would count as a demonstration. For all these reasons, the status of claims and counterclaims about string theory mattered. They were not "just rhetoric" constructed after the fact; they were, in part, struggles over the present and future of physics. Values, in the conjoint moral and technical sense I have in mind, were not, as one historian derisively called them, mere "graffiti." Nor is it the case that these values respected a distinction between "inside" and "outside" science. These were not values in the sense of "propriety," such as whether or not honorary authorship would be countenanced on a group paper. Rather, I will argue that the "values" in debate crosscut through the central, guiding commitments of physics and mathematics into the wider, everyday sense of the term. For this reason, dismissing the role of values in shaping theoretical research makes it impossible to understand both the moral passion behind these debates over demonstrative standards, the participants' own understanding of their distinctive scientific cultures, and ultimately, the scientific persona of the physicist.

The argument will follow this order: section two tracks the reaction by particle physicists to the shift by string theorists away from the constant interplay between theory and experiment. Sections three through five home in on the creation during the late 1980s and early 1990s of a striking "trading zone" between physicists and mathematicians around a variety of developments including that of "mirror symmetry," a remarkable development at once physical and mathematical. Mirror symmetry offered insight both into the shape of string theory after the higher dimensions curled up and provided new ways of understanding not only fundamental features of algebraic geometry, but also reshaped the status of geometry itself. By following the ways mathematicians and physicists saw one another in the episode of mirror symmetry—and the ways each side came to understand aspects of the other's theoretical culture—it becomes possible to characterize the broad features of a new place for theory in the world. Finally, in light of the tremendous impact of this hybrid physics-mathematics, section six analyzes the mathematicians' sometimes conflicted reaction: a response mixing enormous enthusiasm with grave reservations about the loss of rigor that accompanied the mathematicians' collaboration with the physicists. In a sense the physicists' and mathematicians' anxieties mirrored one another: both saw danger in parting from historically established modes of demonstrations that gave identity to their fields. These discussions were hard fought and explicit: in the midst of remarkable new results, claims, and critics, string theorists never had the luxury of being unself-conscious: the purpose, standards, and foundation of their budding discipline were *always* on the table.

At the end of the millennium, string theory resided, both powerfully and precariously, in the hybrid center between fields, bounded on one side by the continent of physics under the flag of experiment and on the other side by the continent of mathematics under the flag of rigor. To understand the aspirations and anxieties of this turbulent territory is to grasp a great deal of where theory stood at the end of the twentieth century.

Theory without experiment

In 1989, David Gross, from Princeton, chose physicists' favorite site for metaphors (the romantic mountains) on which to erect a new relation between theory and experiment, one far different from the close cooperation that had marked the mid-1970s. "One of the important tasks of theorists is to accompany our experimental friends down the road of discovery; walk hand in hand with them, trying to figure out new phenomena and suggesting new things that they might explore."[2] Burnt into the collective memory both of pro- and antistring theorists was the example of the J/psi and other "charm" discoveries of November 1974.[3] During the frenetic days of that late fall, and the months that followed, experimentalists tossed new particles into the ring, so to speak, and theorists worked furiously to explain them; theorists postulated new particles, new effects, and new theories—experimentalists responded with tests that could be prosecuted almost immediately. At the time Gross was writing, in the late 1980s, that highly responsive dialogue had fallen into silence—few experimental results were coming out of the accelerators, and the discoveries that were being reported had a wickedly short life: the neutrino oscillations (indicating that the neutrino might have a mass) came and went; proton decays were reported and retracted; monojets spurted momentarily from CERN, then vanished; the fifth force grabbed attention for a while and then loosened its hold. Under these circumstances some theorists—Gross included—were less and less inclined to theorize furiously after each new sighting. These were no longer the days of new, hot experimental news and papers written on airplanes returning from the accelerator laboratories, of quick phenomenological calculations using a couple of Feynman diagrams and Lie group calculations that could be done on one's fingers. Looking back in 1989, Gross put it this way:

> It used to be that as we were climbing the mountain of nature the experimentalists would lead the way. We lazy theorists would lag behind. Every once in a while they would kick down an experimental stone which would bounce off our heads. Eventually we would get the idea and we would follow the path that was broken by the experimentalists. Once we joined our friends we would explain to them what the view was and how they got there. That was the old and easy way (at least for theorists) to climb the mountain. We all long for the return of those days. But now we theorists might have to take the lead. This is a much more lonely enterprise.[4]

Without knowing the location of the summit, or how far it was, the theorists could promise little reassurance to themselves or to the experimentalists. In the meantime, experimentalists were not only left behind; they were left out altogether.

Not surprisingly, many experimentalists were shocked by the theorists' string ambition, not so much by the idea of theorists leading the way on an uncertain trek up an uncharted mountain, but because the experimentalists did not see how they could even gain a toehold in the foothills. Carlo Rubbia, who only a few years before had taken home a Nobel Prize for his team's 1983 discovery of the intermediate vector bosons, the W and the Z, lamented the loss of contact between experiment and theory at a meeting on supersymmetry and string theory:

> I am afraid I am one of the few experimentalists here. In fact, I can see we are really getting fewer and fewer. I feel like an endangered species in the middle of this theoretical orgy. I am truly amazed. The theories are inventing particle after particle and now for every particle we have there is a particle we do not have, and of course we are supposed to find them. It is like living in a house where half the walls are missing and the floor only half finished.

After the bruising W and Z search, and a contentious struggle with the top quark, Rubbia had little appetite for a zoo of unknown particles as numerous as the known. Even one or two particles were terribly hard to find— Rubbia's UA1 collaboration had employed some 150 physicists for years at a cost of hundreds of millions of dollars to find the W and Z. Now this new breed of theorists was ordering a supersymmetric partner for every known entity: a selectron as partner to the electron, and so on all the way down the line.[5]

Not only was this half-missing world overwhelming in its mandate for experimental discovery, but the very motivations cited by the theorists had moved ever further from the accelerator floor. Gosta Ekspong, a senior European experimentalist who often worked at CERN, addressed the purported aesthetic satisfactions of the string theorists:

> I would like to address the question of truth and beauty; truth being experiment, beauty being theory. . . . The problem is that the latest [superstring] theories are so remote from experiment that they cannot even be tested. Therefore they don't play the same role as Dirac's equation. . . . I hope that this search for beauty does not drive theorists from experiments, because experiment has to be done at low energies, from one accelerator to the next and so on. Big jumps do not seem to be possible.[6]

In the 1970s and 1980s, *theory* and *experiment* were categories (better: subcultures) with an intermediate trading zone of phenomenologists straddling the fence. Ahmed Ali was one of these, having worked at both DESY (*Deutsches Elektronensynchrotron*) and CERN, and he invoked the idiom of the fund-seeking experimentalist when he declaimed, "The present superstring theories are like letters of intent written by a lobby of theoretical physicists. They are very good in intent; but often what is said in the letter of intent and what is measured in the experiment are two very different things. The figure of merit of a theory is its predictive power which could be tested in an experiment in a laboratory."[7]

In a sense, the discomfiture of experimentalists, and those working hand in glove with them, could be expected. New techniques in theory had left experimentalists ill at ease with gauge theories in their early stages, though by 1974 experimentalists had found the gauge theories suggestive of a wide range of predictions, tests, and directions for empirical work. The case of superstrings seemed much worse; there was no clear avenue for the accelerator laboratory to follow, and the theories themselves offered precious little to hold on to in the way of physically "intuitable" entities. Less expected, perhaps, was the vehement reaction against string theory from *within* the theoretical high-energy physics community.

I now turn to that part of the opposition that was certainly *not* grounded on a hostility to wide-ranging claims about unification, nor on a broad opposition to the disproportionate resources allocated to particle physics. (That is, I am not focusing here on criticisms mounted over the 1970s and 1980s by condensed-matter physicists such as Philip W. Anderson, who had in mind both a defense of emergent properties in many-body physics and an argument for a reallocation of human and material backing.)[8] Far from being outsiders to the tradition of "fundamental" physics, Howard Georgi and Sheldon Glashow were as central to the gauge revolution of the 1970s as anyone. No, the dispute hinged on something even deeper, I believe, than the relative fundamentality of physical domains. It circled around differing visions of what theoretical physics should be.

Before the 1984 *annus mirabilis* of strings, Georgi opened the 1983 Fourth Workshop on Grand Unification with a transparency of a recent advertisement he had spotted:

HELP WANTED
Young Particle theorist to work on Lattice Gauge
Theories, Supergravity, and Kaluza-Klein Theories

Here, Georgi asserted, was a telling sign of the times, a position caught between "chemistry" (particle physicist nomenclature for a calculational activity in which fundamental principles were no longer at stake) and "metaphysics and mathematics" (particle physicist jargon for work without experiment).[9] Superstrings had not yet emerged with the force it would a few years later, but the problem of choosing between truth as beauty and truth as experiment had, and Georgi sided squarely with those who demanded accessible measurements as a sine qua non of desirable theory.

The next years polarized the situation further. In 1986, Paul Ginsparg, who had himself contributed to the Alvarez-Gaumé and Witten no-anomaly demonstration in superstrings, collaborated with his Harvard colleague Sheldon Glashow to bemoan the loss in superstrings of the historically productive conflict between experiment and theory:

> In lieu of the traditional confrontation between theory and experiment, superstring theorists pursue an inner harmony where elegance, uniqueness and beauty define truth. The theory depends for its existence upon magical coincidences, miraculous cancellations and relations among seemingly unrelated (and possibly undiscovered) fields of mathematics. Are these properties reasons to accept the reality of superstrings? Do mathematics and aesthetics supplant and transcend mere experiment? Will the mundane phenomenological problems that we know as physics simply come out in the wash in some distant tomorrow? Is further experimental endeavor not only difficult and expensive but unnecessary and irrelevant?[10]

This was an altogether different view than that which Glashow had taken in the euphoric moments after he and Georgi had produced the first grand unified theories (GUTs), theories that while leaving aside gravity, aimed to bring together the strong, the weak, and the electromagnetic forces. GUTs, at least the SU(5) and SO(10) versions, did have very high energies—only a few orders of magnitude less than the Planck scale. And like the string theories, GUTs too forecast a "desert" in which no new experimental results could be found. But, Glashow and Georgi believed, several crucial items differentiated GUTs and superstrings. GUTs forecast a crucial (and measurable) parameter in the electroweak theory that measured the relative strength of the weak and the electromagnetic forces—$\sin\theta_w$. Further, SU(5) and SO(10) both predicted a decay of the proton that ought to have been measurable in deep mine experiments; finally, by construction the new grand unified theories reproduced all of the known phenomenology of both the electroweak and quantum chromodynamic theories. Strings could do neither; that is, they could not make new predictions (such as proton decay and $\sin\theta_w$), nor could they reproduce the known phenomenology of the standard model. Closing the August 1987 Superworld II conference, Glashow remarked that the proliferation of superstring theories undermined claims for a unique "theory of everything." Despite such setbacks, Glashow continued, "stringers enthusiastically pursue their fascination with ever purer mathematics, while some survivors grope towards the baroque to the beat of their superdrums. Perhaps we unstrung and unsung dinosaurs will have the last laugh after all."[11]

By 1989, for Georgi, the proliferation of GUTs, especially their assimilation into the even larger unification schemes of the superstring, was a source of a mighty ambivalence: "I feel about the present state of GUTs as I imagine that Richard Nixon's parents might have felt had they been around during the final days of the Nixon administration. I am very proud that the grand unification idea has become so important [but] I cannot help being very disturbed by the things which GUTs are doing now." GUTs, he insisted, had been motivated by the desire to complete the unification of forces by accounting for the weak mixing angle and explaining the quantization of charge.

> They [GUTs] were certainly not an attempt to emulate Einstein and produce an elegant geometrical unification of all interactions including gravity, despite the parallels which have been drawn in the semipopular press. Einstein's attempts at unification were rearguard actions which ignored the real physics of quantum mechanical interactions between particles in the name of philosophical and mathematical elegance.[12]

Imitating the aging Einstein in his failed quest for a completely unified theory, Georgi contended, was a losing proposition.

As far as Georgi was concerned, at stake was not an incidental question of style or philosophy, but rather the defining quality of the discipline. The nature of physics itself was in contest. "Theorists," Georgi insisted, "are, after all, parasites. Without our experimental friends to do the real work, we might as well be mathematicians or philosophers. When the science is healthy, theoretical and experimental particle physics track along together, each reinforcing the other. These are the exciting times."[13] When experimentalists get ahead, as the bubble chamber experimentalists did in the 1960s, the discipline becomes eclectic, overrun with results without order or explanation. At other times theory gets ahead and the path is strewn with irrelevant speculations out of touch with reality.

The string theorists read history differently; they foresaw a different future, and they found in the new theory not a violation of the defining values of physics, but rather their instantiation. Far from being a decline from "the exciting times" of the past, so they argued, these were the golden days of physics, days like none the discipline had seen since the founding years of quantum mechanics in the mid-1920s. If such struggles over values seemed to go beyond typical debates within physics, it was because more than the choice of the right Lagrangian was in play. In contemplating the fate of string theory, theorists, experimentalists, and mathematicians saw that the constitutive practices of their scientific cultures were at stake.

What theorists want

Why, one might well ask, would any theorist want anything beyond the standard, gauge theory model of the electroweak and strong nuclear forces? For, try as they might, the powerful accelerator laboratories at CERN, DESY, and SLAC had only found ever-greater confirming evidence for the gauge theories—ever-more precise measurements that in every way seemed to celebrate the powerful work accomplished from 1967 through 1974. Contra Kuhn there was no experimental anomaly gnawing at the theorists—no aberrant motion of the perihelion troubled them, no unexpected spectral-line splitting, no failure to find a phenomenon like aether drift. The issues in play were fundamentally intratheoretical. How, theorists asked, could a theory be considered complete when it held (not including the neutrino masses) some nineteen free parameters that were utterly unfixed by basic principles and could only be inserted by hand? How could a theory whose nineteen parameters had to be tuned to a precision of ten or more decimal points be right? Surely this required "fix" hinted at something more natural, more fundamental that lay beneath the surface. How could gravity and the particle theories be reckoned in utterly different ways in disconnected theories? After all, in the highly curved spacetime near a black hole, quantum effects must enter into the picture. It was not just unaesthetic to banish gravity from the theory of matter; it was manifestly inconsistent—and straightforward attempts to make a quantum field theory for gravity led to an inconsistent, nonrenormalizable theory. It seemed as if the infinities that plagued the quantum field theories signaled a pathology in the theory, suggesting that something was wrong with reasoning that allowed lengths to exist down to their vanishing point.

String theory seemed to bridge these gaps. It would start from a single scale—the Planck scale formed of the various fundamental constants of quantum mechanics (h), gravity (G), and relativity (c). This quantity, $(hG/c^2) = 10^{-33}$ centimeters, set the fundamental length, the "Planck length" of the theory. Out of this single parameter, so the hope was, would follow all the parameters of the standard model—including the masses of the quarks and leptons, the coupling constants of the forces. No host of seemingly arbitrary values, no fine tuning of their ratios to get sensible answers that could be reconciled with experiment. Instead of building physics out of point particles, there would be, at the root of all things, *finite* strings of Planck length: all currently known "fundamental" particles of string theory would be no more than the low-lying excitations of these strings. Instead of a Keplerian music of the spheres, forces and masses would be the music of these tiny strings, vibrating under an intrinsic tension of some 10^{39} tons.

Born in the context of strong interaction theory—a long way from unified theories of everything—string theorists got far enough to show that the theories would only exist in 10 or 26 dimensions, far enough, that is, to seem irrelevant for real world 4-dimensional physics. Then, in 1984 three things happened: Michael Green and John Schwarz showed that the theory was probably finite to all orders of perturbation theory—*finite*, not renormalizable.[14] Second, the Princeton "string quartet" of David Gross, Jeffrey Harvey, Emil Martinec, and Ryan Rohm exhibited a particular model that actually seemed a candidate for unifying gravity and known particle physics. And third, Philip Candelas, Gary Horowitz, Andrew Strominger, and Edward Witten provided a picture of what string theory might look like once the "extra" dimensions had curled up (compactified), giving a glimpse of how "low-energy" (accelerator-accessible) particle physics might fit into the theory. String theory exploded. From less than a hundred titles a year between 1974 and 1983, the number skyrocketed to over twelve hundred in 1987.[15]

With Planck-scale physics looking promising in 1985, string-friendly physicists hoped that, with the advances of the high-energy theory, there might be a way to pick out the low-energy consequences of the theory and so meet experiment. The basic strategy for this procedure was this. It was argued that the Planck-scale string theory would "compactify"; that is, 6 of its 10 dimensions would curl up—they had better do so since the world we live in has but 4 dimensions of space and time. After this compactification, the "effective theory" defined on this new space—4 space-time dimensions plus the compactified 6-dimensional real space (equivalent to 3 complex dimensions)—could then be analyzed to determine what particles were predicted to exist and how they should interact.

There are stringent requirements on the nature of this complex 3-dimensional manifold. (An n-dimensional space is a manifold if it can be covered by patches of Euclidean space $\mathbb{R}^n$ or, if complex, patches of $\mathbb{C}^n$.) Most importantly it still had to have a minimal supersymmetry, the pairing of particles like the electrons that obey the Pauli exclusion principle with ones that do not (in the case of an electron, its supersymmetric twin was a postulated entity known as the selectron). Conversely particles (bosons) that tend to bunch together like photons are postulated to have twins that do obey the exclusion principle (the photon's hypothetical twin that *would* obey the exclusion principle was dubbed the photino). The demand for supersymmetry restricted the structure of the 6 "curled-up" dimensions to a particular kind of complex manifold. Not too many years earlier, Eugenio Calabi, a mathematician at the University of Pennsylvania, and Shing-Tung Yau, a mathematician at Harvard, had explored these manifolds and their various properties: in a certain sense flat, closed, and configured so that a vector transported around a closed loop exhibited very special characteristics. Supersymmetry made these mathematicians' speculations the perfect home for curled-up dimensions in string theory. The physicists named them, eponymously, Calabi-Yau manifolds.[16]

At first, everything the physicists knew about these objects came directly from Yau: he had provided a single example in his original paper and in later conversations told Andrew Strominger there were at least four more. For a brief moment, the physicists hoped that all but one would be ruled out, and that the single remaining space would give the true theory. But even before examples had multiplied greatly, Yau began to suspect there were tens of thousands.[17] Still, the requirements the physicists placed on such spaces were stringent and the number of candidates was sure to decrease when those constraints were imposed. In particular, and to the great surprise of many of the string theorists, it turned out that the number of particle generations was tied to a *topological* feature of the Calabi-Yau space. For a 2-dimensional closed surface, the topological Euler characteristic is defined as 2 (1-g) where g is the number of handles in that surface. In higher dimensions, the Euler characteristic is a means of identifying the topological complexity of the manifold. Old-style gauge physics provided no reason for the existence of this multigenerational repetition of particles: electron and electron neutrino, for example, were repeated as the muon (just a heavier version of the electron) and its own associated (muon) neutrino. This same structure repeated a third time with yet a heavier version of the electron known as the tau and its associated (tau) neutrino. Here was a clear example of a way in which the mathematics of algebraic topology fixed a physical feature (number of generations) that had absolutely no constraint in gauge theory other than brute, experimental measurement.[18] The hope, then, was that one day a unique string theory in 10 dimensions would be found such that, when compactified, it would issue in a Calabi-Yau space *predicting* the number of generations to be three. It might then, after all these years, be possible to answer I. I. Rabi's longstanding question about the muon, "Who ordered that?," with an answer: whoever chose the number of holes in the manifold. Or one better: "she who ordered the original 10-dimensional theory that compacted into a Calabi-Yau with Euler number plus or minus 6 was, by doing so, ordering the muon." But for now, absent that final theory, theorists could restrict their attention to those Calabi-Yau spaces with the right Euler number—that is, with the right number of generations.

$$| X | \text{ (the Euler characteristic)} = 2 \text{ x (number of generations)}$$

So to match the known world of three generations, string theorists began hunting for 3-dimensional Calabi-Yau manifolds with Euler characteristic plus or minus 6. (Six divided by two gave the three generations, and it was generally thought that the sign of the characteristic could be conventionally fixed later.) With a more or less well-articulated problem, physicists began calling in the mathematicians. David Morrison, a Harvard-trained mathematician at Duke, found the encounter unnerving:

> Physicists began asking, "Can you algebraic geometers find us a Calabi-Yau with Euler characteristic plus or minus 6? It was a pretty interesting experience being asked this. . . . We were asking questions for internal mathematical reasons. Suddenly some physicist knocks on your door and says: if you can answer this it might be a solution to the problem of the universe. But the communication barriers were immense. A parody of the interaction was this. A physicist asks a mathematician: "Can you find me an X?" The mathematician (after many months): "Here's an X." Then the physicist says, "Oh that. Actually we wanted Y not X." Ad infinitum.[19]

Israeli physicist Doron Gepner was at Princeton from 1987 to 1989, where he was struggling to understand the structure of 2-dimensional quantum field theories—field theories defined by one time and one space dimension. These are well-studied objects, simpler in many ways than full-blown 4-dimensional theories. But for string theorists the 2-dimensional theory also had the virtue of representing the world-sheet swept out by a string. As Gepner examined these 2-dimensional theories, he began to ask an intriguing question. Suppose, as a number of string theorists did at the time, that one divided the 10-dimensional string space-time into a 4-dimensional part and the compacted 6-dimensional part. It is possible to think of the 6-dimensional part as parameterized by six fields, one for each dimension. For a 2-dimensional field theory to represent this 6-dimensional space, it would have to satisfy two constraints. First, it would need to register the six fields with what is called a "central charge" proportional to 6. And second, the 2-dimensional theory would have to be free (or almost free) of anomalies—a quantum effect that spoils the good behavior of the theory. Gepner considered such a minimal model. It was a 2-dimensional, conformal field theory with a weak anomaly and the right central charge.

What Gepner suggested, in particular, was that there was a geometrical interpretation of the particular minimal model he had written down that had an actual, explicit manifold associated with it. In other words, like any quantum field theory, the minimal conformal field theory specified all the configurations the objects it described could take. And the space of those configurations was, for Gepner's minimal conformal field theory, a completely specifiable Calabi-Yau manifold. More precisely, the Calabi-Yau manifolds that Gepner had in mind were spaces defined by solutions to a polynomial in five complex variables, $z_1^5 + z_2^5 + z_3^5 + z_4^5 + z_5^5 = 0$, known as Fermat hypersurfaces. Just as a quadratic equation $x^2 + y^2 = 9$ (with x and y real) defines a one-dimensional curve in 2 dimensions, the Fermat defining equation with z_1 through z_5 picks out a (complex) 4-dimensional hypersurface in 5-dimensional complex space (C^5). Gepner's argument identifying the minimal model and the Fermat hypersurface proceeded in two steps. First, he showed that both had certain quantities in common (the Hodge numbers, which I'll describe in a moment). And second, he pointed out that the Fermat hypersurface and the minimal model exhibited the same discrete symmetries. If, for example, one multiplies, z_1 (or z_2 or z_3 or z_4 or z_5) by a fifth root of 1, call the root α, then since $(\alpha z_1)^5 = z_1^5$ nothing would change in the equation specifying the manifold. Third, he showed that, for certain points in the algebraic formulation of the field theory and certain points in the Calabi-Yau manifold, the spectrum of particles would be the same.

Finally, Gepner went out on a limb, conjecturing that the link between his particular minimal 2-dimensional conformal theory and a Calabi-Yau space was no accident. He simply declared that *every* minimal model would have a corresponding Calabi-Yau space. No proof, just a strongly held hunch.

An aside: it turns out, quite generally, that there was an enormous simplification in mathematical analysis that could be had by moving from a complex space of $n + 1$ dimensions (C^{n+1}) to a "complex projective space" of one fewer dimensions (CP^n) that consists of all the lines through the origin in C^{n+1}.[20] So instead of looking at the Fermat hypersurface in complex 5-space (C^5), Gepner turned to the equivalent problem in the projective space, CP^4. It was clear to both physicists and mathematicians that, while CP^4 by itself is not a Calabi-Yau space, degree 5 hypersurfaces embedded in it would be. Starting with this embedded degree 5 complex space, the Fermat-type polynomial took the problem down one dimension and Gepner could lop off another dimension by posing the analysis in projective 4-space. Together this left a space of 3 complex dimensions— and it was this space that Gepner conjectured was the explicitly given Calabi-Yau manifold in which the compactified 2-dimensional string theory lived. So in the end Gepner had a few promising examples but a larger hope. That hope was for a particular construction of a conformal field theory that would be both physically realistic and uniquely attached to a geometry; in short, there would be a correspondence:[21]

Conformal Field Theory $\Leftrightarrow$ Calabi-Yau.
Geometry might parallel algebra.

Gepner's 1987 claim about geometry and minimal conformal field theories was heard, used, and challenged. Cumrun Vafa, at the time Harvard's lone string theorist, was also after the links between algebra and geometry. He and colleagues Wolfgang Lerche and Nicholas Warner began with the algebraic relations of the 2-dimensional quantum field theory (just as Gepner had) and they too asked, What must the geometry be that would produce these algebraic relations? But what they saw surprised them: there was an ambiguity on the geometry side—that is, there appeared to be *two radically different geometries* either of which seemed as if it could be associated with essentially the same conformal field theory. (Conformal field theories are the quantum field theories used to represent the strings; they are, by definition, left invariant under transformations that preserve the flat [Minkowskian] metric up to a position-dependent rescaling.) This went considerably farther than what Gepner had in mind, and in conversation Gepner criticized Vafa for treating size- and shape-changing parameters as if they were linked. Lance Dixon too began to wonder about this geometrical ambiguity.[22]

We shift scenes now, from Cambridge to Texas. Philip Candelas, based in the physics department at the University of Texas, Austin, had entered the field through astrophysics (having completed his doctoral dissertation on Hawking radiation in 1977), had then begun studying quantum gravity theories, and had turned to string theory in the mid-1980s. By spring 1988, Candelas and his students Rolf Schimmrigk and Monika Lynker were struggling to understand what it was that Gepner had, in fact, done—what exactly was he saying about the relation between conformal field theory and geometry? After giving a seminar to the physics group in Austin during the spring of 1988, Dixon crowded into Candelas's office along with the students to discuss the meaning of Gepner's analysis. Dixon showed the Austin group how in Gepner's scheme labeling a set of particles a generation or an antigeneration was arbitrary. But on the geometrical side, he pointed out, there was a huge distinction between the geometries that would supposedly correspond to the two alternatives. In particular it was puzzling that finding one geometry (with negative Euler numbers) was easy but finding the geometry corresponding to the antigeneration, a supposedly equivalent field theory, was impossible. (The Austin group had practically no relevant manifolds with positive Euler numbers.)[23] Later on, Candelas recalled, "I remember being in seminars where Gepner would say that these Calabi-Yaus came in pairs with opposite Euler numbers. I remember saying: 'Gepner is crazy.' "[24] Crazy because the pairing Gepner conjectured demanded the so-far unseen manifolds with positive Euler numbers. Here it is worth being more precise about the characterization of a manifold.

Since the nineteenth century, mathematicians have had many ways of understanding the topological properties of a manifold—properties such as the genus, the number of holes—that were independent of the particular metric (a rule for calculating distances) imposed on the space. One of the most basic topological characteristics is captured by the Betti number, that is, the number of independent cycles of various dimensions that could be defined on the space. Intuitively (by-passing the precise definition of cycle) think of a torus. It has two closed curves that can be neither smoothly contracted to a point nor deformed to one another: the cycle that goes the long way around a donut and the cycle that goes around the circular cross-section. Since the Betti number is a topological characteristic, so are any linear combinations of Betti numbers, and, in particular, the Euler characteristic can be defined as the alternating sum of Betti numbers. For a real 3-dimensional manifold, for example,

$$X = b_0 - b_1 + b_2 - b_3.$$

Generalizing to the case of complex manifolds, mathematicians defined the generalized Betti number (known as the Hodge number, $h^{p,q}$) to be a count of the number, p, of complex cycles and the number, q, of complex conjugate cycles. For the class of manifolds considered here (Kähler), the relation between the Hodge numbers and the Betti numbers is just one of addition: b_2, for example, is just a sum of all the Hodge numbers such that the total number of cycles is 2, that is, where $p + q = 2$. So here $b_2 = h^{1,1} + h^{2,0} + h^{0,2}$.

In general, one could say a great deal about a 3-dimensional complex manifold by writing down all its Hodge numbers, a task conveniently displayed in the Hodge Diamond:

$$
\begin{array}{ccccccc}
&&& h^{0,0} &&& \\
&& h^{1,0} && h^{0,1} && \\
& h^{2,0} && h^{1,1} && h^{0,2} & \\
h^{3,0} && h^{2,1} && h^{1,2} && h^{0,3} \\
& h^{3,1} && h^{2,2} && h^{1,3} & \\
&& h^{2,3} && h^{3,2} && \\
&&& h^{3,3} &&&
\end{array}
$$

Simple complex conjugation takes a complex variable into its conjugate (as in $x + iy$ goes to $x - iy$). Accordingly there is a symmetry around the vertical axis: $h^{1,0}$ must equal $h^{0,1}$, for example. Other symmetries enforce an identity of terms flipped over the horizontal axis, for example, $h^{2,0} = h^{3,1}$. The conditions that define a Calabi-Yau manifold set certain Hodge numbers equal to 0 and others to 1. When the dust settled, there were only two surviving, independent terms ($h^{1,1}$ and $h^{1,2} = h^{2,1}$) in the Hodge Diamond of a Calabi-Yau manifold:

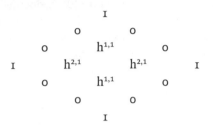

Roughly speaking, the Hodge number $h^{1,1}$ counts the number of non-trivial 2-surfaces on the Calabi-Yau manifold (that is, the number of 2-dimensional surfaces); $h^{1,1}$ also counts the number of parameters that rescale the manifold without changing its shape. By contrast, $h^{2,1}$ leads easily to the number of 3-surfaces (the number of 3-surfaces is 2 $h^{2,1}$ + 2); $h^{2,1}$ also gives the number of parameters that change the shape (complex structure of the manifold) without altering its topology. Summing up the Hodge numbers in an appropriate way gave the Euler number and, as discussed above, the Euler number is directly proportional to the number of generations. For the 3-dimensional Calabi-Yaus

$$|X| = 2 \,|\, (h^{1,1} - h^{2,1}) \,| = 2 \text{ (number of generations).}$$

So by late spring 1988, Candelas and his group understood Gepner's conjecture as posing a well-defined puzzle: if a family of particles was defined on a Calabi-Yau of Euler number X, then the antifamily would be defined by a manifold of Euler number $-X$. If this was right, Calabi-Yau manifolds that corresponded to conformal field theories came in pairs—one with X and one with $-X$—both of which essentially corresponded to the same (conformal) string theory.

"You see," Candelas mused, "the mathematicians were never after thousands of examples of something like Calabi-Yau manifolds—they knew half a dozen. You don't get promoted in math for such things."[25] By contrast, the physicists *were* after thousands of these manifolds; they were precisely interested in the grubby details of their internal geometric structure—for in one of those thousands of manifolds, in some set of geometrical relations (so they hoped), was more than a mathematical example. Somewhere among these imagined structures was the one manifold that would yield an effective theory corresponding on one side to a compactification of a still-to-be found Planck-scale string theory, and on the other side to the zoo of observed particles that came flying out of particle colliders. Somewhere in the panoply of manifolds might lie the one solution to the theory of everything.

Brian Greene was one of the physicists knocking on mathematicians' doors. Greene had been an undergraduate at Harvard, then pursued a doctorate at Oxford with Roger Penrose, finishing in 1986. With a National Science Foundation (NSF) postdoc in hand, he then returned to Harvard to work with Vafa in the hopes of doing something "more physical" than the cosmological work he had followed in England. Greene sought to exhibit, explicitly, the manifold pairings that Gepner, Dixon, and Vafa thought existed. By sewing together pieces of manifolds in a suitable way, Greene and a graduate student, M. Ronen Plesser, found explicitly the missing manifold of opposite Euler number to the ones that Gepner had explored.

Remarkably, one member of the pair $h^{1,2}$ corresponded to the $h^{1,1}$ of the other (and vice versa)—a mirror flip, if you will, across a diagonal axis on the Hodge Diamond. And since this flip interchanged those two quantities, $X = 2 (h^{1,1} - h^{2,1})$ switched sign. For the well known Calabi-Yau with $X = -200$, Greene and Plesser now could offer one with $X = +200$; for the one with $X = -88$, there now was a twin with $X = +88$.[26] And this sewing and gluing process would, by construction, leave the basic physics (the conformal field theory) unchanged in all its predictions despite the fact that seemingly very different properties of the twin manifolds had switched roles. In short,

Calabi-Yau ‗ Conformal Field Theory $\Leftrightarrow$ Mirror Calabi-Yau

This was a remarkable state of affairs: two different manifolds—different in their very topology—were indistinguishable in terms of physical predictions. While $h^{1,1}$ counted the ways in which size could be altered (rescalings of the metric), $h^{2,1}$ did something completely different—it altered the *shape* of the manifold by changing its complex structure. Nothing in general relativity prepared the physicists for this twinning: in general relativity, to have a radically different geometry meant a different physical situation.

Meanwhile, unknown to Greene and Plesser, Philip Candelas was exploring the same territory, but using very different lines of reasoning. Candelas developed a way of constructing large classes of Calabi-Yaus and finding their Hodge numbers, and from that, extracting their Euler numbers. Up to that point most of the known Calabi-Yau manifolds had negative Euler numbers, so it seemed highly unlikely that they actually existed in pairs. But in the spring of 1989, to check the conjecture Candelas, along with Lynker and Schimmrigk, did what physicists do when encountering a question of large numbers of entities. They tried lots of examples. More specifically, they generated some six thousand examples of these newly discovered Calabi-Yaus on the computer and printed out a scatter plot. "We did not believe they came in pairs," Candelas noted. "We wanted to know if we could find Calabi-Yaus with positive Euler numbers. Could there be a large number of Calabi-Yaus with positive X? Could the computer do the job in less than the age of the universe?" As it turned out, yes and yes.

When Candelas and his collaborators saw the Euler number printout of their various Calabi-Yaus, they were floored. Roughly speaking, the number of positive and negative Xs were equal. Moreover, there were some twenty-five hypersurfaces with Euler number 6 (or -6) meaning there would be, as needed, three generations. Candelas: "When I got this graph, I brought it to Brian Greene and said, 'You're going to fall off your chair when you see this.' He said 'we know that—we know they [Calabi-Yau manifolds] come in pairs.'"[27]

Greene, as it turned out, remained upright in his chair because he, too, had been grappling with Gepner's wild idea. Having taken a smaller class of surfaces than Candelas and Xenia de la Ossa had considered (Greene and Plesser were using the equation $z_1^5 + z_2^5 + z_3^5 + z_4^5 + z_5^5 = 0$, with no cross terms between the different z's), the two Harvard postdocs had begun building up conformal field theories on these Calabi Yau manifolds and discovered, to their immense surprise, that they indeed could get pairs with opposite Euler numbers. So when Candelas came to his pairing conclusion on the wider set, Greene was ready to believe. Now there were two pieces of evidence for what Greene had begun calling "mirror symmetry." From Candelas came a rough argument (same number of positive and negative Euler numbers for a wide class of manifolds) and from Greene a precise argument restricted to a narrower class of manifolds: Greene and Plesser could show exactly that one and the same conformal string theory could sit on Calabi-Yau manifolds with opposite Euler numbers.

With this concordance in hand, Candelas and Greene began showing their results to both physicists and mathematicians. Not surprisingly, Vafa liked the result—it confirmed his earlier conjecture. But when Greene walked his result the fifty yards or so from the physics department to Shing-Tung Yau in the Harvard mathematics department, the reception was quite different. As far as Yau was concerned there simply was no reason for this pairing; he was sure they had simply made a mistake. Candelas too found the mathematicians highly dubious. Yau just did not think that his manifolds were twinned. "I was beginning to believe that they came in pairs," Candelas said. "But the mathematicians were not having any of it. We published the diagram and I began phoning up the mathematicians—we got no response. There just wasn't any reason why they *should* come in pairs. The first time I mentioned it to Yau, he said it must be bullshit."[28] To other mathematicians the arguments by Greene, Candelas, and their collaborators seemed dreamlike, incomprehensible. For the mathematicians around 1990, the two Calabi-Yaus seemed utterly unrelated. First, the conformal field theory itself meant nothing to them— the fact that the same theory could be defined equivalently on the two manifolds simply did not move them one way or the other. Second, almost all the Calabi-Yaus known had negative Euler numbers, so the idea of each Calabi-Yau having a twin of opposite Euler number seemed preposterous. Third, the manifolds themselves seemed unrelated. The moduli space (the space defined by the parameters that define the vacua of the theory) on one side involved deformation of the complex structure of the Calabi-Yau and had a special geometry (variation in shape). On the other side, the moduli space was parameterized by deformations in the Kähler structure (variation in size).

For the physicists around 1990, the mirror symmetry conjecture seemed puzzling for completely different reasons. In the moduli space defined by the parameters that varied complex structure, describing quantum corrections to scattering of strings was particularly easy—it could be shown that these were not quantum corrected. The geometry was exact. On the other side (the moduli space built on the size-changing, "Kähler" deformations), the scattering amplitude was corrected. Because of the quantum corrections, the geometrical picture, it seemed, would be lost. So how, the physicists asked, could these two different realizations of the theory possibly be physically equivalent?

To understand why the result was, nonetheless (physics notwithstanding), so entrancing to mathematicians, we need to turn away from the physics of strings, away from conformal field theories and scattering amplitudes, and toward the mathematicians' own world of algebraic and enumerative geometry.

When mathematicians approach systems of algebraic equations in several variables, they are after the structure of the solutions. One way to get that is to consider the space of solutions to the equations as a topological manifold, a manifold in which one does not assume the existence of a distance but only the topological properties such as the number of holes in a surface. Algebraic geometers, by contrast, consider the equations as having coefficients in (an arbitrary field) and the solutions to these equations lie (in its algebraic closure). As one textbook put it, "the arguments used are geometric, and are supplemented by as much algebra as the taste of the geometer will allow."[29]

There are high-brow problems in algebraic geometry—problems of classification of manifolds, for example. But it was not the high flyers that made first contact with the physicists but rather the mathematicians who aimed at high-level classifications by studying specific examples. Herbert Clemens loved examples. Having taken his doctorate in complex geometry with Philip Griffiths in 1966, Clemens was after variations of the complex structure of 3 (complex) dimensional manifolds. For odd dimensions a simplification exists because the local spatial structure (cohomology) has a simple extra symmetry. And for 3-folds there is a further reduction in difficulty if the Hodge numbers at the far left and far right of the Hodge Diamond are zero. That is, if the Hodge numbers $h^{3,0} = h^{0,3} = 0$, then, roughly speaking, the 3-dimensional manifold is called "of Fano type" and behaves in certain respects as if it were a *one*-dimensional object. A particularly simple Fano-type 3-fold is the cubic hypersurface in projective 4-space. And it is there that Clemens did his work. An old conjecture stipulated that this hypersurface was equivalent to an affine space (a space without a metric but where distance could be defined along parallel lines). Clemens and Griffiths proved it was not.

Over the years, Clemens continued studying these Fano spaces, exploring the structures of these simplified 3-dimensional entities and how they related to the one-dimensional case. "I much preferred working with concrete examples," he said, "and then, after many years, trying to see generality." The holy grail of people in the field was to prove Hodge's conjecture that, for Fano-type 3-folds, translates to the assertion that the homology—the set of nonequivalent cycles that specify the topology of the manifold—would be parameterized by closed loops of algebraic curves. "There are many out there who would sacrifice an arm and a leg and probably more if they could prove the Hodge Conjecture, even in a restricted setting close to the Fano 3-fold case."[30]

One way for Clemens to pursue this purely mathematical goal was to examine the simplest case where one could *not* move the curves around. And the Calabi-Yaus defined by a quintic hypersurface in projective 4-space seemed to fit that bill: it was the lowest degree hypersurface that was not nearly Fano and it certainly had a nontrivial homology. If Hodge was right, then the rational curves (the special category of curves that might parameterize the general curves) would have to be rigid for a general quintic—that is, not distendable into one another.

So for the physicists, Calabi-Yau 3-manifolds were the conceptual site on which they expected to describe a realistic, three-generation string theory at low energies. But for the mathematicians, Calabi-Yau manifolds were to be a site for exploring the Hodge Conjecture in the particular case of Fano-like 3-folds. Now, if the rational curves really were rigid—and Clemens had shown that some were rigid in arbitrarily high degree—then a homogeneous coordinate z_i could be restricted to a function of arbitrarily high degree on a rigid curve. So one ought to be able to count curves of fixed degree.[31] While there would still be much else to understand, a determination of their number would say at least *something* about these curves. So Clemens encouraged Oklahoma mathematician Sheldon Katz to take a look.

Counting geometrical objects like this—known as enumerative geometry—was a modest corner of algebraic geometry. Katz vividly remembered advice he got soon after graduate school at Princeton: building a reputation as an enumerative geometer might kill his budding career.[32] Although mathematicians had long admired the tantalizing numbers of enumerative geometry, these numbers were by no means the most highly prized, most abstract results of algebraic geometry. One old result, due to the magician-like Hermann Schubert in the nineteenth century, showed that exactly 27 lines could be drawn on a general cubic surface.[33] Next on the totem pole of such counting problems was the task of finding the number of rational curves of higher degree that could be put on a quintic 3-fold. The simplest case, lines (curves of degree one), had been calculated in 1979; it turns out, as Joe Harris showed, that there are 2,875 of them. The next case, that of conics (curves of degree two), fell to Sheldon Katz in 1986, who had met Clemens's challenge: there were 609,250 of them.[34] What Katz really wanted to know was what lay behind these numbers: "I wanted to know what makes Schubert's ideas work. What deep unexpected relationships were there on higher-dimensional manifolds?"[35] By the end of the 1980s, it was well known among enumerative geometers that the next case, the tally of curves of degree three, was going to be vastly harder to compute. Two Norwegian mathematicians were hard at work on it; I will return to them in a moment.

Suddenly, in late 1990, Candelas and his collaborators Xenia de la Ossa, Paul Green, and Linda Parkes (COGP) saw a way to use mirror symmetry to barge into the geometers' garden.[36] From Brian Greene and Ronen Plesser, Candelas knew there was, in principle, an equation that allowed one to pass from calculations of string interactions on a manifold to the same calculation on the mirror. Greene and Plesser thought actually computing some key quantities would simply be too hard—Candelas, who liked calculating things, thought maybe it was in fact tractable using some algebra and a home computer. Candelas and his younger colleagues reasoned this way. Suppose three strings interact. Since the conformal theories are essentially the same whether described in terms of a Calabi-Yau manifold, M, or its mirror, W, the amplitude for the string collision must be the same whether computed on M or W. On one manifold this expression turned out to be relatively easy to evaluate. On the mirror manifold, the expression for the amplitude amounted to the number of rational curves. So suddenly the computation of a seemingly impossible quantity became nearly trivial.

This is worth explaining in more detail. A string traveling through space-time carves out a surface; a closed string depicts a cylindrical surface (world sheet or surfaces of higher genus). Quantum mechanically, a particle takes all possible paths, as Richard Feynman taught back in the 1940s. So in computing an amplitude—the probability of a three-string interaction—one must consider all the appropriate world-sheet embeddings in the Calabi-Yau manifold. Now this manifold could well have uncontractible holes in it (hollowed out spheres), and so the interaction among strings must be corrected by terms corresponding to the string world-sheet wrapping, in a minimal and smooth way, around these spheres one, two, three, or more times. Such uncontractible topological defects are known as instanton corrections, and increasing the number of wrappings corresponds to making higher order corrections. From the mathematicians' perspective, each winding number around a defect corresponds to a particular degree curve on the quintic hypersurface. So what for the physicists was an expansion yielding the quantum (instanton) corrections was for the mathematicians a series giving the number of curves from lines to conics, to cubics, and so forth.[37] But this identification while interesting was not yet news. Neither physicists nor the mathematicians could compute the series.

Mirror symmetry broke the computational blockade. Suddenly the computation on the size-changing (Kähler) manifold could be equated to the shape-changing (complex structure) mirror manifold. All at once Candelas and his collaborators had an answer to the enumerative geometers' dream: Even if the problem was intractable on the quintic hypersurface, any conformal field theory calculation there could be converted to the mirror manifold that could then be deformed in shape as desired. But just this shape invariance kept calculations on the mirror from being quantum corrected: calculations on that side of the mirror equation could be done easily, and explicitly, with some algebra and a Mac.[38]

Astonishingly, at the end of a brief calculation, Candelas and crew had the next member of the series, the till-then elusive number n_3, that they assessed at 317,206,375. But the mathematicians were dubious—the methods the physicists used corresponded to nothing remotely comprehensible to the mathematicians. Conformal field theories, Feynman path integrals— as far as the algebraic geometers were concerned, these were ill-defined concepts plucked like a clutch of rabbits out of thin air. So COGP went back to their Mac and offered the mathematicians the next number, n_4, and the next after that, n_5, all the way up the series to n_{10}. To the mathematicians the whole procedure, from start to finish, seemed dubious indeed.

Meanwhile, the Norwegian mathematicians, Geir Ellingsrud and Stein Arild Strømme, were struggling with a direct, mathematically grounded, geometrical computation of the number of rational curves. Both had taken mathematics degrees in Oslo, both had completed doctorates with the same advisor (Olav Arnfinn Laudal), and they had been collaborating since 1978 on a range of mathematical problems surrounding the moduli spaces for various geometric structures and in particular on cohomology or intersection theory. By 1987, they had cranked up a crude desktop computer, a Sinclair Spectrum, to help with their computations, and, over the course of the next years, they studied the space of twisted (rational) cubic curves. Then in 1989 Ellingsrud and Strømme met Clemens who, fresh from his success in motivating Katz to count the curves of n_2, now gave similar encouragement to the Norwegians to reckon n_3. Ellingsrud and Strømme thought they could do it—after all, having completed years of work on twisted cubics and their mobilized Spectrum, they thought they could find intersections of the twisted cubics with just about anything. It was equally obvious that the calculation of n_3 would be far too difficult to undertake by hand. And so, though both were relative neophytes on the computer, they pounded out a new computer package, using an algebraic program known as Maple. For almost a year they worked at it until, in June 1990, the computer displayed the fruits of its labor: $n_3 = 2,682,549,425$.

Instantly, they shot the ten-digit number over the Internet to Sheldon Katz, Herb Clemens, and others; Katz electronically relayed the news to Candelas:[39]

> Dear Sheldon Katz, June 6, 1990
> We finally got a number for the number of twisted cubics on the general quintic threefold. I know that Geir saw you not long ago, but he is out of reach for the moment so I ask you directly this way what you know about the number. We get 2 682 549 425 = $5^\wedge 2 * 17 * 6311881$. I get the impression from your 1986 papers that you only claim that the number is divisible by 5, but Herb Clemens thought perhaps it should be $3 * 5^\wedge 3$. Obviously there is the possibility of a programming error on our part (we used Mathematica, Maple and Macaulay, and MPW to shuffle results back and forth between these), but it would be most interesting if you can say right away that this number has to be wrong. . . .
> Best regards,
> Stein Arild Stromme

Now Candelas found himself in a complicated position. On the positive side, he had a computer program that had produced a scatter plot strongly indicating that mirror symmetry should work over a large class of manifolds and the mirror symmetry hypothesis squared with the older conjectures and the more recent constructions by Greene and Plesser. On the negative side, he had a bevy of very dubious mathematicians. On the positive side again, he could reproduce the mathematicians' n_1 (2,875) and even the three-year-old result of Sheldon Katz for n_2 (609,250). All this was remarkable. But the third member of the series, the long-sought n_3, clashed directly. The Norwegians posted 2,682,549,425, and that was not the same as physicists' 317,206,375 no matter how you sliced it. These were numbers that had to be the same, but they were not. Candelas and his collaborators went over the calculation again and again. There would be no middle ground: either the physicists or the mathematicians were wrong.

Between physics and mathematics

In part propelled by the clash over n_3, a corps of mathematicians and physicists scheduled a workshop at Berkeley's Mathematical Science Research Institute for 6–8 May 1991. From the physics side were Greene, Plesser, Candelas and his collaborators, Vafa, and Witten among others; among those from the mathematics side Yau, Katz, Ellingsrud, Strømme, and Shy-Shyr Roan. A lingua franca did not come easily, and each day's lectures were followed by intensive question sessions during the evenings. As Yau put it, both mathematicians and physicists "each attempted to grasp the vantage point and conceptual framework of the other." Language as well as specific expertise made communication exceedingly difficult. "As with any important new development which straddles traditional academic disciplines, two pervasive obstacles facing prospective adherents are the differences in language and assumed knowledges between the respective fields."[40] The "language gap" was echoed throughout the papers, as Greene and Plesser stressed in their contribution: "At present, mirror symmetry finds its most potent expression in the language of string theory," for it was principally the fact that both manifolds contained identical physical theories—the same conformal field theory—that vouchsafed (for physicists) the equivalence of the underlying manifolds.[41] Mathematicians, by contrast, simply took "string" to be a "mnemonic" for a more precise definition.[42]

Despite—or perhaps because of—the numerical clash, David Morrison, a mathematician from Duke University, saw in Candelas's form of argumentation something important for mathematicians. Exactly what was harder to say. "The language problem," as Candelas put it, "is a very difficult barrier to surmount."[43] Some of the problem revolved around specific concepts, as Candelas noted:

> There were aspects of this that had been terribly mysterious—very hard to think of modularity space and how you moved around it. Monodromies—to a physicist a complex structure was to be understood by

hypergeometric functions represented by integrals. The really fundamental thing would always be these integrals, though as the calculations got more complex, one encountered higher-dimensional versions of these hypergeometric functions. For the physicists the fact
that if you walked around a singularity, it was mysterious that quantities altered in certain ways.[44]

Mathematicians saw not mystery in the movement of these points around
the surfaces but straightforward geometry. Monodromies—well known
geometric entities—were quantities that were locally single-valued but
changed values if one took them around a nontrivial closed path. Conversely, for the mathematicians the existence of mirror symmetry itself was
mysterious. For the physicists, however, it was not—it came out of "natural" assumptions about the conformal field theory—and the key quantitative features, coefficients of various quantities, were merely factors that had
to be introduced to keep track of coordinate changes.

For the algebraic geometers the idea of treating key coefficients as if they
were a coordinate transformation appeared unrigorous, even arbitrary, as
Morrison made clear in a talk during July 1991, a talk in which he deliberately tried to place the physicists' result in a language familiar to mathematicians from standard techniques in number theory:

> By focusing on this q-expansion principle, we place the computation
> of [COGP] in a mathematically natural framework. Although there
> remain certain dependencies on a choice of coordinates, the coordi-
> nates used for calculation are canonically determined by the mono-
> dromy of the periods, which is itself intrinsic. On the other hand, we
> have removed some of the physical arguments which were used in the
> original paper to help choose the coordinates appropriately. The re-
> sult may be that our presentation is less convincing to physicists.[45]

Indeed, while to a physicist the invocation of a patchwork of plausibility
arguments was entirely reasonable, to the mathematicians, the physicists'
formulations looked ill-defined, practically uninterpretable. Morrison continued his quotation from COGP to show the mathematicians just how
cryptic it looked in the original: "To most pairs (X,S), including almost all of
interest in physics, there should be associated . . . ." What, Morrison asked,
was the scope of "almost all," especially when applied to a category picked
out by the (mathematically) incomprehensible property "of interest in physics"? To this he added: "To be presented with a conjecture which has been
only vaguely formulated is unsettling to many mathematicians. Nevertheless, the mirror symmetry phenomenon appears to be quite widespread, so
it seems important to make further efforts to find a precise formulation."[46]

Again, Morrison took an extract of the COGP paper, and while emphasizing throughout his argument that the physicists' conjecture was of great interest, he concluded that the formulation struck mathematicians as completely unrigorous. From COGP, he reproduced, verbatim, the justification for their formula that counted rational curves on quintic threefolds:

> These numbers provide compelling evidence that our assumption about the form of the prefactor is in fact correct. The evidence is not so much that we obtain in this way the correct values for n_1 and n_2, but rather that the coefficients in equation [$K_{ttt} = 5 + 2875\ e^{2\pi i\ t\ n} + 4876875\ e^{4\pi i\ t} + \ldots$] have remarkable divisibility properties. For example the assertion that the second coefficient 4,876,875 is of the form $2^3\ n_2 + n_1$ requires that the result of subtracting n_1 from the coefficient yields an integer that is divisible by 2^3. Similarly, the result of subtracting n_1 from the third coefficient must yield an integer divisible by 3^3. These conditions become increasingly intricate for large K. It is therefore remarkable that the n_k calculated in this way turn out to be integers.[47]

"These arguments," Morrison added, "have a rather numerological flavor" reminiscent of the physicists' speculations about the "monster group" that mathematicians had at first labeled "moonshine."[48] Maybe this case of mirror symmetry would turn out to be more than moonshine, too. But, as Morrison noted, there remained the stubborn numerical obstacle presented by the Norwegian mathematicians: "Unfortunately, there seem to be difficulties with n_3."

As Ellingsrud and Strømme mentally stepped, line by line, through their program, a glitch suddenly leapt out at them. Two subroutines, "logg" and "expp" figured in the calculation, routines related to ordinary logarithmic and exponential functions. Just as in ordinary logarithms on a slide rule, these little programs sped up the manipulation of truncated power series expressions, reducing multiplication and division to simple additions and subtractions. But where, ordinarily, these functions were applied to terms that had constant terms equal to 1 (where the log of 1 is 0) in this application, the constant term, W1, was not unity. And since "logg" ignored the constant terms, any information contained in W1 was lost. The next line's routine, "expp" could not retrieve it. The instant that logg hit W1, the calculation was doomed.

Swapping out the faulty lines, they reran the program. And on Thursday, 31 July 1991, the mathematicians' barrier vanished: their computer too spat out $n_3 = 317{,}206{,}375$. Herb Clemens received an email from the Norwegians and immediately forwarded it to Candelas with its white-flag header: "Physics wins!"

Date: Wed, 31 Jul 91 11:06:34 MDT
From: Herb Clemens ⟨xxx@xxx.edu⟩
To: candelas@YYY.edu
Subject: Physics wins!
Message-Id: ⟨AAA@xxx.edu⟩

To: Herb Clemens ⟨xxx@xxx.edu⟩
From: stromme@ZZZ (Stein Arild Stromme)
Dear Herb,
we just discovered the mistake in one of the computer programs.
Once that was corrected, our answer is the same as that of Candelas &
Co. I am glad we found it at last!
Best regards,
Stein Arild

Returning to the electronic preprint file, Morrison drew a line through the words "Unfortunately, there seem to be difficulties with n_3," and inscribed, "Not any more!"[49] To Joe Harris at Harvard, Geir Ellingsrud shot a similar electronic concession, probably also that same day:

We found an error in the program we use computing the number of twisted cubics on quintics a few days ago. After having fixed it, we now get the same number as the physicists. I feel a little bad about not having discovered the error before, but that's life and for mathematics I think the outcome is the best. Please tell Yau and the others about it. Best regards, Geir.[50]

It was a hard turn of events for Ellingsrud and Strømme. Harris offered some consolation in a return email: "Don't feel bad about the miscalculation, though—it seems to me that the point of all this is not the number but the ideas and techniques, and those are if anything vindicated. Joe."[51]

During the summer of 1991, while Morrison was struggling with the gap between mathematicians' and physicists' views, Sheldon Katz arrived at Duke for a previously scheduled, year-long visit. The two of them began interrogating the physics graduate students and inscribed lists of terms on the blackboard, a list beginning "conformal field theory," "correlation functions," and continuing on from there. The next year, Morrison shared an office with Brian Greene at the Institute for Advanced Study, and they set aside an hour each day to lecture to each other. By the time Greene and Yau had assembled a second major mirror symmetry volume in 1997,[52] the joining of mathematics and physics was legible in the author list: now instead of explaining the physicists' work to the mathematicians, Morrison and Greene, among others, were coauthoring papers. "One of my roles," Morrison commented a few years later, "is as an interpreter between the two communities. They are after different things and I've tried hard to maintain the distinctions. Mathematicians want to know which parts of this stuff are proven rigorously and which parts are conjectural. Physicists don't see that—they don't like to be told something is not a theorem. They have an argument and think it is so. Different standards."[53]

The borderland prospered. Physicists and mathematicians alike began talking about geometry in a radically new way. Ordinary geometry—the geometry built up out of points—held a special relation with a physics predicated on point particles. But now that strings were beginning to take over, it was becoming apparent that the point-based geometry was only a limit, just the way at distances large compared to the Planck scale of 10^{-33} cm, space "looked" as if it were made up of points. So it was imagined to be in geometry: another geometry, the hidden half, so to speak.[54] String theory, on this interpretation, offered that generalized geometry and reduced in an appropriate limit to classical geometry. The situation was analogous to a noncommutative algebra of quantum operators that reduced to the commutative case as the Planck constant h headed to zero.

Mirror symmetry was but one of the boundary regions increasingly populated by both string theorists and mathematicians. Edward Witten used the conformal field so essential to string theory to prove novel results in the theory of knots, and string theory led to a new understanding of an enormous simple finite discrete group dubbed the Monster and a host of new insights into, inter alia, Donaldson theory of 4-manifolds and a new proof of the Atiyah-Singer index theorem.[55] Writing to the National Science Foundation in 1994, Witten and mathematician Pierre Deligne, his colleague at the Institute of Advanced study, outlined a plan designed to capitalize on this new domain. "For most of the past hundred years," they wrote, "the role of theoretical physics has been to explain known experimental or observational phenomena and to make predictions that then lead to further experimentation." Such ventures covered a broad range of phenomena from quantum mechanics to general relativity, all of it borrowed from already extant mathematics, including Riemannian geometry. "In such significant historic examples, the mathematics did not drive the physics, but it was ready at hand and utilized by the physicists with little need for reference back to the mathematical foundations."[56] Now that was changing, the authors argued. With physicists digging deeper into string theory, it became apparent that the required mathematics did not exist— though isolated pieces of recent math had proved useful.

Because the larger part of what they required was not available, they have pushed the mathematics, frequently on an ad hoc basis, leading to startling predictions. Physicists' arguments do not automatically translate into mathematical proofs but have yielded striking new mathematical ideas and results. These results have usually involved what physicists perceive as manifestations of the unknown new conceptual-geometical framework of string theory.

The structure of this common cause would involve senior mathematicians who "can listen to the physicists and communicate their ideas to [other] mathematicians in a way that captures the physical context of their thought." The geographical scope of possible mathematical recruits included MIT, IHES, Tohoku, Oxford, and Cambridge. Physicists would come from Harvard, UCLA, the University of Chicago, Tokyo, Texas A&M, and UCSB. Four to six "younger researchers" would enter the program to "develop in ways noticeably different from those of their colleagues who are more traditionally focused, either in mathematics or physics [the new breed of researcher] should feel equally at home in both worlds."[57]

Contested boundaries

The National Science Foundation replied, with regrets. It is of great interest to understand why. One referee began by lauding the contributions of the principal investigators, and celebrated the increased use of physics by mathematicians. But "I do not think that this activity [on the border between mathematics and physics] should result in the production of a large number of physics students who would work in this field." Others reiterated that sentiment: excellent investigators but the NSF should pause before further "populating" the border region. "I fear," another wrote, "that in this case the results may be analogous to those obtained by axiomatic field theory in the past—which did not further physics understanding of field theory in a substantial way." The harshest critic agreed in his praise for the leaders (Witten "is probably the best person in the entire galaxy to lead the proposed program") but the program itself raised fundamental questions about the direction of physics. "My conscience would not rest if I did not record those doubts here, even though I am fully aware that my opinion is highly contrarian." The referee continued:

> I tend to think that the most conspicuous development of the last decade is the training of a generation of very bright young theorists who know and care more for geometry and topology than for the standard model and current experimental efforts to discover the next step beyond it. Since I am convinced that the key advances in physics emerge from physical rather than mathematical insight, I must view this as a negative development. I think that theoretical physics would be in better shape if this group of very capable people had been taught

to practice research with better balance between physical fact and mathematical intuition.[58]

Ultimately, this evaluator's greatest concern was not for the mathematicians but for the physicists, especially young ones, whom the program "would tend to subvert." Mathematics, the referee continued, was a tool, but one that must be secondary to the concerns of a fundamentally physical nature. "The main goal of theoretical physics is to understand the laws of nature, and for most of the 20th century this has involved a closer connection between the most capable theorists and experiment than exists today in particle physics."[59] What was needed, this reviewer argued, was an amplification, not a diminution of that bond between laboratory and blackboard.

After revision, Witten and Deligne resubmitted the proposal, armed with a more detailed exposition of new results—this time successfully. Again they aimed to create an environment that would allow mathematicians to explore links between previously disconnected mathematical domains, fields that, to the physicists, appeared manifestly linked: "before the current period, mathematicians have tended to try to treat each idea coming from physics as a separate, isolated phenomenon, with proofs to be given in each case in an *ad hoc* fashion, unrelated to the context in which the ideas appeared." It was necessary, Witten and Deligne argued, for the mathematicians to step beyond such a piecemeal approach, to see the physicists' problematic in its "natural context," not in vitro. For the string physicists over the two years prior (1993–95), it had become common wisdom to see the various different string theories as asymptotic versions of some unknown, underlying theory. And grasping the mathematics of this theory—in the absence of the fundamental symmetries, variables, and geometrical ideas governing it—was exceedingly difficult and would surely involve both new physical ideas and "mysterious new mathematical structures." At stake, the authors contended, was the future: "The extent of success in understanding what that theory really is will very likely shape the fate of physics in our times."[60] The referees concurred; funding followed.

Once approved, prospective applicants found descriptions of the planned math-physics collaboration on the Web, including one posted on 15 December 1995 that sought to differentiate the current collaboration from previous uses of physics by mathematics:

> It is not planned to treat except peripherally the magnificent new applications of field theory. . . . Nor is the plan to consider fundamental new constructions within mathematics that were inspired by physics, such as quantum groups or vertex operator algebras. Nor is the aim to discuss how to provide mathematical rigor for physical

theories. Rather, the goal is to develop the sort of intuition common among physicists for those who are used to thought processes stemming from geometry and algebra.[61]

I have used the IAS as an example but it was not alone in its search to create a new kind of scientist, a new personhood in science, if you will, one not only with particular problems and procedures but with a hybrid way of thinking. At stake were not only "thought processes" and "intuitions" but ultimately identity. Small wonder that the move met resistance. Some particle theorists—several of whom had been coauthors with some of the principal string theorists—took the withdrawal of string theory from experiment as the harbinger of a dark age of speculation. Charges of "theology" echoed off the walls, and the battles were fought over positions from graduate student through postdoc, junior, and senior faculty positions. The first round of NSF referees' responses to the Institute for Advanced Study's "Integrated Program" was but one site that revealed these tensions; similar battles erupted in physics departments across the country, precipitating a far-reaching debate over the nature and meaning of theoretical physics.

If the string theorists were to use mathematics as their new constraint structure, they had to find a modus vivendi with the mathematicians. And here the results were contradictory, informatively contradictory, forcing to the surface long-dormant resentments and ambitions. That they would need the mathematicians, however, was clear. Precipitating this stage of the debate was an article published by Arthur Jaffe from Harvard and Frank Quinn from Virginia Polytechnical Institute—both mathematical physicists heading mathematics departments in the early 1990s. Titling their July 1993 piece, " 'Theoretical Mathematics': Toward a Cultural Synthesis of Mathematics and Theoretical Physics," in the *Bulletin of the American Mathematical Society*, they unleashed a torrent of response by public and private email and in print—and in the process surfaced views about the nature and importance of theoretical and mathematical culture. Jaffe and Quinn began by recounting how the string theorists had lost their historical tie to experiment and then continue:

> But these physicists are not in fact isolated. They have found a new "experimental community": mathematicians. It is now mathematicians who provide them with reliable new information about the structures they study. Often it is to mathematicians that they address their speculations to stimulate new "experimental" work. And the great successes are new insights into mathematics, not into physics. What emerges is not a new particle but a description of a representation of the "monster" sporadic group using vertex operators in Kac-

Moody algebras. What is produced is not a new physical field theory but a new view of polynomial invariants of knots and links in 3-manifolds using Feynman path integrals or representations of quantum groups.[62]

Here we have the crux of the issue. Physicists were using their standard set of tools (such as Feynman path integrals, vertex operators, and representations of quantum groups) to solve mathematicians' problems—and not trivial ones, either: representations of the Monster sporadic group, polynomial invariants of knots and links in 3-manifolds. Suddenly, the hard-won theorems of mathematics were being exceeded by methods the mathematicians found utterly lacking in rigor.

Rather than reject this incursion into mathematical territory out of hand, Jaffe and Quinn wanted a "dignified" name for the activity that would nonetheless isolate it from the mainline of rigorous mathematics. Borrowing from physics itself, they chose the name "theoretical mathematics": "The initial stages of mathematical discovery—namely, the intuitive and conjectural work, like theoretical work in the sciences—involves speculations on the nature of reality beyond established knowledge. Thus we borrow our name 'theoretical' from this use in physics."[63]

By employing the term *theoretical mathematics* in this way, they deliberately displaced two older, competing notions of *theoretical*. On one hand, they refused to identify "theoretical" in mathematics with the "pure" in contrast to "applied" mathematics. On the other hand, they refused to employ "experimental" mathematics to designate computer simulations—for in fact they took such speculative explorations to be under their categorization, "theoretical."[64]

So far, just a redefinition. But words alone could not efface the different ways in which the two groups treated some of the same sets of symbols. Both mathematicians and physicists might want to characterize the knots and kinks in 3-manifolds, but the methods that each group used led to a direct confrontation:

> Theoretical physics and mathematical physics have rather different cultures, and there is often a tension between them. Theoretical work in physics does not need to contain verification or proof, as contact with reality can be left to experiment. Thus the sociology of physics tends to denigrate proof as an unnecessary part of the theoretical process. Richard Feynman used to delight in teasing mathematicians about their reluctance to use methods that "worked" but that could not be rigorously justified. He felt it was quite satisfactory to test mathematical statements by verifying a few well-chosen cases.[65]

Complementing Feynman's views, Glashow lambasted overly mathematical string theorists: "Until the string people can interpret perceived properties of the real world, they simply are not doing physics. Should they be paid by universities and be permitted to pervert impressionable students?" Conversely—and Jaffe and Quinn did not hesitate to raise the point—mathematicians had no great respect for the weightiness of physicists' contributions to knowledge. In one anecdote that resonates on a gendered as well as epistemological level, they likened the physicist's proof to the woman who traced her ancestry to William, the Conqueror . . . with only two gaps.

Exaggerating for emphasis, many mid–twentieth-century physicists thought that mathematicians were supererogatory and mathematicians thought that physicists were superficial. But by 1992, neither side could so lightly dismiss the other; in the past their fiefdoms had been at sufficient remove that each could polemicize at a distance. Now they overlapped on territory vital to both.

Branches of algebraic geometry had become a trading zone—with each side contributing to it, each interpreting joint results differently. For the physicists, mirror symmetry along with other dualities promised to become some of the most powerful theoretical tools they had available. It showed how some of the string theories might be further reduced by demonstrating their equivalence—and even offered a deep geometrical understanding all the way down to Lagrangian quantum field theory and classical electrodynamics. On the mathematical side, new forms of calculation corrected their own work and extended it in certain cases infinitely beyond their previous capacities.

Stepping between the fields was delicate work, as Morrison indicated in his contribution to the 1997 mirror symmetry volume. He argued that the enumerative geometry results gained through the new techniques were physically powerful, but that they should be used with mathematical caution: "The calculations in question can often be formulated in purely mathematical terms, but it should be borne in mind that the arguments in favor of the equivalence of the answers . . . rely upon path integral methods which have not yet been made mathematically rigorous. For this reason, mathematicians currently regard these calculations as *predicting* rather than *establishing* the results."[66]

One mathematical response to the unlocking of these mirror equivalents and similar string theory successes was a full-tilt emulation by some senior mathematicians of physicists' style of work. It was a route, Jaffe and Quinn cautioned, that was a minefield, littered with danger. Imitation had happened without the evolution of the community norms and standards for behavior which are required to make the new structure stable. Without rapid development and adoption of such "family values" the new relationship between mathematics and physics may well collapse. Physicists will go back to their traditional partners; rigorous mathematicians will be left with a mess to clean up; and mathematicians lured into a more theoretical mode by the physicists' example will be ignored as a result of the backlash.[67]

Only "truth in advertising" could avoid the disciplinary rendition of this (unconventional) family romance: when proofs were nowhere in sight, practitioners ought to label their wares "theoretical," and both "theoretical" and "experimental," proof-governed mathematics ought be credited by the community.

Reaction to Jaffe and Quinn was swift, some of it laudatory and some condemnatory. In a flurry of email to the journal in early 1994, some of the most senior mathematicians and physicists responded. Michael Atiyah, a senior mathematician and master of Trinity College, took up the question of values. He pointed out that this admixture of algebraic geometry and string theory (as opposed to previous ones) engaged "front-line ideas" in both areas. This was no sideshow, setting foundational questions straight for already-accepted physical theory. No, these new developments "might well come to dominate the mathematics of the 21st century." Still, he castigated Jaffe and Quinn for emphasizing the danger to mathematicians of being "led astray": "I think most geometers find this attitude a little patronizing: we feel we are perfectly capable of defending our virtue."[68] Others were less sure of the mathematicians' capacity to defend their reputations, especially when it came to "the young" and "the impressionable students."

Throughout their piece, Jaffe and Quinn came back again and again to the necessity of imparting *values* as part of the educational process. At the first level, this meant physicists had to accord mathematics the same respect, the same necessity that experiment had previously: "students in physics are generally indoctrinated with anti-mathematical notions; and if they become involved in mathematical questions, they tend not only to be theoretical but often to deny that their work is incomplete."[69] Mathematicians, in turn, had to ward off the undue temptations that the purely speculative would have for their young. Jaffe and Quinn: "We suggest it is a dangerous thing for a student who does not understand the special character of speculative work to set out to emulate it."[70] The consequences of ignoring these dangers were catastrophic: "theoretical [mathematical] work" could lack feedback; further work could be discouraged and confused because unreliable; "dead areas" could be created when all credit fell to theorizers; and more generally, the young could be "misled."[71]

Sounding more like sociologists of science than the chairs of two major math departments, Jaffe and Quinn underscored the inseparability of values and cognitive content:

> Mathematicians tend to focus on intellectual content and neglect the importance of social issues and the community. But we are a community and often form opinions even on technical issues by social interactions rather than directly from the literature. Socially accepted conventions are vital in our understanding of what we read. Behavior is important, and the community of mathematicians is vulnerable to damage from inappropriate behavior.[72]

No one was more sympathetic to Jaffe and Quinn's moral crusade to guard against such "damage" than Steven Krantz, a fiercely harsh critic of Yale's Benoit Mandelbrot and his fractal enthusiasts. In an email shot to Jaffe on 16 November 1992—before the article had even officially appeared in print, he recalled an incident from his own early days in mathematics: "When I was just starting out in this profession Richard Arens came up to me at a party and said 'Young man, do you want to be famous?' Of course I said 'Yes.' He replied, 'Well, then, go [mess] up a subject.' Truer words were never spoken." His rage at what he perceived as the mathematical sloppiness of some physically oriented geometers was redoubled precisely because of the fame it had brought them. "The people who do fractal geometry," he continued, "are not just confused. They are amoral. If you could get one of them to sit still long enough to read your article, they would not have a clue what you are talking about. Their speculation is uninformed, pointless, and has contributed nothing to the pool of scientific knowledge. Much of it is poached."[73] Even the softened version of Krantz's objections that Jaffe and Quinn cited in their "two cultures" piece relit the fire. Mandelbrot promptly retorted to Jaffe and Quinn that he found their unasked-for labeling of some mathematicians as "theoretical" to be a kind of "police state within Charles mathematics [mathematics centered around the Charles Street (Providence, Rhode Island) headquarters of the AMS] and a world cop beyond its borders."[74]

Here then was the bottom-line issue: a struggle over the very meaning of theory that was inseparably about technicalities, rigor, credit, pedagogy, and professional identity. Saunders Mac Lane, one of the deans of algebraic topology, retired from the University of Chicago, vehemently protested the notion of theory proposed by Jaffe and Quinn:

[Jaffe's and Quinn's] comparison of proofs in mathematics with experiments in physics is clearly faulty. Experiments may check up on a theory, but they may not be final; they depend on instrumentation, and they may even be fudged. The proof that there are infinitely many primes—and also in suitable infinite progressions—is always there. We need not sell mathematics short, even to please the ghost of Feynman.

Since World War II, he contended, physics had played the dominant role in American science—but the discipline was itself now in trouble. Mathematicians need not, ought not, pine after the methods of this crepuscular science. "Mathematics does not need to copy the style of experimental physics. Mathematics rests on proof—and proof is eternal."[75]

Tied to pedagogy, credit, and epistemic security, ultimately the culture war over theory had consequences for "reality," as Morris W. Hirsch, the Berkeley algebraist and differential geometer, made clear. His claim was that however much Jaffe and Quinn protested that they wanted to eschew terminology per se, they still wrongly spoke about "mathematical reality": "It is important to note at the outset that their use of 'theoretical' is tied to a controversial philosophical position: that mathematics is about the 'nature of reality,' later qualified as 'mathematical reality,' apparently distinct from 'physical reality.' They suggest 'Mathematicians may have even better access to mathematical reality than the laboratory sciences have to physical reality.' " On Hirsch's view, mathematics was not a theory of anything, and certainly not of a special branch of reality. Neither poems nor novels "referred," and mathematics was no different. True enough, Hirsch said, physicists used mathematics as a tool with which to construct "narratives" (his term). They might be telling a story about how a certain system worked and would use the concepts of mathematics to continue, as for example they characteristically would do when invoking the assumption of equilibrium, nonzero determinates, or the independence of random variables. But the uses of mathematics (on Hirsch's view) spoke not a word about the intrinsic referentiality of mathematics itself, and if, as he believed, "mathematical reality" was idle chatter, then the very category "theoretical mathematics" ought cede to the ontologically neutral one: speculative mathematics.[76] For many, including James Glimm from Stony Brook, more was at stake in this struggle than the referential structure of mathematics:

> It bears repeating that the correct standards for interdisciplinary work consist not of the intersection, but the union of the standards from the two disciplines. Specifically, speculative theoretical reasoning in physics is usually strongly constrained by experimental data. If mathematics is going to contemplate a serious expansion in the amount of

speculation [it will] have a serious and complementary need for the admission of new objective sources of data, going beyond rigorously proven theorems, and including computer experiments, laboratory experiments and field data. [The] absolute standard of logically correct reasoning was developed and tested in the crucible of history.

Such standards, Glimm concluded, had to be preserved and defended in the rapid expansion of mathematical horizons.[77]

On the cultures of theory

String theorists, prominently among them Edward Witten, took their history from the young Einstein—the Einstein who constructed general relativity with the barest of experimental ties, such as the precession of the perihelion of Mercury. Georgi and Glashow, arguing against string theory, called to the stand a different historical Einstein—the aging hermit chasing after the illusion of unification, self-blinded to the worlds of Lagrangian quantum field theory, meson exchange, and new particles.

Mathematicians and mathematical physicists including Quinn, Jaffe, Mac Lane, Glimm, and Atiyah also battled over histories—the collapse of a brilliant but in some mathematicians' view too speculative Italian algebraic geometry versus the path-breaking formulae of Euler's divergent series or Ramanujan's number-theoretical insights. These wars over the past were tightly coded interventions aimed more at shaping the future than on chronicling the past. Would physics students learn about cross sections, Lagrangians, and particle lifetimes? Or would they train in Calabi-Yau manifolds, knot theory, and topological invariants? Would mathematicians learn to think in terms of physics categories like vertex operators, conformal field theories, and Feynman integrals? Or would they guard the proof structure of Bourbakian morality, "family values," and disciplinary traditions? These prophesies and evaluations were not "outside" the creolized physicomathematics of strings—they helped, in no small measure, to constitute it. Through shared institutions and intuitions, mathematicians and physicists are constructing a conjoint field of inquiry, whether one calls this trading zone "quantum geometry" or situates it within string theory or algebraic geometry. Joint appointments, common conferences, research collaborations, and training methods all have created a substantive region of overlap in practices and values. From the perspective of this account, it is perhaps not surprising that by 1999 the mathematics department at Columbia began demanding that its graduate students take a course in quantum field theory. Even ten years earlier, such a requirement would have been unthinkable.

By creating such a substantial border region, more than results have shifted. As the "contrarians" rightly noted, these alterations signaled a shift or perhaps an expansion in the meaning of theoretical physics and algebraic geometry. And with these changes, what it means to be a geometer or a theoretical physicist altered as well. The next generation of mathematicians and physicists would know a different world from their elders: in addition to physical sites (such as the Institute for Advanced Study at Princeton), they would also frequent virtual places, such as the joint mathematics and physics "Calabi-Yau Home Page" that, by the end of the twentieth century were already joining activities and techniques in the border zone. Trained differently from physicists in the 1970s or 1980s, working to different standards with different tools, it became possible for a young investigator to say: "I don't know whether I am doing physics or mathematics," an utterance either unthinkable or unacceptable even a few years earlier. With the new sense of theoretical physicist and geometer came also a new object of inquiry: in its present form not quite mathematical and not quite physical, either. One day pieces of such entities may be folded back into physics or into geometry, but at century's end they were *conceptual objects*, hugely productive and yet seen with discomfort by purists in both camps.

In the late twentieth century, understanding the shifting cultures of theory was not just of abstract significance. It was a problem of urgent concern to leading physicists and mathematicians; these debates would shape central features of their disciplines into the next century. Because this dynamic so thoroughly mixed constitutive values with the disciplinary identity of the practitioner and the allowed objects of inquiry, the story of this hybridized mathematics and physics raises equally pressing concerns for historians, sociologists, and philosophers of science. Our histories and our shifting present are always reconstituting the triad: persons, values, and objects.

Notes

1   On trading zones as sites of local coordination between different scientific and technological languages, see Peter Galison, *Image and Logic: A Material Culture of Microphysics* (Chicago: University of Chicago Press, 1997).

2   Antonio Zichichi (ed.), *The Superworld I* (New York: Plenum Publishing, 1990), 237.

3   Galison, *Image and Logic*, chapter 6; Galison, "Pure and Hybrid Detectors: Mark I and the Psi," in Lillian Hoddeson, Laurie Brown, Michael Riordan, and Max Dresden (eds.), *The Rise of the Standard Model* (Cambridge: Cambridge University Press, 1997), 308–337.

4   David Gross, "Superstrings and Unification," in R. Kotthaus and J. H. Kühn (eds.), *Proceedings of the XXIV International Conference on High Energy Physics: Munich, Federal Republic of Germany, August 4–10, 1988* (Berlin: Springer-Verlag, 1989), 328.

5   Zichichi, *Superworld I*, 235.

6   Ibid.

7   Ibid.

8   Philip W. Anderson, "More Is Different: Broken Symmetry and the Nature of the Hier-
    archical Structure of Science," *Science* 177 (1972): 393–396; reprinted in Anderson, *A
    Career in Theoretical Physics* (Singapore: World Scientific, 1994), 1–4. For a good discussion
    of various condensed matter theorists' views on unity and disunity in physics, see Jordi
    Cat, "The Physicists' Debates on Unification in Physics at the End of the 20th Century,"
    Historical Studies in the Physical and Biological Sciences 28 (1998): 253–299.

9   P. J. Steinhardt (ed.), *1983 Fourth Workshop on Grand Unification* (Boston; Birkhäuser Bos-
    ton, 1983), 3.

10 Paul Ginsparg and Sheldon Glashow, "Desperately Seeking Superstrings?" *Physics Today*
    39 (1986): 7.

11 Glashow, "Closing Lecture: The Great LEP Forward." in Antonio Zichichi (ed.), *The
    Superworld II* (New York: Plenum Press, 1990), 540.

12 Howard Georgi, "Effective Field Theories," in Paul Davies (ed.), *The New Physics* (Cam-
    bridge: Cambridge University Press, 1989), 446.

13 Ibid., 452.

14 Michael Green and John Schwarz, *Physics Letters* B 149 (1984): 117–122.

15 Gross, J. Harvey, E. Martinec, and R. Rohm, "Heterotic String Theory," *Nuclear Physics* B
    256 (1985): 253–284; P. Candelas, G. T. Horowitz, A. Strominger, and E. Witten, "Vac-
    uum Configurations for Superstrings," *Nuclear Physics* B 258 (1985): 46–74.

16 Candelas et al., "Vacuum Configuration for Superstrings," 46–74.

17 Shing-Tung Yau interview, 20 August 1998; Strominger interview, 2 April 1999.

18 Candelas et al., "Vacuum Configurations for Superstrings," 46–74.

19 David Morrison interview.

20 A simpler, visualizable analogue to projective space can be imagined if we start with R3,
    real 3-dimensional space. Then RP2, the real projective 2-space, is defined by identifying
    (x_1, x_2, x_3) with $\lambda(x_1, x_2, x_3)$ where λ is a real number. The elements of this real projective
    2-space—the lines through the origin—can be represented by the intersection of such
    lines with a real 2-sphere around the origin. Each such pair of points (located opposite
    each other on the sphere) picks out a single element of the space, namely a line. The goal
    in shifting attention to projective space is that now many limiting processes contain the
    asymptotic limit within the space rather than convergent to a point *outside* the space. That
    is, the space is *compact*. If x,y has a point (1,3) in it, then (2,6) is too, as is (x, 3x) in general
    for any real number x. So suppose we start with (1,x), a point that looks like it will head
    off to (1,•) for x growing indefinitely. By the rule that scaling does not change the
    identity of the point, we can always identify (1,x) with $1/x(1,x) = (1/x, 1)$. And 1/x, 1
    behaves just fine as x gets really large. It is this feature of containing its limit—being
    compact—that makes projective space so useful to mathematicians.

21 D. Gepner, "Exactly Solvable String Compactifications on Manifolds of SU(N) Holon-
    omy," *Physics Letters* B 199 (1987): 380–388; and "Space-Time Supersymmetry in Compac-
    tified String Theory and Superconformal Models," *Nuclear Physics* B 296 (1988): 757–778.
    Gepner's conjecture was later proven by B. Greene, C. Vafa, and N. P. Warner, "Calabi-
    Yau Manifolds and Renormalization Group Flows," *Nuclear Physics* B 324 (1989): 371–
    390.

22 W. Lerche, C. Vafa, and N. P. Warner, "Chiral Rings in N = 2 Superconformal Theories,"
    Nuclear Physics B 324 (1989): 427–474; L. Dixon, lectures at the 1987 ICTP Summer
    Workshop in High Energy Physics and Cosmology; see also "Some World-sheet Proper-
    ties of Superstring Compactifications, On Orbifolds and Otherwise," in G. Furlan et
    al. (eds.), *Superstrings, Unified Theories, and Cosmology 1987* (Singapore: World Scientific,
    1988), 67–126.

23 Email message, Rolf Schimmrigk to author, 17 March 1999.

24 Candelas interview, 29 May 1998.

25 Ibid.

26 B. R. Greene and M. R. Plesser, "Duality in Calabi-Yau Moduli Space," HUTP-89/A043, printed as Nuclear Physics B 338 (1990): 15–37.

27 Candelas interview, 29 May 1998; also Candelas, M. Lynker, and R. Schimmrigk, "Calabi-Yau Manifolds in Weighted P_4," Nuclear Physics B 341 (1990): 383–402.

28 Candelas interview, 29 May 1998; also Yau interview, 20 August 1998.

29 Serge Lang, Introduction to Algebraic Topology (Reading, Mass.: Addison-Wesley, 1972), v.

30 Both citations from Herbert Clemens interview, 31 October 1998.

31 H. Clemens, Some Results on Abel-Jacobi Mappings, in Topics in Transcendental Algebraic Geometry (Princeton, N.J.: Princeton University Press, 1984).

32 Sheldon Katz interview, 30 October 1998.

33 See, e.g., Phillip Griffiths and Joseph Harris, Principles of Algebraic Geometry (New York: John Wiley, 1978, 1994), 480–489; Steven L. Kleiman, "Problem 15. Rigorous Foundation of Schubert's Enumerative Calculus," in Proceedings of the Symposium of Pure Mathematics 28 (1976), part II, 445–482. Cf. Hermann Schubert, Kalkuel der Abzaehlenden Geometrie (Leipzig: B. G. Teubner, 1879).

34 S. Katz, "On the Finiteness of Rational Curves on Quintic Threefolds," Composition Mathematics 60 (1986): 151–162; For n_3 see J. Harris, "Galois Groups of Enumerative Problems," Duke Mathematical Journal 46 (1979): 685–724.

35 Sheldon Katz interview, 30 October 1998.

36 P. Candelas, X. C. de la Ossa, P. S. Green, and L. Parkes, "An Exactly Soluble Superconformal Theory from a Mirror Pair of Calabi-Yau Manifolds," Physics Letters B 258 (1991): 118–126, hereafter cited as COGP-1; for the first ten n_k, see ibid., "A Pair of Calabi-Yau Manifolds as an Exactly Soluble Superconformal Theory," in Shing-Tung Yau (ed.), Essays on Mirror Manifolds (Hong Kong: International Press, 1992), 31–92.

37 The degree of a map is different from the degree of a curve. Because of this, extracting the number of fixed-degree curves from string theory computations is quite difficult. For example, the multiple winding of string world-sheets contributes multiple covers of lower degree curves—and these count in the enumeration of higher-degree maps.

38 Why is the complex structure manifold not quantum corrected? In string theory there is only one free parameter, α', the string tension. So a string expansion on the complex side would necessarily involve α'/R^2, the R^2 entering to make the expansion parameter dimensionless. But R is a parameter that changes the size of the Calabi-Yau manifold, and supersymmetry guarantees that the parameters that change size and the parameters that change shape will not mix—so R cannot occur on the shape-changing side of the equation. Therefore, no string corrections exist for the shape-changing (complex structure) side of the mirror pair.

39 Stein Arild Strømme, email message to Sheldon Katz, 6 June 1990.

40 Yau, "Introduction," Essays on Mirror Manifolds, iv.

41 B. R. Greene and M. R. Plesser, "An Introduction to Mirror Manifolds," in Yau, Essays on Mirror Manifolds, 2–3.

42 P. S. Aspinwall and C. A. Luetken, "A New Geometry from Superstring Theory," in Yau, Essays on Mirror Manifolds, 318.

43 Candelas interview, 29 May 1998.

44 Ibid.

45 David R. Morrison, "Mirror Symmetry and Rational Curves on Quintic Threefolds: A Guide for Mathematicians," DUK-M-91-01; July 1991, alg-geom/9202004 10 Feb 92.

Note that this citation can also be found in Xenia de la Ossa, "Quantum Calabi-Yau Manifolds and Mirror Symmetry" (Ph.D. diss., University of Texas, Austin, 1990), 2.

46   Morrison, "Mirror Symmetry," 13.

47   Ibid.

48   Ibid., 18.

49   Ibid., 17.

50   Geir Ellingsrud, email message to Joe Harris, n.d. [probably 31 July 1991].

51   Email message, Harris to Ellingsrud, 1 August 1991, 18:07 PDT.

52   B. Greene and S.-T. Yau (eds). *Mirror Symmetry II* (Providence, R.I.: American Mathematical Society, International Press, 1997).

53   Morrison interview, 4 November 1998.

54   See, e.g., B. Greene and H. Ooguri, "Geometry and Quantum Field Theory," in Greene and Yau, *Mirror Symmetry II*, 26.

55   Some relevant papers by Witten include "Quantum Field Theory and the Jones Polynomial," *Communications in Mathematical Physics* 121 (1989): 351–399; "Fional Gauge Theories Revisited," *Journal of Geometry and Physics* 9 (1992): 303–368; "Supersymmetry and Morse Theory," *Journal of Differential Geometry* 17 (1982): 661–92.

56   E. Witten and P. Deligne, "Mathematical Sciences: An Integrated Program in Mathematics and Theoretical Physics," NSF-DMS 9505939, submitted 18 November 1994.

57   Ibid.

58   Referee reports for Witten and Deligne, "Mathematical Sciences."

59   Ibid.

60   E. Witten and P. Deligne, "An Interdisciplinary Program in Mathematics and Theoretical Physics," NSF DMS 96-27351.

61   Robert MacPherson, 15 December 1995, http://www.math.ias.edu/QFT/fall/letter.txt.

62   Arthur Jaffe and Frank Quinn, " 'Theoretical Mathematics': Toward a Cultural Synthesis of Mathematics and Theoretical Physics," *Bulletin of the American Mathematical Society* 29(1) (1993): 3.

63   Ibid., 2.

64   Ibid.

65   Ibid., 5.

66   Morrison, "Making Enumerative Predictions By Means of Mirror Symmetry," in Greene and Yau, *Mirror Symmetry II*, 457.

67   Jaffe and Quinn, " 'Theoretical Mathematics,' " 4.

68   Michael Atiyah, *Bulletin of the American Mathematical Society* 30 (1994): 179.

69   Jaffe and Quinn, " 'Theoretical Mathematics,' " 5.

70   A. Jaffe and F. Quinn, "Response to Comments on 'Theoretical Mathematics,' " *Bulletin of the American Mathematical Society* 30 (1994): 209.

71   Jaffe and Quinn, " 'Theoretical Mathematics,' " 9.

72   Ibid., 10.

73   Steven G. Krantz, email to A. Jaffe, 16 November 1992, 15:06 CST.

74   Benoit Mandelbrot, *Bulletin of the American Mathematical Society* 30 (1994): 193–194.

75   Saunders Mac Lane, *Bulletin of the American Mathematical Society* 30 (1994): 193. Armand Borel of the IAS also demurred from the Jaffe-Quinn category of theoretical mathematics, though on the grounds that experiment ought to correspond to special cases, reserving theory for the purely mathematical notion of general theorems; see *Bulletin of the American Mathematical Society* 30 (1994): 179–180.

76   Morris Hirsch, *Bulletin of the American Mathematical Society* 30 (1994): 186.

77   James Glimm, *Bulletin of the American Mathematical Society* 30 (1994): 184.

Nonlinear Dynamics and Chaos

2 Chaos, disorder, and mixing: a new fin-de-siècle image of science?

Amy Dahan Dalmedico

Chaos and the sciences of disorder or mixing concern particular disciplinary sectors (e.g., mathematics, fluid mechanics, physics, and engineering science), but they also form disciplinary crossroads, which symbolize a new mode of conceiving and practicing science. Moreover, they go hand in hand with a discourse proclaiming a new episteme or a new paradigm—the idea of a third scientific revolution has even been mentioned. To characterize this new area of science is a delicate matter since it involves both specific scientific practices and theories and diffuse representations of them among audiences more remote from practice. Such a representation of science is always partial; even when it has a hegemonic ambition, it competes with other representations, built by other scientific subgroups, who regard it as partly ideological. But since my aim is to reflect on a certain *air du temps* in contemporary science rather than to establish a Kuhnian paradigm or a Foucauldian episteme, I will deliberately use somewhat vague terms such as *images* and *representations* of science.

A few historical remarks on representations that epitomize an air du temps may be useful. Our historiographical tradition has constructed an image of seventeenth-century science as being about a world run by clockwork, whose harmony was mathematical. Nature was an automaton exhibiting repetitive phenomena ordained by a God who was likened to an architect (or a clockmaker). Several recent historical studies, however, have shown that this representation did not exclude others coexisting with it, even within one individual (Newton for example).[1] Moreover, this representation differed significantly for different individuals and groups, for example, between continental mechanical philosophers (Descartes, Pascal, or Huygens) and English natural philosophers (Boyle and Newton).[2] Perforce, it varied considerably among groups with more remote interests and practices: the astronomers of the Academy of Sciences in Paris and the Royal Society in London had little in common with Baconian naturalists or the practitioners who experimented with thermometers and barometers or tested the stiffness of wood and glass.[3] In spite of this heterogeneous variety, however, the mechanical image retains some validity; it no doubt conveys something essential about the seventeenth century.

As a second example, consider the common representation of Laplacian science at the end of the eighteenth century. In a recent study, I have shown how Laplace's vision has been demonized and turned into a Manichean opposition between a Laplacian totalitarian universe and the freedom authorized by contemporary theories of chaos. In fact, Laplace merely presupposed in principle the possibility of total prescience and complete predictability, but he was perfectly conscious that their practical possibility was out of reach. For this precise reason, he judged it necessary to forge the tools of probability theory, allowing for a statistical description of some processes. We can, nevertheless, claim once again that Laplacian determinism expressed a "general conviction of the time," that is, a belief in the causal structure of the world, a faith in the mathematical intelligibility of laws which the scientist has to discover or to approach as nearly as possible.[4]

In both of these cases, we meet constructed but incomplete images expressing some important characteristic of the air du temps.[5] Let me now summarize some key elements that characterize the representation of contemporary science:

> —The logic of scientific reasoning, that is, the relation between causes and effects, has changed: in brief, "the world is nonlinear," and a small detail can bring forth a catastrophe.
> —Complexity was formerly believed to be decomposable into elementary units, yet only a global approach is pertinent.
> —Sand heaps and water droplets reveal the profound, physical nature of matter's behavior, which cannot be revealed by atomic physics.
> —Reductionism has reached its limits, and its end has been announced for the near future.
> —The scientific method associated with Galileo and Newton and once believed immutable has now been supplanted by a historical point of view which has become dominant in the geosciences and in the life and social sciences.

In the representation following from the above theses, the hierarchy of the disciplines has been upset. In particular, mathematics and theoretical physics have lost their dominant position, and the biological and earth sciences have been promoted in the new pantheon of science.

Within the confines of this article, it will not be possible to discuss in detail either the many texts that have adopted these theses (and sometimes loosely extended them to economic, political, or social domains) nor the history of their development over the past twenty or thirty years (even fifty years in the case of cybernetics). Instead, I will review a few recurrent themes constituting this new representation of science and, for each, examine what is at stake—in general, to show that everything is more complicated and subtle than is often supposed. Since the themes are intimately linked, to separate them as if their evolution were independent from one another is rather artificial, but it will enhance the intelligibility of the text. Finally, concerning intelligibility, we should recognize that the subtlety of the themes lies in the fact that most of the expressions involved, such as determinism, predictability, randomness, and order, assume a precise meaning only in the framework of a mathematical formalism. But in this remark, far from aiming at discouraging unavoidable, and probably desirable, nonformal discussions, I intend to underscore the difficulties of conducting a wide debate on the issues without exhibiting an excessive scientistic authoritarianism.[6]

Order and disorder

When people have talked about the science of disorder (revenge of the God Chaos), it has been to underscore an opposition to classical science, construed as the discovery of order and regularities in nature and the restitution of the world's transparent intelligibility beyond confused, opaque appearances. In this opposition, two aspects may be distinguished, which I will treat in succession: (1) nonlinearity and the butterfly effect, and (2) determinism and randomness.

Nonlinearity and the butterfly effect

Associated with the classical conception of science was the principle of proportionality of causes and effects, which played a paradigmatic role for (linear) causality. In several seventeenth-century propositions, the idea was explicit: for example, Newton's law of cooling, Hooke's law for springs, and Mariotte's (and Boyle's) pressure-volume law. In mechanics, Varignon invoked this principle to argue in favor of a law of proportionality between force and velocity, on which he wished to rely. In the eighteenth century, the same principle underlay the controversy about the foundations of mechanics, involving Euler, d'Alembert, and Maupertuis, among others. At the turn of the twentieth century, although scientists were by then beyond the proportionality of causes and effects, the linear harmonic oscillator remained an omnipresent model in the domain of electromagnetism.

And the founders of quantum mechanics came back to it whenever they faced important difficulties, as evidenced by the Bohr-Kramer-Slater theory in 1924.[7]

Now one often hears, "The world is nonlinear." This affirmation is older than is often said. As early as the 1930s, for Soviet mathematicians and physicists at Moscow, Kiev, and Gorki (Mandelstam, Andronov, Bogoliubov, Krylov, Pontryagin, among others), the world was already nonlinear, and what they then called "self-oscillations" (or self-sustained oscillations) became a universal model replacing the harmonic oscillator. More generally, conceptual tools popularized by the "sciences of disorder," whether mathematical (dynamical systems theory), or information-theoretical (Shannon and Kolmogorov), are rather old.[8]

What is relatively recent and has stamped the development we are witnessing is the following:

> —A new credo and interest not so much for disorder, but rather for a renewed appreciation of very particular, subtle combinations of order and disorder, which lends specificity to the science of chaos.
> —A different technological environment, which concerns both tools and social demands. New tools have appeared: above all the computer (an omnipresent tool for calculation, algorithms, images, construction of attractors, simulations, etc.) but also lasers and ultrasound devices to probe matter, its defects, irregularities, and asymmetries, and to exhibit transitions to turbulence (for example, in the experiments of Gollub and Swinney in the 1970s). At the same time there has been a strong social demand for new technologies and materials, and for expertise on problems of vibration and stability, transmission, signal theory, and many others.

Ever since the nineteenth century, disorder and its quantification via entropy have attracted attention, but this was statistical disorder, which, using the law of large numbers, can produce average values that obey simple laws. Although results obtained by Poincaré and Birkhoff, among others, showed that this opposition between order and disorder was not so clear-cut, until the late 1950s only statistical methods seemed adequate to deal with "disordered" phenomena. These were the methods privileged by physics, meteorology, hydrodynamics, and mechanics. A significant departure from these methods is the numerical experiment on nuclear chain reactions started by Fermi, Pasta, and Ulam at Los Alamos in the 1940s and picked up again at the beginning of the 1950s. While ergodicity (which was expected) would have justified a statistical treatment, the semiorder revealed by the experiment resisted any mathematical formulation and remained very hard to interpret.[9]

The same happened when Kolmogorov addressed the International Congress of Mathematicians in 1954 in Amsterdam and introduced what would soon be famously known as the Kolmogorov-Arnold-Möser (KAM) theorem. At least for mechanics and probably elsewhere too, this theorem helped to change the conception of the order/disorder relation. The theorem showed that order was much more powerful than usually assumed, that it resisted perturbations, and that disorder, conceived in terms of ergodicity, contradicted the stability observed in several physical systems. In his work on celestial mechanics, Poincaré had exhibited the homoclinic tangle, a mesh of intertwined curves picturing the "chaotic" nature of a multitude of possible solutions and the infinity of allowed scenarios. Afterward, mathematicians concluded that no solution to the problem of perturbed Hamiltonian systems was a smooth curve, or that stable orbits did not exist. But this was shown to be erroneous. When perturbations are sufficiently small, the KAM theorem shows that a majority of orbits are stable; although nonperiodic, they never stray far from the periodic orbits of the unperturbed system. Such orbits are said to be quasiperiodic. Others are chaotic and unpredictable. Still others are caught in islets of stability in an ocean of chaos. Thus it could be said that the "KAM theorem was order without law"![10] For that matter, the methods of the KAM theorem have become "paradigmatic," and are today applied in contexts far from their origins.[11] The succession of names given to this domain indicate the successive understandings of order, disorder, and law. While at first the terms *disorder* and *chaos* were used almost interchangeably,[12] by the end of the 1970s the habit developed of associating order *and* chaos to designate the domain. In 1986, the publication of Bergé, Pomeau, and Vidal's *L'Ordre dans le chaos* exhibited, at least in France and in the scientific community, the desire of specialists to distance themselves from confusions about the meaning of chaos evident among a few intellectuals and writers.[13] Henceforth, the terminology *deterministic chaos* was established.

Concerning the question of causality, in the 1950s the theme of feedback and circular causality became very important, especially as a result of Wiener's work (much preoccupied with holistic conceptions, where interactions among a whole set of elements were present) and the first wave of cybernetics. The first conference on cybernetics was titled: "Conference on Circular Causal and Feed-Back Mechanisms in Biological and Social Systems." At that time, simple-return action was already familiar to control engineers and was later developed in automatism, but cyberneticists envisioned more, namely, a retroactive action on the very model of action.

From this point of view, the work of meteorologist Edward Lorenz in the 1960s constituted a crucial moment. If indeed sensitivity to initial conditions and the existence of systems whose trajectories diverge exponentially were already known to Poincaré and Hadamard, the impact of Lorenz's work would be crucial on two counts. On the one hand, Lorenz discovered that, to give rise to chaotic behavior, it sufficed to have only three variables; that is, a nonlinear system with three degrees of freedom—a very simple formal system indeed—could exhibit a very complex dynamical behavior. This result would profoundly upset the usual understanding of the relation between the simple and the complex. On the other hand, the intervention of the computer in Lorenz's work is in itself crucial at two levels: (1) the sensitivity to initial conditions—later called the butterfly effect—was revealed through numerical instability; (2) on his computer screen Lorenz also exhibited a surprising image of the "Lorenz attractor," which he succeeded in representing graphically as a two-dimensional projection. While Poincaré had long before, in a few complicated sentences, imagined and described the homoclinic tangle, Lorenz could explain his construction by means of iterative processes and, moreover, could visualize it.[14]

Some fifteen years—roughly from 1963 to 1976—would be needed for these results to be assimilated by various groups of scientists, from meteorologists to mathematicians, from astronomers to physicists and population biologists.[15] Through many kinds of numerical explorations, each group could then reclaim the domain as its own: logistic equations, Hénon's mapping, Metropolis, Stein, and Stein's iterations, Feigenbaum's bifurcations, and so on. Initially, this exploration was not performed on computers, but above all on simple calculators, which were much more common than computers in the late 1970s. Interestingly, in their attempt to counter the craze for chaos by underscoring the rather old age of most fundamental results, pure mathematicians specializing in the abstract topological methods of dynamical systems and focusing on their great classification programs often missed this striking new computational dimension.

This explains the great a posteriori theoretical importance of demonstrating the "shadowing lemma." When the trajectory of a dynamical system is computed for a given initial position, numerical instabilities make it probable that this trajectory is false due to the noise resulting from an accumulation of rounding errors. The shadowing lemma, however, ensures that there exists a "true" trajectory—that is, an "exact" trajectory corresponding to an initial position close to the original—which follows the computed trajectory with any desired degree of precision. This means that the accumulation of rounding errors is counterbalanced by a simple translation of the initial condition. Thus, the property of structural stability legitimates the computation of trajectories for dynamical systems: computers indeed plot "real" trajectories. For some fifteen years, however, a few mathematicians had done away with this justification in order to explore chaotic systems numerically.[16] One may add that the mathematical properties of the Lorenz attractor were only established in 1998, according to the discipline's traditional requirements for rigor. This underscores the specific, experimental character of the science of chaos, in which mathematics partly intervenes in an a posteriori manner, that is, to justify methods, results, or processes explored by means of numerical simulations on the computer.[17]

The new articulation of randomness and nonrandomness and speculations about determinism

It must be recognized at the outset that, at least in France, the status of randomness and probability theory has considerably changed in the course of the last decades. Until the 1970s, probability theory sat at the bottom of the hierarchy of mathematical branches. Outside pure mathematics, it was not even judged useful enough to be taught in preparatory classes. Even though the situation was not the same everywhere and even though statistics and probability received considerable development in the twentieth century, the field often remained outside the official boundary inscribed in the mathematical community's institutions.[18] Only Kolmogorov's work in the Soviet Union, as early as the 1930s, unfolded over two domains: mathematics and mechanics on the one hand, and probability and information theories on the other, with understanding the essence of order and chaos as his supreme objective. In its ambition—but not in spirit, which remained radically different—this monumental program can only be compared to Hilbert's and to the Bourbakist movement.

Characteristic of this domain is a blurring of borders—between order and disorder, as we have already seen, but also between randomness and nonrandomness, between what is deterministic and what is not. In an experimental system, even one with a small number of degrees of freedom, such as Lorenz's model, when the parameters are unknown, one may be facing either a chaotic dynamical system for some values of the parameters, or a stochastic perturbation, a noise, for other values of the parameters. A choice is allowed between two models, one that is deterministic and the other stochastic. Taken by assuming an a priori deterministic system, measurements themselves will enable the observer to distinguish deterministic systems from a whole set of systems, not all of which may be deterministic. Moreover, even if deterministic, a system about which information is partial may appear stochastic. The baker's transformation whose initial condition is "hidden" exemplifies this situation. Lastly, the fact that, following Kolmogorov's work in particular, an algorithmic definition of randomness is now available has sensibly changed the understanding of the notion of randomness and its mathematical handling through computer-generated processes (random sequences of numbers, random walks, etc.). To say that the trajectory of a system is random amounts to saying that it cannot be defined by an algorithm simpler than the one continuously going over the trajectory and that information concerning this trajectory cannot be compressed (by means of a formula, a process, etc.).

In a chaotic dynamical system, after a long evolution with respect to its Lyapunov time—which locally measures an exponential divergence of trajectories—there is a loss of memory about the initial state. Consequently, the trajectory no longer seems a pertinent idealization. One needs a statistical description of a probabilistic type that is incompatible with the notion of a trajectory. This is what Sinai has termed the "randomness of the nonrandom" while showing that a true dialectics is established between the instability of a chaotic dynamical system and its structural stability (in this case, one is on an attractor that exhibits a large degree of structural stability). Sensitivity to initial conditions, memory loss, and algorithmic complexity become organically linked with each other and refer back to local instability. This is a leitmotiv in the work of Prigogine especially, who claims that dynamical chaotic systems should be described in terms of probability distributions. While for trajectories the Lyapunov time is indeed an element of instability, in terms of probability distributions it becomes an element of stability: the larger the Lyapunov time, the faster the damping and convergence to uniformity. On this view, irreversibility is due to the very formulation of unstable dynamical systems' dynamics, and one has: instability (or chaos) ⇒ probability distributions ⇒ irreversibility. For Prigogine, probability distributions and irreversibility are intimately linked with one another and therefore require the arrow of time.

In the 1980s, deterministic chaos and this new dialectics between randomness and nonrandomness brought back fantastic speculations about determinism. The need to revert to statistical methods for the study of deterministic chaotic systems raised a profound question: Did the description in terms of probability distributions have an intrinsic character, or did it derive principally from our ignorance? Symbolized by the figures of René Thom and Ilya Prigogine, a polemic arose—slightly losing steam today—which showed that the problem belonged to metaphysics: nature's global, ontological determinism is neither falsifiable nor provable but that does not imply that the philosophical options have no consequence for heuristic choices. On this matter, Sinai's demonstration (at the end of the 1960s) of the mixing character of plane billiards (i.e., the stochastic character of a billiard ball's behavior) bestowed legitimacy on the application of deterministic chaos to every conception of the physical universe. A few specialists, such as Ruelle and Ekeland, even believed they could spot in deterministic chaos a scientific, existential virtue for reconciling human free will with determinism.[19] Again, the tension emerges between the precision implied by the use of specific terminology and extreme idealizations.

Is everything chaos?

By raising this question, we also touch upon the legitimacy of analogies derived from chaos theory and their audacious (or aggressive) transfer to other fields. As understood by practitioners, chaos theory is restricted to a specific framework: the theory of deterministic (generally nonlinear) dynamical systems whose behaviors are studied in phase space with particular attention to final states and their global aspects (hence the attempts at classifying final states and attractors). A priori, the theory does not deal with anything outside this frame. This offers the advantage of abstraction since the theory gives no information about nor any interpretation of phase space coordinates. As a consequence, the universality of the theory is greatly enhanced. But this hardly means that everything is chaos or that the theory is everywhere applicable.

The quantitative study of chaos in a system requires the quantitative understanding of its dynamics; if time-evolution equations are well known, they can be integrated on the computer, which is the case for solar-system astronomy, hydrodynamics, and even meteorology. In the case of oscillating chemical reactions, evolution equations are not known, but precise recording of experimental data over a long time period, called time series, can be obtained and, since they are rather simple, they can be used to reconstruct dynamical equations. But this is not true in other areas. There is indeed an important problem: When one observes a physical, econometric, or biological system whose differential equations are not known and only measured information is available, what can be done? Sometimes even the number of variables in the system is unknown and only one is measured.

Practitioners often try to reconstruct the attractor of the system, and then the question is to know which observables are accessible through such reconstructions. They regularly emphasize that possible artifacts should be avoided by checking the independence of results from the choices made during a reconstruction. Here, the property of ergodicity becomes decisive (but very hard to show rigorously), since an average over a large number of initial conditions would then essentially reproduce the same image with the same density as a single typical orbit observed over a long period. Several observables are sought. The more widely pursued is (grossly speaking) the dimension of the attractor, that is, the number of coordinates necessary to locate a point on the attractor (the Grassberger-Procaccia algorithm then proves very useful), or else the Kolmogorov-Sinai entropy, that is, the number of information bits gained by extending the duration of the observation by one unit of time. There also exist algorithmic procedures to obtain the Lyapunov exponent (which measures local instability), but with the express condition that the properties of the system remain unchanged during the course of the measurement. Hence there are great difficulties in applying chaos theory to the biological or social sciences. In these domains, not only is it difficult to obtain long time series with good precision, and not only are the dynamics generally quite complicated, but often the system "learns" over time and its nature changes. By its formulation the problem no longer belongs to dynamical systems theory.[20] Ruelle writes: "For such systems (ecology, econometrics, the social sciences), the impact of chaos stays at the level of scientific philosophy rather than at that of quantitative science."[21]

Controlling chaos

In engineering science, one starts from the observation of a finite series of discrete states. To characterize a chaotic behavior and to distinguish it from noisy effects then constitutes a difficult, fundamental problem. Large classes of discrete dynamical systems give rise to models given in terms of non–one-to-one functions: this is the case in control engineering, where systems use sampled data, pulsed modulations, neural networks, and other models; in nonlinear electronics and radiophysics, where various feedback devices are used; and in numerical simulations, signal theory, and other areas. As P. Bergé and M. Dubois wrote, because the notion of chaos is inseparable from the phenomenon of transition to chaos, various "roads" to chaos and several "scenarios"—by period doubling, by intermittency, or by quasiperiodicity—were studied, exhibited, and explored as early as the 1970s.

Despite those who lamented a technical, utilitarian orientation in contemporary science and dreamed of having found their revenge in a mainly qualitative, morphological mode of understanding with no possibility of action on reality,[22] engineering science today seeks to control and master chaos.[23] The extreme sensitivity of chaotic systems to subtle perturbations can be used both to stabilize irregular dynamical behaviors (stabilization of lasers or of oscillatory states) and to direct chaotic evolution toward a desired state. The application of a small, correctly chosen perturbation to an available parameter in a system can enhance its flexibility and performance. Various applications have been studied in signal theory (synchronization of chaotic signals), in cryptography, and even in automatic vision (with the help of a method of pattern recognition inspired by Lyapunov's theory). In brief, applications in the engineering sciences are highly promising.

For the social sciences, consider briefly the case of economics. Here chaos broadens the spectrum of available models mainly by introducing models with internal mechanisms explaining change in activity cycles (irregularities, disorders, etc.), which earlier were associated with external random shocks.[24] Having expressed the classic theory of general equilibrium in terms of equilibrium cycles, neoclassical econometric theory integrated motion, first under the rather simple form of growth, then through fluctuation mechanisms associated with external shocks and perturbations. Economists defending evolutionist approaches need to go beyond this fundamentally static, ahistorical viewpoint, because it tends to produce an essentially stable view of the economy with only exogenous stochastic perturbations.[25] In this view, changes, if they occur at all, would happen according to universal mechanisms and within structurally invariable frameworks. According to common epistemologies, history is indeed dialectically opposed to science, the ideographic mode of the former being incompatible with the nomological ambition of the latter. In a way, the structuralist episode has shown how the social sciences internalized this principle, according to which the scientific is constructed by excluding history.[26]

The evolutionist movement, however, uses dynamical systems theory—which according to S. Smale's definition truly constitutes the "mathematics of time"—to introduce a time dimension in its starkest possible form: history. In particular, the concept of bifurcation could be used to introduce structural changes and transformations of capitalism in a particularly radical form, that is, structural historicity, which more or less corresponds to the regulationist position (insisting on structural transformations of capitalism). I shall simply list here some of the ways history has been introduced into economic dynamics: multiple equilibrium situations with different basins of attraction, path-dependence, the concept of hysteresis (by analogy with physics), the role of historical contingency, modeling with positive feedback (Polya's urns), bifurcations (passages through critical points) as the phenomenology associated with great crises, the articulation of dynamics with different time scales, and many others. This effort in economics does not seek to free itself from the quantitative substrate of chaos theory to retain only a qualitative rhetoric; on the contrary, following a long tradition of mathematical modeling in economics, it mobilizes sophisticated, properly mathematical results from dynamical systems theory.[27]

The transposition of chaos to the social and political domains has also gone on freely. In these cases, however, it has mostly been confined to a metaphorical point of view, even in the discourses of some technical specialists. Starting from Sinai's billiards or the butterfly effect, philosophers and politicians have speculated on the profound difficulties of practical rationality or the conflict between rationality and democracy.[28] Concerning the conduct of complex, evolving, and potentially chaotic systems, David Ruelle has, with a slight irony, suggested that politicians' random action might be for the best. But is it possible to accept this transposition of modern society taken as a whole into a chaotic dynamical system that should be directed? Whatever the answer, these discursive tropes themselves indicate a contemporary air du temps.

Complexity, reductionism, and mixing

As noted earlier, discovering the Lorenz system and recognizing its stakes challenged the traditional understanding of the relation between simplicity and complexity. One can find complexity in simple systems and simplicity in complex ones. Richard Feynman's classic image—the world as a gigantic chess game in which each move taken in isolation is simple and in which complexity (and irreversibility) only comes from the very large number of elements in the game—is no longer valid.[29] While there was, formerly, a tendency to associate complexity with an extrinsic, accidental character linked to a multiplicity of causes, complexity can now appear as an intrinsic property of a system.

The erosion of the reductionist program

Underlying the development of physics until the beginning of the twentieth century, the reductionist program found early formulations in Galileo's *Assayer* (1623)[30] and in a famous passage of Locke's *Essay Concerning Human Understanding* (1690): both sought to discover the ultimate elements of reality, the "atoms" of matter bearing primary qualities (solidity, form, extension, motion, and number), and to show how their combinations and interactions with our senses explain secondary qualities (color, flavor, sound, and diverse other effects).[31] Nearly two centuries later, chemistry would reduce the apparent variety of substances to combinations of a restricted number of atoms, and physicists would quite naturally exalt this program as a norm. No matter what difficulty physicists might face, from then on they would seek to resolve the problems of reductionism by reiterating the same process: atomic physics, then nuclear physics, then (elementary?) particle physics.

Long before the advent of mixing physics, some physicists stressed the limitations of reductionism, as well as some of its failures to ascend to synthesis. On balance, the type of explanation that accounts for phenomena at one level in terms of properties of constituents belonging to the next lower level is far from satisfactory. It seems impossible, for example, to predict the very complex properties of water on the sole basis of the composition of its molecule H_2O. Similarly, the discovery of superconductors was a total surprise which could not be explained theoretically. Recent investigations of the impenetrability of solids, density questions, and sand-heap physics raise enormous difficulties for the explanation of macroscopic properties, without mentioning the fact that the "reduction of complicated visible things to simple invisible ones," as Locke said, supposes a conception of simplicity which is now being challenged. At the quantum level, the "simplicity" of invisible things seems to be relying on the introduction of new ideas that remain counterintuitive.[32]

Driven by the reductionist program and the search for a grand unification, fundamental physics probably witnessed the apex of its heroic age with the "standard model."[33] Despite its nineteen parameters, it represented a grand accomplishment for the unification of natural forces. But the golden age may now have lost this image. Ironically, the very concepts on which the model was built have eroded the bases of the program whose accomplishment it was, namely the concept of symmetry-breaking and the renormalization group.

Concerning critical phenomena, as well as the understanding of the elementary world, Kenneth Wilson's works have great importance.[34] Above all, however, they seem to mark a decisive epistemological turn. A particle physicist, and a Gell-Mann student who was also close to Hans Bethe, Wilson nevertheless promoted a conception of physics very different from the dominant one, which sought a theory a priori valid for all scales, from the astronomically large to the inconceivably small, meaning at unattainably high energies. Wilson thought that at energies higher than the accessible, unknown phenomena must intervene (quantum fluctuations, unification of forces, etc.). Starting from the unknown at short distances and, by successive iterations, eliminating unobservable fluctuations with too short a wavelength, he sought to construct a theory at accessible scales. Ultimately, Wilson ended up with a theory coinciding with the one constructed by the founders of quantum electrodynamics. The lasting image of his accomplishment was of "a theory valid not because it is susceptible of being applied down to the shortest distances but because it is an effective theory at great distances, a universal one resulting from a hidden world that was even more microscopic."[35] The same program would be developed in the domain of critical phenomena and condensed matter, where the "physics of mixing" emerged.

Experimental as well as theoretical tools contributed to the physics of mixing. In the 1960s, new experimental means served as a way both to get closer to critical points and to explore the nature of local order and its fluctuations. The diffusion of neutrons—sensitive to magnetic order—and that of laser light changed the experimental landscape. For P.-G. de Gennes, the 1991 Nobel laureate in physics, for example, the same tool—neutrons and their interactions—served both for the exploration of condensed matter and the study of critical phenomena and phase transitions. His earliest work on disordered, mixed materials (in particular, alloys of magnetic and nonmagnetic materials, and then superconductors) led him to the concept of "percolation" and to consideration of characteristic length scales, large with respect to interatomic distances. K. Wilson's works on phase transitions, involving geometrical studies of self-similar systems, showed that scaling explained their universal characteristics and in 1972 led to the famous theorem permitting the application of renormalization-group methods to the study of polymers. Once more, the cancellation of short-scale quantum fluctuations led to a simple large-scale image in which scaling laws and phenomena are said to be universal, independent of specific systems. In short, even in complex situations, simple effective models can capture large-scale physics. De Gennes then used this method for tangled polymer chains, dealt with liquid crystals, and tackled percolation in heterogeneous media, moistening, colloidal suspensions, and elsewhere.

Interested in the analogy between phase and hydrodynamic transitions, theoreticians (F. Dyson or D. Ruelle in statistical physics, for example) as well as experimenters (Ahlers at Bell Labs, Gollub and Swinney at New York University, Bergé and Dubois at the Commissariat à l'énergie atomique, Saclay) went from the study of critical fluctuations to that of instabilities. Wiener's cybernetics already contained the idea that deep connections existed between a large variety of disciplines, like statistical mechanics, information theory, biology, and a few others. By the 1970s, however, a group of physicists (K. Wilson, Kadanoff, P.-G. de Gennes, I. Prigogine, and others) had forged the tools necessary for studying the way systems pass from a mechanical or thermodynamic equilibrium to an unstable state with emergent phenomena (phase transitions, critical phenomena, onset of turbulence, crystal growth, solid breaking, etc.), and their ideas resonated with researchers who dealt with complex or chaotic systems (Ruelle, Feigenbaum, Libchaber, Pomeau). For example, M. Feigenbaum's observations about the appearance of chaos in simple deterministic systems by period-doubling cascades owed much to the fact that he had been introduced to K. Wilson's ideas at Los Alamos. Perhaps the idea of analyzing the appearance of a singular threshold in terms of scaling laws and of showing its universality would never have occurred to pure mathematicians specializing in the study of dynamical systems.

In this interaction between critical phenomena and complex chaotic systems, relatively old conceptual tools such as entropy, Brownian motion, random walks (for diffusion phenomena), and Ising models were revived along with the new tools that played a crucial role, such as fractals and self-similarity, multiscale processes, and percolation (abnormal random walks on a badly connected disordered network). Geometry, on the one hand, and statistical methods, on the other, increasingly entered this "physics of disorder."

Scaling laws truly constituted an epistemological turn, namely, toward the indispensable joint consideration of atomic, molecular, mesoscopic, and macroscopic levels with all levels being regarded as equally fundamental and no level regarded as reducible to the properties of the others. This "solidarity among scales," which is characteristic of mixing physics, is opposed not only to the former reductionist faith but also to the basic epistemology of theoretical physics developed during the nineteenth century. The new physical theories of the nineteenth century—kinetic theory of gases, electromagnetic theory, thermodynamics—were as distinct from classical mechanics as they were from each other and were accompanied by the view that the scientific method consisted in separating out relatively distinct systems in the universe while leaving room for the unknown. Since these systems were characterized by qualitatively different levels of organization or pertained to different phenomenological classes, distinct physical theories, with limited domains of validity, had to be conjugated.[36] Two further conceptual outcomes of the classical scientific method appeared inescapable: (1) the concept of scale, characterized by the smallest volume inside of which everything remained uniform and by the shortest time interval within which everything remained constant, and (2) the concept of observation domain, characterized by the largest volume and the longest duration to which investigations were extended. In contrast, the theoretical study of chaotic or complex systems as well as the experimental study of phase transitions and of disordered and mixing media have installed a potent new dialectic of the local and the global, hence breaking away from the former classical epistemology of bounded domains.

For a mundane science

The spectacular rise of mixing physics over the last twenty years has been associated above all with two elements. First, cultural change became perceptible as early as the late 1970s among various groups of physicists: on the one hand, an aspiration to a more mundane science and a return to the concrete and, on the other, a saturation, indeed a rejection, of both highly theoretical, abstract particle physics and the rigid forms of organization imposed by the needs of large, high-energy physics laboratories.[37] Today, research programs in elementary particle physics invoke length scales below 10^{-31} centimeters, requiring energy levels several billion times higher than those reached by contemporary particle accelerators. Conceivable experiments are exceptionally expensive as well as dangerous. Concerning the horizon of usable applications, this scale has so little relation with the human scale that many people seem to envisage either abandoning this line of research or at least recognizing its necessarily marginal character.[38] Meanwhile, mixing physics deals with mundane objects and natural phenomena encountered in everyday life.

Second, the rise of increasingly effective and, increasingly omnipresent, technologies and tools has favored the study of mixing. First, lasers and ultrasound instruments transformed mixes into "transparent boxes" whose local concentrations or velocity fields could be observed and measured without being destroyed or perturbed. Second, numerical modeling and simulations became determinative as a means of exploring how various media such as fluids, colloids, and granulated media could be mixed, separated, and operated upon, even before delicate physical experiments could be undertaken.

As recalled by E. Guyon and J.-P. Hulin, the Hebrew verb in the Holy Scriptures usually translated as "to create" literally means "to separate."[39] In recounting Creation, the Bible tells the process of separating or sorting out all things, which was also a process of organization: "God said: 'Let there be light' and there was light . . . God then separated the light from the darkness. . . . Then God said: 'Let there be a dome in the middle of the waters, to separate one body of water from the other.' And so it happened. God made the dome, and it separated the water above the dome from the water below. . . . Then God separated the great sea monsters and all kinds of swimming creatures with which the water teems." This text was written thousands of years ago, but since the start of the modern era, the metaphysical quest for the pure and the simple has gone hand in hand with a degraded representation of mixing. This historical relation now seems to be undergoing inversion.[40] In fact, the apparently opposed notions of separation and mixing often are quite close to one another and a slight evolution of the environment or small changes in experimental parameters can blur their border and invert the processes (which does not mean that they are reversible). This is the case, for example, when the constituents of a mixture maintain their identity in the form of droplets or particles (unstable emulsions, granular mixtures with aggregates and sediments, etc.).

Concerning mixing, the focus has recently moved from the study of static composite materials to that of dynamical processes leading to mixing. In this change, mixing physics has become a key element of process engineering.[41] Indeed, if we call process engineering the whole set of industrial processes in fields such as chemistry, farming, and civil engineering that involve manipulations, conditionings, moldings, and other transformations, we realize that mixing steps in these processes is often crucial at the level of efficiency and cost. From this point of view, mixing physics constitutes a fundamentally interdisciplinary domain that remains close to the engineering sciences. But mixing physics also concerns geophysics. All soils and, more generally, the terrestrial crust, are particular static mixtures—called "fixed mixtures"—built out of the porous piling up of small-size grains, whose efficiency, one most frequently hopes (especially in the case of industrial pollution), should be small. Finally, the study of mechanisms linking mixing with convective motions—whether they originate thermally in composition or density variations—is essential for understanding motion in the atmosphere and in the ocean, as well as in the terrestrial mantle.

True, mixing physics and the study of complex systems employs theoretical tools that predate them: mathematical techniques coming from dynamical systems theory, results from turbulent-fluid studies, and concepts from the physics of critical phenomena. As a new domain at the disciplinary crossroads, however, mixing physics was not constituted as a field where a priori theoretical research predominated and applications followed; on the contrary, it started with the problem of resolving "real" problems that sought to obtain precise mixtures (homogeneity or carefully balanced heterogeneity), to stabilize mixtures, to carry out antimixing processes (preventing pollution or diffusion in soils), and to master the construction of complex systems, among other results. Favored by the return to the macroscopic character of phenomena, the interaction between theory and application very quickly went back and forth.

Historicity and narrative in the sciences

Consider now the question of historicity and of the arrow of time with respect to physics. The nineteenth century has bequeathed a dual heritage. On the one hand, we have the classical laws of microscopic physics, epitomized by Newton's laws and their successors. They are deterministic and ahistorical, and they deal with certainties in the sense that they univocally link one physical magnitude to another. Moreover, they are symmetric with respect to time; past and future play the same role in their formulation.[42] On the other hand, through the second law of thermodynamics, which expresses the increase of entropy over time and thereby introduces the arrow of time, we have the vision of an evolving universe. These two points of view (symmetrical microscopic descriptions and macroscopic irreversibility) have been reconciled by considering irreversibility as a result of statistical approximations introduced in the application of fundamental laws to highly complicated systems consisting of many particles. According to Prigogine, the existence of chaos (unstable dynamical systems) makes this interpretation through approximation untenable;[43] in particular, one cannot rely on complicated systems because chaos can occur in very simple systems with a few degrees of freedom. From this fact, Prigogine concludes that instability and irreversibility are part of a fundamental, intrinsic description of nature. In this (controversial) interpretation, chaos would force researchers to revise the way they conceive the laws of nature, which would express what is possible—not certain—in a manner analogous to the more purely natural-historical disciplines like geology, climatology, and evolutionary biology.

In all phenomena perceptible at the human scale, whether in physics, chemistry, biology, or the human and social sciences, past and future play different roles. While classical physics dealt with repeatable phenomena, today's physics is much more concerned with phenomena that cannot be identically repeated and with singular processes. In molecular diffusion, dispersion in porous media, or in turbulent chaotic mixtures, irreversibility of mixing is crucial. History, it can be said, is entering the physical sciences in these unique "narratives." Especially through bifurcations, history enters the systems of chemistry, hydrodynamics, and engineering science: at each bifurcation, a "choice" governed by probabilistic processes emerges between solutions and in this sense, with its series of bifurcations, each chaotic evolution is truly a "singular" history. Hence the domain of chaos emphasizes the notion of "scenario," in the sense of a "possible road" to chaos. Titling his book *Of Clouds and Clocks* in 1965, K. Popper had already used vivid terms to express this opposition between the two kinds of physics.

In physics as well as in biology, an extreme sensitivity to perturbations and parameter variations seems to be a specific trait in the formation of complex mixing systems and in the spontaneous emergence of complex structures. For example, the role of small intrinsic effects or microscopic thermal fluctuations in crystal anisotropy is well known, though this does not usually enable the prediction or control of equilibrium forms, as in the cases of dendrite growth (snowflakes), microstructure formation (alloys), or fracture dynamics (solid or terrestrial crust).[44] Confronting this extraordinary proliferation of emergent forms and dynamics, the theorist J. S. Langer writes that it is not a certainty whether complexity physics can be successfully reduced to a small number of universality classes. The prospects for complexity science, nevertheless, seem excellent, he concludes, even though we may have to accept both the infinite variety of phenomena and the idea that we may never find simple unifying principles. Observed, described, and simulated on computers, the evolving behavior of unstable complex systems enriches our knowledge of what is possible as well as our understanding of emergence and the mechanisms of self-organization. Without necessarily leading to the formulation of general laws, it can also contribute to our understanding of why complexity emerges so easily in nature.

In summary, not only does chaotic and complex systems science install a new dialectics between the local and the global, it also forces us to rethink the relationship between the individual and the universal, and between the singular and the generic. In an article published in 1980, Carlo Ginzburg has distinguished two great modes of exploring and interpreting phenomena in the social sciences.[45] Inspired by the Galilean natural sciences, physics and astronomy, and aiming at conquering universality, the first mode is Ginzburg's Galilean paradigm. The second mode, the paradigm of clues, is concerned with the barely visible detail, the trace, the revealing symptom of a hidden reality. Associated with these two paradigms are two distinct methodologies: the former hypothetical-deductive, the latter inductive.[46] Even if we still do not know exactly how, the sciences of complex systems, including biology and physics, will soon have to come to terms with this duality for themselves.

An image shift in mathematics

The promotion of what I have called a fin de siècle image of science sometimes goes hand in hand with an aggressive, confused dispute over the place of mathematics in the general configuration of knowledge. From various sources we have heard critiques aiming either to end a status for mathematics deemed too prestigious, or to characterize its role as much less important for disciplines on the rise today, or to contest its overly abstract, overly theoretical representation. In these critiques, several levels interfere: epistemological, conceptual, political, and institutional (notably with respect to questions concerning teaching and training). Here again, the polemic raging in France is perhaps more lively and more "overdetermined" than elsewhere. It would be superfluous to show how ridiculous is the claim that mathematics is not important in computer science, macroscopic mixing physics, modeling, or any such scientific practice. I want instead to discuss the context and stakes of contemporary controversies about mathematics. These debates cannot be understood without considering what has been a true "image war" about what mathematics is, what it deals with, and how. Triggered by the end of World War II, the conflict raged with great intensity until the 1980s. It was decisively stirred up by the French mathematical school, and especially the prestigious Bourbaki group. Spurred by the mathematical community itself, the battle mainly focused on the dichotomy between pure and applied mathematics.[47] Let us look back a little to see more precisely what this was about.

Recent scholarship has emphasized that until the nineteenth century the dichotomy between pure science and applied science did not exist in its current form, with the former motivated by the disinterested pursuit of the laws of nature and the latter by the need for technological mastery of things and processes in a context of markets, powers, and applications. In fact, it now seems that these two types of activities were intimately interwoven through complex links and through overlapping networks of actors.[48]

For example, sixteenth- and seventeenth-century mathematicians were concerned with problems of artillery, fortification, surveying, astronomy, cartography, navigation, and instrumentation, which also left room for purely philosophical debates, especially about the "certainty" of mixed mathematics.[49] In the eighteenth century, men such as the Bernoullis, Euler, Lagrange, Monge, and Laplace were still concerned with a variety of problems, mathematical as well as mechanical, navigational, astronomical, and engineering. They proposed laws for the whole corpus of analysis (Euler, Lagrange), provided profound conceptual reorganizations and new foundations (Lagrange), and developed important abstract research in number theory. But they did all this without constructing a value hierarchy over the whole collection of studies. Moreover, these men were invested with various political and institutional responsibilities; during the revolutionary period, for example, Monge and Laplace appeared as true scientific coordinators.

During the mid-nineteenth century, mathematics' center of gravity clearly moved from Paris to Berlin, where many factors contributed to separating that part of mathematics concerned with rigor (Weierstrass), generality, and proof from that part mainly occupied with providing tools for physics and engineering. The rise of the German university system, with its ideal of pure research, the practice of seminars which was then established, and the development of disciplines such as number theory and abstract algebra (Kummer, Kronecker) favored the figure of the academic mathematician, rather isolated from other disciplines and the rest of society, and engaged in research motivated above all by the internal dynamics of mathematical problems. But this image of the mathematician, represented in the early twentieth century by David Hilbert at Göttingen, co-existed with clearly distinct images, such as that of Felix Klein, who played the role of a true *Wissenschaftspolitiker* at Göttingen from 1893 to 1914. Diverging in their mathematical inclinations and styles, Klein and Hilbert jointly established Göttingen as the world's mathematical center. Only gradually did the Hilbertian ideal of conceiving mathematics axiomatically impose an ordered, hierarchical architecture on the whole mathematical corpus. Originally developed by such students of Hilbert as E. Noether and B. van der Waerden for the sole domain of algebra, the ideal of an axiomatic, structural, abstract mathematics would become the single privileged image of modern mathematics, especially in the hands of the group called Bourbaki.[50] By the audacity of their joint enterprise of rewriting all of mathematics as well as by their individual research, Bourbaki's founding members (H. Cartan, Chevalley, Dieudonné, Weil) systematically promoted the structural ideal. On a personal level, they projected an image of strong, elitist, and variously gifted virtuosi (notably in music and ancient languages), mathematicians above the common lot who needed only their brains to rework the edifice of mathematical knowledge.[51] Their imperial objective was immense in its ambitions.

World War II played a complex and ambivalent role in the evolution of mathematics. To start with, the Göttingen school was brutally destroyed when Hitler seized power. Then, applied mathematics went through major developments in the United States, redefining disciplinary boundaries and reshaping the figure of the mathematician.[52] Benefiting from an intense cooperation with the military in the collective war effort, the new domains included the study of partial differential equations associated with wave propagation, the theory of explosions and shock waves, and probability theory and statistics (prediction theory, Monte Carlo methods, etc.). Game theory and decision-making mathematics and operations research, which would soon become systems analysis, also emerged at this time. Although not alone in his efforts, John von Neumann is often associated with this mutation. Socially engaged and intervening in the technological and political choices of the United States, von Neumann blurred the borders between pure and applied mathematics, between what concerned mathematics and what previously had come under various other disciplinary domains (mechanics, engineering science, physics). In particular, he strongly associated hydrodynamics with computer and numerical analysis.

Despite these considerable developments, however, the international mathematical community did not care much about applied mathematics. From roughly 1950 to 1970, pure mathematicians succeeded in maintaining a cultural hegemony over their discipline. Clearly, they privileged problems stemming from internal interfaces between branches of mathematics. The more valued branches have consistently been the more structural, more abstract ones: algebraic and differential geometry, algebraic topology, number theory. These areas constituted the profound part of mathematics, to which the best students were directed. Simultaneously, more applied branches (such as differential equations, probability theory, statistics, and numerical analysis) were devalued in higher education and research as well as in the institutions of the professional community. Thirty years later, Peter Lax commented on the American situation in the 1950s: "the predominant view in American mathematical circles was the same as Bourbaki's: mathematics is an autonomous subject, *with no need of any input from the real world, with its own criteria of depth and beauty,* and with an *internal compass* for guiding further growth. Applications come later by accident; mathematical ideas filter down to the sciences and engineering."[53] The philosophy of mathematics that informed this conception is clearly expressed in a famous text signed by Bourbaki: "In the axiomatic conception, mathematics appears as a reservoir of abstract forms, the mathematical structures; and it happens—without one knowing quite why—that certain aspects of experimental reality are cast in certain of these forms, as though by a kind of preadaptation."[54] Thus did mathematicians find a legitimacy for neglecting the world. Their view, it must be said, was not independent of the role of structuralism as a dominant mode of thought in the 1960s: structures of language, mental structures, structures of kinship, structures of matter, structures of society. The ambition of virtually all disciplines was to discover fundamental structures and mathematics was the science of structure par excellence, providing a universal key for the intelligibility of all knowledge.[55]

In brief, two concurrent images of mathematics were clashing. On the one side was pure mathematics developed "for the honor of the human spirit,"[56] whose paradigmatic methodology was axiomatic and structural. It progressed through internal dynamics at the interfaces of several branches of mathematics and sought through set theory to reduce mathematics to a structurally unified corpus, to which, as for works of art, one applied a rhetoric of esthetics and elegance. On the other side was the image of applied mathematics stemming from the study of nature, from technological problems, and from human affairs (numerical analysis, approximations, modeling). It was less noble and less universal, because dependent on material interests and societal conflicts. Pure mathematicians, clearly more prestigious until the end of the 1970s, had put their stamp on this opposition.

In the course of the 1980s, within the new economic, technological, and cultural contexts of contemporary societies, the general landscape of mathematics began to change. Domains left dormant for decades were rejuvenated and new domains opened, linked in particular with the computer and experimental mathematics. Distressed by their isolation and concerned to improve their image in society,[57] mathematicians now promoted an open ideal of mathematics, mathematics in interaction with other disciplines, the world, and human needs. What P. Lax called "the tide of purity" receded. Ideological representations of the purity of mathematics now share the stage with other representations which promote different values: the pragmatic and operational character of results, links with state power and corporate wealth, and entrepreneurial dynamism.

If *structure* was the term emblematic of the 1960s, *model* was the term for the 1990s. The practice of mathematical model building (in the physical sciences and climatology, in engineering science, in economics) has progressively been extended over the last decades. Today its range is immense and it is almost always accompanied by experimentation and numerical simulation. In some parts of the mathematical community, it also produces distress. What theorems have those mathematicians precisely and clearly demonstrated who study, with the help of the computer, supersonic fluid dynamics, plasmas in fusion, or shock waves? Or those who model a nuclear reaction or a human heart in order to test, respectively, an explosion velocity or the viability of an artificial heart? Do they share the same profession with traditional mathematicians? In August 1998, at the Berlin International Congress of Mathematicians, the old opposition between pure and applied mathematics was expressed differently: "mathematicians making models versus those proving theorems."[58] But the respect formerly enjoyed by the theorem provers is now generally shared by the modelers.

The applied, concrete, procedural, and useful versus the pure, abstract, structural and fundamental: this set of oppositions expressing the shift in values and hierarchies that has occurred in mathematics resonates with that described earlier: disordered, complex, mixed, macroscopic, and narrative versus ordered, simple, elementary, microscopic, structural. It appears that the shift in mathematics is contributing strongly to the formation of a general fin de siècle image of science.

Notes

Translated by David Aubin. The text benefited from critiques by David Aubin, Dominique Pestre, and M. Norton Wise. I thank them warmly.

1   For a very complete bibliographical essay on historical studies of seventeenth century science, see S. Shapin, *The Scientific Revolution* (Chicago: University of Chicago Press, 1996).

2   Sophie Roux, *La Philosophie mécanique (1630–1690)*, 2 vols. (Ph.D. diss., Ecole des Hautes Etudes en Sciences Sociales, Paris, 1996).

3   C. Licoppe, *La Formation de la pratique scientifique. Les discours de l'expérience en France et en Angleterre, 1630–1820* (Paris: La Découverte, 1996).

4   A. Dahan, "Le déterminisme de Pierre-Simon Laplace et le déterminisme aujourd'hui," *Chaos et déterminisme*, ed. A. Dahan, K. Chemla, and J.-L. Chabert (Paris: Seuil, 1992). I attributed this demonizing particularly to I. Ekeland in *Le Calcul, l'imprévu* ((Paris: Seuil, 1984), *Calculus and the Unexpected*, trans. by the author (Chicago: The University of Chicago Press, 1988), and *Le Chaos* (Paris: Flammarion, 1995) where he writes: "C'est de cette chappe étouffante, de cet univers clos [l'univers de Laplace] où il ne peut rien se passer, où il n'y a ni inconnu ni nouveau que nous délivre la théorie du chaos" (100–101).

5 Another much-discussed example appears in Paul Forman, "Weimar Culture, Causality, and Quantum Theory, 1918–1927: Adaptation by German Physicists and Mathematicians to a Hostile Intellectual Environment," *Historical Studies in the Physical Sciences* 3 (1971): 1–115.

6 On this matter, see A. Dahan and D. Pestre, "Comment parler des sciences aujourd'hui?," in B. Jurdant (ed.), *Les Impostures scientifiques* (Paris: Seuil, 1998), 77–105.

7 Cf. "The Quantum Theory of Radiation," *Philosophical Magazine* 47 (1924): 758–802. Indeed, at the height of the crisis in the old quantum theory (following the display of the Compton effect in 1923), Bohr and his colleagues proposed to return to a continuous theory of radiation in the form of a "virtual" field carrying no energy but giving rise to transition probabilities. For this purpose, they associated the atom with a set of virtual harmonic oscillators. Six months later, this theoretical attempt collapsed as a result of Bothe and Geiger's experiment. Cf. Catherine Chevalley's introduction and notes in her edition of N. Bohr, *Physique atomique et connaissance humaine* (Paris: Folio-Gallimard, 1991).

8 See S. Diner, "Les voies du chaos déterministes dans l'école russe," in A. Dahan, K. Chemla, and J.-L. Chabert (eds.), *Chaos et déterminisme*, 331–370; A. Dahan, "La renaissance des systèmes dynamiques aux États-Unis après la deuxième guerre mondiale; l'action de Solomon Lefschetz," *Supplemento ai Rendiconti del Circolo di Palermo* (II) 34 (1994): 133–166; A. Dahan, "Le difficile héritage de Henri Poincaré en systèmes dynamiques," in F.-L. Greffe, G. Heinzmann, and K. Lorenz (eds.), *Henri Poincaré, science et philosophie* (Paris: A. Blanchard, 1996), 13–33; D. Aubin and A. Dahan, "Writing the History of Dynamical Systems and Chaos: *Longue Durée* and Revolution, Disciplines and Culture," *Historia Mathematica* 29 (2002): 273–339; and D. Aubin, "A Cultural History of Catastrophes and Chaos: Around the Institut des Hautes Études Scientifiques, France" (Ph.D. diss., Princeton University, 1998), UMI #9817022.

9 The article "Studies on Non Linear Problems" (Los Alamos Document, 1940) by E. Fermi, J. Pasta, and S. Ulam, was reprinted in S. Ulam, *Sets, Numbers, and Universes: Selected Works*, ed. W. A. Beyer, J. Mycielski, and G.-C. Rota (Cambridge: MIT Press, 1974), 977–988. In 1955, Ulam devoted the Gibbs Lecture of the American Mathematical Society to this topic.

10 Möser writes in 1975: "The development of statistical mechanics had led to the belief that most mechanical systems—at least those composed of numerous particles—were ergodic, that is to say, that after a given amount of time their behavior becomes entirely independent of initial conditions. This is, however, in flagrant contradiction with stability. In fact, physicists in the nineteenth century, who started off from this perspective, attempted to prove that almost all mechanical systems will have an unstable behavior if given enough time. That this was not, after all, the case for many realistic systems has not been thoroughly demonstrated." We should note that the theorem states that the difference between stability and instability is subtly linked to a question of number theory—the approximation of irrational numbers by rational numbers.

11 Aside from those dealing with celestial mechanics or mathematics, physicists only took notice of the KAM theorem in the 1970s; the article by G. H. Walker and J. Ford, "Amplitude Instability and Ergodic Behavior for Conservative Non-Linear Oscillator Systems," *Physical Review* 188 (1969): 416, helped to diffuse the theorem.

12 The first occurrence of *chaos* in the sense of homogeneous disorder in statistics was in N. Wiener, "The Homogeneous Chaos," *American Journal of Mathematics* 60 (1938): 897–936.

13 Among intellectuals, we may cite E. Morin, M. Serres, H. Atlan, and others. See K. Pomian (ed.), *La Querelle du déterminisme* (Paris: Le Débat-Gallimard, 1990). An exceptional bestseller in France, James Gleick's book *Chaos: Making a New Science* (New York: Viking,

1987; translated as *La théorie du chaos* (Paris: Albin Michel, 1989) produced a certain annoyance because it was perceived as having glossed over non-American contributions in the history of chaos.

14 Cf. A. Dahan, "History and Epistemology of Models: Meteorology (1946–1963) as a Case Study," *Archive for History of Exact Sciences* 55 (2001): 395–422.

15 Cf. Tien-Yien Li and James A. Yorke, "Period Three Implies Chaos," *American Mathematical Monthly* 82 (1975): 985–992; and Robert May, "Simple Mathematical Models with Complicated Dynamics," *Nature* 261 (1976): 459–467. These and several other seminal articles are reprinted in Pedrag Cvitanovic (ed.), *Universality in Chaos* (Bristol: Adam Hilger, 1989); and Hao Bai-Lin (ed.), *Chaos* (Singapore: World Scientific, 1984), and *Chaos II* (Singapore: World Scientific, 1990).

16 See A. Douady, "Déterminisme et indéterminisme dans un modèle mathématique," in A. Dahan, J.-L. Chabert, and K. Chemla, *Chaos et déterminisme*, 11–18.

17 See W. Tucker, "The Lorenz Attractor Exists," *Comptes Rendus de l'Académie des Sciences*, series I, 328 (1999): 1197–1202.

18 Although schematic, this description nevertheless fairly characterizes the international community of mathematicians, which never awarded a Fields Medal to a probability theorist. In the United States, an excellent school for probability theory was developed by three prominent personalities: J. L. Doob (general theory of martingales), W. Feller (Markov process theory), and M. Kac (spectral methods in probability theory), the last of whom stood at the interface with physicists.

19 This is David Ruelle's case. "As deterministic automata, our thoughts and our actions are rigidly determined while we have the impression that many of our choices are free or 'randomly' made. What can be said is that the constraints imposed by physics on 'deterministic automata' are rather weak and allow for a large portion of fantasy. Natural laws grant us a behavior which is unpredictable in practice, and there is no manifest contradiction between determinism and freewill." See Ruelle, "Le problème de la prédictibilité," in K. Pomian, *La Querelle du déterminisme*, 160. I. Ekeland is more assertive and even more lyrical: "What an admirable, subtle dosage of chance and necessity! . . . Here are at once resolved a whole army of false problems concerning human freedom in a deterministic universe. No more do we see, like Laplace, a sunny sky opened on an infinite horizon, so clear as to give the impression of being able to touch it. Neither do we see a cloudy sky swamped in a fog, which conceals the horizon to our sight. What we see is both together, like a rainy sky, where gusty winds set up a few bursts of faraway horizons laden with sunshine" (*Le Chaos*, 104).

20 See J.-P. Eckmann and D. Ruelle's work, and especially their review article, "Ergodic theory of Chaos and Strange Attractors," *Reviews of Modern Physics* 57 (1985): 617–656.

21 D. Ruelle, *Hasard et chaos* (Paris: O. Jacob, 1991), 105.

22 In particular, this is true for René Thom's case (see his book *Prédire n'est pas expliquer* [interviews by E. Noël] [Paris: Eshel, 1991]) and also for some of his disciples. Cf. A. Boutot, "La philosophie du chaos," *Revue philosophique* 2 (1991).

23 Concerning engineering science, many references are to be found in C. Mira, "Some Historical Aspects of Nonlinear Dynamics: Possible Trends for the Future," Visions of Nonlinear Science in the Twenty-first Century, Seville, 26 June 1996.

24 I do not wish to criticize the global epistemological pertinence of the mathematical modeling approach in economics, a domain in which I am not sufficiently competent.

25 I am relying above all on F. Lordon's work, but he is obviously not alone. One could mention a whole group of regulationist economists whose main representative in France is Robert Boyer.

26 Lordon recalls that the inclusion of history in the understanding of economic dynamics leads to one of the most important rifts in the field, created by the clash between two worldviews. On the one side, the exigency of scientificity relies on object invariance and the exclusion of history (universality of structures); on the other side, in the name of an elementary realism, there is a refusal wholly to reject historical transformations; cf. F. Lordon, "Formaliser la dynamique économique historique," Économie appliquée 44(1) (1996): 55–84.

27 This does not mean that an epistemological critique is impossible, but that it should be concerned with the fundamental level of the concepts and principles adopted for modeling.

28 See D. Parrochia, Les Grandes Révolutions scientifiques du XXème siècle (Paris: Presses Universitaires France, 1997); also N. Katherine Hayles (ed.), Chaos and Order: Complex Dynamics in Literature and Science (Chicago: University of Chicago Press, 1991).

29 Cf. R. Feynman, The Character of Physical Laws (Cambridge, Mass.: MIT Press, 1967), especially chapter 5.

30 Galilée, L'Essayeur, trans. C. Chauviré (Paris: Les Belles Lettres, 1980), 239.

31 J. Locke, Essay Concerning Human Understanding, book 2, chapter 8.

32 Cf. the examples given by J.-M. Lévy-Leblond, "Une matière sans qualités? Grandeur et limites du réductionisme physique," conference at the Collège de France, 1989. See also J.-M. Lévy-Leblond, Aux Contraires (Paris: Gallimard, 1996).

33 S. Schweber, "The Metaphysics of Physics at the End of a Heroic Age," forthcoming.

34 Wilson obviously was not alone and notably used Kadanoff's work on spin glass.

35 E. Brézin, "Bref survol des phénomènes critiques," Des phénomènes critiques au chaos. Actes du colloque scientifique à la mémoire de Pierre Bergé, CEA-Saclay, March 1998, 23–29.

36 According to Pierre Duhem, a distinct physical theory is what ordains, in a coherent formalized whole, the representation of a class of phenomena; see La Théorie physique, son objet, sa structure (1906; Paris: Vrin, 1981).

37 James Gleick's book Chaos is quite instructive on this respect as it reports numerous interviews with Los Alamos physicists.

38 See the 100th meeting of the American Physics Society in Atlanta in 1999 and the rather contradictory interventions of Weinberg and Laughlin (the 1998 Nobel Prize winner for work on the quantum Hall effect).

39 E. Guyon and J.-P. Hulin, Granites et fumées. Un peu d'ordre dans le mélange (Paris: O. Jacob, 1997), preface and p. 10.

40 B. Bensaude-Vincent, Éloge du mixte. Matériaux nouveaux et philosophie ancienne (Paris: Hachette, 1998).

41 See E. Guyon and J.-P. Hulin, Granites et fumées.

42 Still, without even raising problems associated with thermodynamics, we should note that modern physics (relativity, quantum mechanics) has modified the conception of time as a given framework and neutral receptacle for phenomena (just as it has modified the conception of space, but this is better understood). Time is now being conceptualized as structured by phenomena themselves. Michel Paty writes: "we see time take a certain 'material' consistency, which it seems to lack in Newton's conception of an absolute mathematical time. The fact that phenomena now governed the properties of time, rather than the reverse, led to a physical time in which one can discern a certain thickness, or consistency, but not that of experienced history [histoire vécue]." Cf. "Sur l'histoire de la flèche du temps," in E. Klein and M. Spiro (eds.), Le Temps et sa flèche (Paris: Frontières, 1994), 52. G. Cohen-Tannoudgi goes even further when he states that "matter is all that gives its arrow to time." "Le temps des processus élémentaires," in ibid., 130.

43 A very clear expression of his positions is to be found in I. Prigogine, *Les Lois du chaos* (Paris: Flammarion, 1994).

44 See the lucid article by J. S. Langer, "Nonequilibrium Physics and the Origins of Complexity in Nature," in V. L. Finch, D. R. Marlow, and M. A. Dementi (eds.), *Critical Problems in Physics* (Princeton, N.J.: Princeton University Press, 1997), 11–27. In particular, he writes: "Note that my scale of goodness is different from the tradition of twentieth century physics, which insists on grand unification and underlying simplicity. I am perfectly comfortable with the idea that we may never find a set of unifying principles to guide us, in the manner of Gibbsian equilibrium statistical mechanics, to solutions of problems in the nonequilibrium physics of complex systems" (21).

45 C. Ginzburg, "Signes, traces, pistes. Racine d'un paradigme de l'indice," *Le Débat* 6 (1980): 3–44.

46 With great pertinence, A. Desrosières has used this distinction in the case of social statistics. See his "Du singulier au général. L'argument statistique entre science et état," in B. Conein and L. Thévenot (eds.), *Cognition et information en société* (Paris: Editions de l'EHESS, 1997; special issue of *Raisons pratiques* 8), 267–282.

47 For a more detailed account of this "image war," see A. Dahan Dalmedico, "Pur versus appliqué? Un point de vue d'historien sur une 'guerre d'images,'" *Gazette des mathématiciens* no. 80 (April 1999): 31–46; and "An Image Conflict in Mathematics after 1945," in U. Bottazzini and A. Dahan Dalmedico (eds.), *Changing Images in Mathematics. From the French Revolution to the New Millennium* (London: Routledge, 2001), 223–253.

48 See D. Pestre, "La production des savoirs entre académies et marchés. Une relecture historique du livre 'The New Production of Knowledge' édité par M. Gibbons et al.," *Revue d'économie industrielle* no. 79 (1997): 163–174; and "Regimes of Knowledge Production in Society: Towards a More Political and Social Reading," *Minerva* 41(3) (2003): 245–261.

49 See Jim Bennet and Stephen Johnston, catalog of the exhibit *The Geometry of War, 1500–1750*, Oxford, Museum of the History of Sciences, 1996; also *The Measurers: A Flemish Image of Mathematics in the Sixteenth Century*, Oxford, Museum of the History of Sciences, 1995.

50 See D. Rowe, "Felix Klein as Wissenschaftspolitiker," and L. Corry, "Mathematical Structures from Hilbert to Bourbaki: The Evolution of an Image of Mathematics," both in Bottazzini and Dahan Dalmedico, *Changing Images in Mathematics*, 69–91 and 167–185, respectively.

51 André Weil wrote: "Qu'un autre hante les antichambres pour se faire accorder le coûteux appareillage sans lequel il n'est guère de Prix Nobel: un crayon et du papier, c'est tout ce qu'il faut au mathématicien; encore peut-il s'en passer à l'occasion." See "L'architecture des mathématiques," in F. Le Lionnais (ed.), *Les grands courants de la pensée mathématiques* (Paris: A. Blanchard, 1948), 308; *Great Currents of Mathematical Thought*, trans. R. A. Hall and Howard G. Bergman (New York: Dover, 1971).

52 A. Dahan Dalmedico, "L'essor des mathématiques appliquées aux États-Unis: l'impact de la Seconde Guerre mondiale," *Revue d'histoire des mathématiques* 2 (1996): 149–213; and "Axiomatiser, modéliser, calculer: les mathématiques, instrument universel et polymorphe d'action," in A. Dahan and D. Pestre (eds.), *Les sciences pour la guerre (1940–60)* (Paris: Presses de l'EHESS, 2004), 33–55.

53 P. Lax, "The Flowering of Applied Mathematics in America," in P. W. Duren, et al. (eds.), *A Century of Mathematics in America*, 3 vols. (Providence, R.I.: American Mathematical Society, 1988–89), 2:455–466.

54 N. Bourbaki, "L'architecture des mathématiques," in Le Lionnais (ed.), *Les grands courants* (*Great Currents of Mathematical Thought*, trans. A. Dresden [New York: Dover, 1971], 321–336).

55 D. Aubin, "The Withering Immortality of Nicolas Bourbaki: A Cultural Connector at the Confluence of Mathematics, Structuralism, and the Oulipo in France," *Science in Context* 10 (1997): 297–342; M. Armatte, "Mathématiques 'modernes' et sciences humaines," in B. Belhoste, H. Gispert, and N. Hulin (eds.), *Les Sciences au lycée* (Paris: Vuibert, 1996), 77–88; J-P. Kahane, "Les mathématiques, hier et demain," in *Les Sciences au lycée*, 89–98.

56 The first occurrence of this expression was in a letter in French from Jacobi to Legendre, 2 July 1830; see C. Jacobi, *Gesammelte Werke* (Berlin: Reimer, 1884), 1:454. It reappears in André Weil, "L'avenir des mathématiques," written in 1943 and published in Le Lionnais (ed.), *Les grands courants*. In the 1960s and 1970s, it was often used by various authors and taken up by Jean Dieudonné as the title of his bestseller, *Pour l'honneur de l'esprit humain. Les mathématiques aujourd'hui* (Paris: Hachette, 1987).

57 In the United States, see in particular the efforts of the American Mathematical Society following the publication of the David report and in France, a conference jointly organized in 1985 by the Société mathématique de France and the Société de mathématiques appliquées et industrielles.

58 Cf. David Mumford, "Trends in the Profession of Mathematics," *Berlin Intelligencer*, International Congress of Mathematicians, Berlin, August 1998.

3 Forms of explanation in the catastrophe theory of René Thom: topology, morphogenesis, and structuralism

David Aubin

Voici maintenant qu'après l'âge des denrées et des matières, après celui de l'énergie, nous avons commencé à vivre celui de la forme.—Pierre Auger, *Proceedings of the First International Conference on Cybernetics*

Peut-être personne n'est plus capable que le mathématicien de suivre une question de forme pure.—George D. Birkhoff, "Quelques éléments mathématiques de l'art"

"Science, some say, is in crisis." When the French mathematician René Thom wrote this in 1975, the catastrophe theory he had imagined during the previous decade had already made him famous. Thom nonetheless agreed that science faced an unprecedented crisis. Decrease of governmental support, students' disaffection from scientific careers, accumulation of trash, and poisoning of the earth: albeit highly visible, these signs merely pointed to a deeper malaise within science itself. Indeed, hidden behind triumphant proclamations of progress and success, Thom saw a "manifest stagnation of scientific thought vis-à-vis the central problems affecting our knowledge of reality." At bottom, he contended, this stagnation was due to the fact that "science [had sunk] into the futile hope of exhaustively describing reality, while forbidding itself to 'understand' it."[1]

Understanding—this was science's "prime vocation," and the way out of the present crisis. Inspired by Kuhnian theses, Thom believed the crisis presaged an important paradigm shift. Science "must come back to this essential goal [which is] to understand reality, to *simulate nature*. . . . If, as I wish to believe, this necessary mutation is to be accomplished, will we not then be able to say of science that it remains man's hope?"[2] The only solution to the problem of contemporary science was more science. This would not be the old science, but a new one, which would endeavor to provide explanations rather than mere descriptions or predictions.

His own catastrophe theory, of course, was for Thom a prime example of this new type of science.[3] Slowly appreciated when it was introduced in the

late 1960s, catastrophe theory was propelled on a wave of hype and enthusiasm during the mid-1970s only to die out in bitter controversies by the
end of the decade. Caught in fierce debates, the movement nearly vanished
from the scene of science. True, the theorems that Thom and his collaborators proved, have survived as "a beautiful, intriguing field of pure mathematics."[4] Even the concepts they introduced lived on in other guises, as
Thom was clearly aware: "Sociologically speaking, it can be said that this
theory is a shipwreck. But in some sense, it is a subtle wreck, because the
ideas that I have introduced gained ground. In fact, they are now incorporated in everyday language. . . . The notions [of catastrophe theory] have
become part of the ordinary baggage of modelers. Therefore, it is true that,
in a sense, the *ambitions* of the theory failed, but in *practice*, the theory
has succeeded."[5]

This is especially true of chaos theory, with which catastrophe theory had
important interactions. Both inspired by topology, these theories shared
some of their mathematical concepts, their modeling practices, their practitioners and institutional locations, their modes of explanation, and their
general aims. Only in the second half of the 1970s did they definitely part
from each other.[6]

Moreover, catastrophe theory offered types of explanation that were directly inspired not only by mathematics, but also by biology and structuralism. The new explanations were perceived as subverting dominant ideolo-
gies in all of these disciplines. Centered on problems of structure and
form, they aimed at providing accounts for the emergence and destruction
of morphologies, based not on underlying forces but on mathematical
principles.[7] "Thus, we are catching a glimpse of the possibility of creating
a dynamic structuralism," Thom declared, while proposing explanations
that grew out of his interest in embryology.[8] We shall see, for example, how
the important concept of an "attractor" emerged from his understanding
of embryologist Conrad Hal Waddington's epigenetic landscapes.

Finally, because of its interdisciplinary character, catastrophe theory
was, for people with widely differing agendas, a cultural connector linking
mathematics, biology, the social sciences, and philosophy. It represented
an incomplete transition from explanations in terms of a few simple, stable
forms to an understanding of nature in terms of the complex, the fluid, and
the multiple. Indeed, it may be possible to see catastrophe theory as a
crucial transitional stage from structuralism to poststructuralism, perhaps
even from modernism to postmodernism.[9]

Although the global ambitions of catastrophe theory dimmed markedly
in the late 1970s, it remains of interest to determine what has survived "in
practice." We can begin by understanding the specific context from which
catastrophe theory emerged. More specifically, we need to reconstitute and

reinterpret the type of explanations Thom proposed. Today his project is often obscured by the settlement that took place in the latter part of the 1970s, effectively dividing the world between stable and chaotic systems in the course of opening new paths for understanding the sources of complexity in the world.[10]

This essay will explore the new forms of scientific explanation that Thom offered as an alternative to what he denigrated as the traditional "reductionist approach." Indeed, his philosophy of science could be summarized as follows: first one classifies a phenomenology by describing its morphologies, then one strives for explanations, which, as Thom believed, could be achieved by following one of two philosophically distinct approaches, the *reductionist* or the *structural*. The former accounted for morphogenesis in terms of other morphologies; the latter eschewed such attempts and looked for autonomous, intrinsic explanations that did not depend on other levels of reality.

None of these steps is self-evident. Their actualization depends on the modeling practices that scientists have deployed. Indeed, morphologies, like explanations, are not imposed on observers, but fashioned by the lens they choose to wear. This is what is meant by modeling practices: the actual processes by which scientists transform, using some specific means, a given raw material, selected by them, into a product which they hope will be considered knowledge about natural phenomena.[11] Together with other mathematicians—Ralph Abraham, Steve Smale, and Christopher Zeeman, who often visited him at the Institut des hautes études scientifiques (IHÉS) in Bures-sur-Yvette, France—René Thom proposed radically new modeling practices. With catastrophe theory, he wished to redefine what it meant to build a mathematical model. His experience in mathematics suggested new means of knowledge production. His forays into embryology shaped his views on what should be sound raw material for modeling. And he found in the French intellectual context—in structuralism—an inspiration for his interpretation of product-knowledge. In each case (mathematics, embryology, and structuralism), Thom used well established technical achievements to go beyond and subvert the original framework to which they belonged. The final part of this essay will examine Thom's philosophy of science as he described it around 1975, that is, after the main tenets of catastrophe theory had been well publicized but just before harsh critiques and the emergence of chaos made him retreat deeper into philosophy.

Sociologically speaking, a mathematician

A bold and comprehensive theory aimed at explaining the dynamics of shapes in the everyday world, catastrophe theory has often been narrowly construed as a mathematical approach able to deal with abrupt, discontinuous changes in nature—a rubber band that breaks, for example. For Thom, however, it was always much more than this.

What was catastrophe theory?

From 1964 to 1968, on his own account, Thom worked on an ambitious book, a manifesto, titled *Structural Stability and Morphogenesis*, which was not published until 1972, due to its publisher's financial trouble.[12] For this reason, catastrophe theory was first presented in two articles, both published in 1968. To the proceedings of a theoretical biology symposium, Thom contributed "A Dynamical Theory of Morphogenesis," and for the French journal *L'Âge de la science*, he wrote "Topology and Meaning."[13] The first article was concerned with biology, the second with semiotics. Not content with introducing a new mathematical language and exploring its consequences in some areas of science, Thom also conceived of his book, and both of these articles, as exposés of an original philosophy of science, indeed a true "natural philosophy."[14] The subtitle of his book, "An Outline of a General Theory of Models," revealed the extent of his ambitions.

A striking paradox raised by Thom may illustrate his epistemological concerns.[15] Consider an eroding cliff and the developing egg of a frog. In the former case, suppose that later microclimatic conditions and the geological nature of the soil are known, then knowledge of the physical and chemical forces at play will be excellent. Nevertheless, it is impossible to predict the future shape of the cliff. As for the egg, Thom contended, although knowledge of the substrate and developmental mechanisms is sketchy, we can still be pretty sure that it will end up as a frog! In his view, this paradox showed that blind reliance on reductionist arguments obscured problems of forms. Clearly, a new method was needed that would focus on shapes, account for their stability, and explain their creation and destruction.

For Thom, catastrophe theory supplied this method. In summary, its goal was to understand natural phenomena by approaching them directly, rather than relying on traditional reductionism. Its main concern was the creation and destruction of forms, but more precisely, as they arise at the mundane level of everyday life. Catastrophe theory posited the existence of a mathematically defined structure responsible for the stability of these forms, which he called the logos of the form. Consequently, he rejected the idea that the universe was governed by chaos or chance. The models built with the help of catastrophe theory were inherently qualitative—not quantitative—which meant that they were not suited for action or prediction, but rather aimed at describing, and intelligibly understanding, natural phenomena. Finally, Thom recognized that catastrophe theory was not a proper scientific theory, but rather a method or a language that could not be tested experimentally and therefore was not falsifiable in the sense of Karl Popper.

Mathematical styles: Bourbaki against intuition

At the source of catastrophe theory, we find a man who still "sociologically" defines himself as a mathematician. Born in 1923, René Thom recalls a "decisive encounter with Euclidean geometry" during his lycée years, when he fell for "the geometric mode of thought and type of proof."[16] However, his geometric, intuitive vision of mathematics was opposed to the dominant trend. In 1943, Thom experienced at the École normale supérieure "the excitement born with Bourbakist ideas."[17] Some of the Bourbaki group, already important members of the French mathematical community, were among Thom's professors. Bourbaki "was a symbol . . . of the triumph of abstraction over application, of formalism over intuition."[18] The Bourbaki group did not reject geometry as much as the intuitive approach to Euclidean geometry, upon which Thom's mathematical intuition and philosophy were built. Thom's opinion of Bourbaki was thus quite ambivalent. As one of Bourbaki's most successful students, Thom praised his introduction into France of the mathematics of Göttingen. But as David Hilbert himself once wrote, two tendencies were present in mathematics: "On the one hand, the tendency toward abstraction, [seeking] to crystallize the logical relation inherent in the maze of material that is being studied, and to correlate the material in a systematic and orderly manner. On the other hand, the tendency toward intuitive understanding, [fostering] a more immediate grasp of the objects one studies, a live rapport with them, so to speak."[19]

For Thom, Bourbaki had clearly chosen the first path, thus failing to keep Hilbert's mathematics alive. "It is a bit as if, at the time of Vesaleus, when the method of dissection eventually imposed itself, one had wanted to identify the study of human beings with the analysis of cadavers."[20] Bourbaki's ascetic formalism killed mathematics.

Thom knew Bourbaki very well. He was once almost recruited by them but says that he literally fell asleep during the lectures.[21] Nevertheless, he was learning. Thom's early achievement was to reconcile his powerful geometric intuition with Bourbaki's arsenal. In 1946, he moved to Strasbourg with his mentor Henri Cartan, who had oriented him toward differential topology. This, in part, motivated Thom's ambiguous assessment of Bourbaki. Multidimensional spaces—which topologists mostly study—are difficult to visualize. So, a systematic, formal mode of thought, however boring and counterintuitive it might be, is then incomparably useful.

1. René Thom lecturing on catastrophe theory at Memorial University, Canada, in the early 1970s, with a section of the cusp behind him. Source: Photo Section, ETV Centre, Memorial University.

Indeed, Thom mastered the techniques offered by Bourbaki's edifice well enough to obtain results, which, according to Jean Dieudonné, marked "the modern rise of differential topology."[22] In 1958, he was awarded the highest distinction for a mathematician—the Fields Medal.

On this occasion, Heinz Hopf identified Thom's strengths. This was a time when topology was in a "stage of vigorous . . . algebraicization." Not only had algebra been found to provide "a means to treat topological problems," but also "it rather appears that most of [these] problems themselves possess an explicitly algebraic side." Still, for Hopf, there lurked the danger of "totally ignoring the geometrical content of topological problems. "In regard to this danger, I find that Thom's accomplishments have something that is extraordinarily encouraging and pleasing. While Thom masters and naturally uses modern mathematical methods and while he sees the algebraic side of his problems, his fundamental ideas . . . are of a perfectly geometric-*anschaulich* nature."[23] Thom was able to use Bourbakist algebraic methods in solving topological problems without losing sight of their anschaulich, or intuitive, character. Tim Poston, a later catastrophist, vividly contrasted Thom's style with a traditional approach *à la* Bourbaki. "Some mathematicians go at their work like engineers building a six-lane highway through the jungle, laying out surveying lines, clearing the under-brush, and so on. But Thom is like some creature of the mathematical jungle, blazing a trail and leaving just a few marks on his way to the next beautiful clearing."[24] Indeed, Thom came to view rigor in mathematics as counterintuitive and counterproductive. "Absolute rigor is only possible in and by insignificance." True to his preference for meaningful wholes over insignificant details, he held that rigor hid the essential. In mathematical research, it should always come second. "Rigor, in mathematics, is essentially a question of housekeeping [*intendance*]."[25]

Mathematical interlude I: Thom's cobordism theory

Thom's work on cobordism, for which he was awarded the Fields Medal, clearly illustrates his intuitive approach as allied with the profound knowledge of Bourbakist methods that guided most of his mathematical work.[26] As Hopf testified, cobordism was important because of the way it mixed topological and algebraic approaches in the classification of manifolds. In the following, the definition of a few concepts will be recalled. Briefly, Thom's cobordism theory enabled him to construct groups Ω^n out of equivalence classes of manifolds of dimension n, and to classify these groups.

Topology is a generalization of geometry that studies spaces with the degree of generality appropriate to a specific problem. One central concern of topology is to study the properties of spaces that do not change under a continuous transformation, that is, translation, rotation, and stretching without tearing. One such property is expressed by the concept of *dimension*: a curve is one-dimensional; a surface has two dimensions; ordinary space, three; and the space-time of general relativity, four.

Mathematicians faced with the problem of characterizing a space locally isomorphic to a Euclidean space use the notion of *manifold*. An n-dimensional manifold is a space M, such that a neighborhood V exists around each point p of M in one-to-one correspondence with a subset W of the n-dimensional Euclidean space $\mathbf{R}^n$. The study of manifolds is called differential geometry, and the classification of all manifolds of a given dimension is an important problem of topology. It is also possible to define

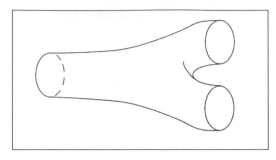

2. The manifold composed of two circles is cobording with the manifold consisting of a single circle because there is a "pant-shaped" smooth surface joining them. Since this is true for manifolds combining any number of circles, the group Ω^1 is the one-element trivial group.

manifolds with edges. If the manifold with edges has $n+1$ dimensions, then the edges are n-dimensional manifolds. For example, a sheet of paper rolled into a cylinder has two circles as edges. A manifold with three circles as edges is pictured in figure 2.

Let us also define equivalence relations and equivalence classes. An equivalence relation, symbolized by $\sim$, over a set S is defined so that, for all a, b, and c in S, the three following properties are satisfied: (1) reflexivity: $a \sim a$; (2) symmetry: if $a \sim b$ then $b \sim a$; and (3) transitivity: if $a \sim b$ and $b \sim c$, then $a \sim c$. The equivalence class $[a]$ of an element a of S is the subset of S that contains all the elements b that are equivalent to a, that is, all b's in S such that $b \sim a$.

Thom defined two manifolds M and N, both of dimension n, to be cobording (in French, *cobordantes*, from *bord* or "edge") if there was a manifold P of dimension $n+1$ so that M and N formed its edge. He then showed that cobording manifolds formed an equivalence class. For example, one circle is cobording with the manifolds formed by the nonintersecting union of two circles, because it is possible to unite them with a two-dimensional manifold with edges (figure 2).

Thom realized that the set Ω^n of all these equivalence classes formed a group, the group operation being defined as the nonintersecting union of manifolds. Exploiting modern formalism with the help of Jean-Pierre Serre, Thom identified the structure of those groups. He found that

$$\Omega^0 = Z; \ \Omega^1 = \Omega^2 = \Omega^3 = 0; \ \Omega^4 = Z; \ \Omega^5 = Z_2; \ \Omega^6 = \Omega^7 = 0.$$

(He also provided partial results for higher dimensions.)

It is worthwhile to note that if M is cobording with N, then it is possible for M to evolve in time and become N. Thus cobordism can be seen as the study of possible continuous transformations of a given shape. Retrospectively, Thom also saw it this way: "The problem of cobordism . . . is of knowing when two manifolds can be deformed one into the other without encountering a singularity in the resulting space, at any *moment* in this deformation."[27] The example of a circle becoming two circles can, very crudely of course, model cell division (figure 2).

The mathematical background of catastrophe theory

When Thom moved to Strasbourg in 1946, it hardly corresponded to the provincial exile that successful French professors often had to endure before they could trek back to Paris. In addition to the presence there of Thom's thesis director Henri Cartan, the Bourbakist Charles Ehresmann directed a topology seminar, where in 1950 Thom heard Hassler Whitney describe his work on singularities of mappings from the plane to the plane.[28] Thom also became acquainted with Morse theory concerning the relation between the topology of spaces and the singularities of real functions defined on them.

From his stay in Strasbourg, Thom drew resources congenial to his attack on the problems of singularity theory, which he founded with Morse and Whitney. Just like "living beings," Paul Montel wrote in 1930, "functions are characterized by their singularities."[29] Trying to make sense of multidimensional spaces, Thom considered singular points a blessing. He once discussed "a philosophical aspect" motivating the emphasis put on singularities, thus revealing his topological intuition. "A space is a rather complex thing that is difficult to perceive globally." To study its structure, one may however project it on the real line. "In this flattening operation, the space resists: it reacts by creating singularities for the function. The singularities of the function are in some sense the vestiges of the topology that was killed: . . . its screams."[30] Publishing in 1955 his first article on singularities, Thom knew that he had found a great topic: "There is hardly any doubt . . . that the study of the local properties of singularities of differential mappings opens the door to an extremely rich domain."[31] His work on singularities provided him crucial mathematical tools for catastrophe theory: the concepts of *genericity* and of *structural stability*, as well as a classification of singularities later to become a list of the seven *elementary catastrophes*.

As an intuitive way of saying that some properties were much more common than others, the concept of genericity had been loosely used by Italian algebraic geometers since the beginning of the century. After a "memorable discussion" with the Bourbakist Claude Chevalley at Columbia in 1952, Thom had the idea of extending its use to other domains. "I quickly perceived that this phenomenon of 'genericity' was an essential source for our present worldview."[32] As for structural stability, it had been introduced from Russia (where it was known as roughness) by Princeton topologist Solomon Lefschetz, who since World War II had been reviving the qualitative study of ordinary differential equations and whom Thom visited in the early 1960s.[33] Structural stability was the assumption that, in order to be physically useful, systems had to exhibit similar behavior when slightly perturbed. That this concept was pivotal for catastrophe theory is reflected in the title of Thom's book, *Structural Stability and Morphogenesis*. Central to Stephen Smale's contemporary development of modern dynamical systems theory, the conjunction of genericity and structural stability likewise guided Thom's research program in singularity theory. Smale wished to show that structurally stable systems were generic; Thom, that structurally stable mappings were generic.[34]

Mathematical interlude II: singularity theory

The "screaming" projection that René Thom described to show the importance of singularities was called a Morse function. It was a smooth mapping f from an n-dimensional manifold M to the real line $\mathbf{R}$. As Thom described it, one of Morse's crucial results allowed "the determination of the relations between the topological characteristics" of M and the singular points of f.[35] Consider a smooth differentiable mapping f from $\mathbf{R}^m$ to $\mathbf{R}^n$, or more generally from an m-dimensional manifold M to an n-dimensional manifold N. Then, a point p in M was a *singular point* of f if there was a direction along which the derivative of f at p vanished.

The name of the game then was, as often in modern mathematics, to classify and characterize singularities. For an arbitrary mapping f and arbitrary manifolds M and N, this was a very hard problem.[36] Thom focused on low-dimensional spaces and on structurally stable mappings, that is, those whose topological character was preserved under small perturbations. He hoped that structurally stable mappings would prove to be very common, so that every mapping was either stable or, in a topological sense, very close to one that was: in mathematical parlance, they were *generic*.

In the above example of real functions, a generic singular point p was such that the second derivative of f at p was nonzero: $f'(p) \neq 0$. Morse theory showed that using an appropriate change of variable $x \rightarrow y(x)$, such that $y(p) = 0$, then f could be written as $f(y) = \pm y^2$ in a small neighborhood. This completely classified the generic singular points for real functions: there was, essentially, only one kind of singularity that could occur, soon to be identified, in Thom's language, with the catastrophe called a *fold*.

Whitney completely classified the singularities that "a good approxima-

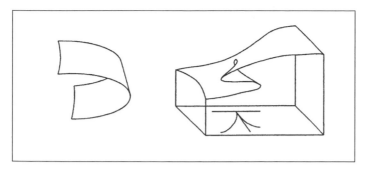

3. The fold and the cusp catastrophes.

tion" of any mapping from the plane to the plane was allowed to have.[37] He considered a surface S (a sheet, for example) projected on a plane underneath. The surface S was just a different parametrization of the plane. Often, there was no problem; there was a one-to-one correspondence between the points of S and those below. But it might happen that there was a fold, close to which two points from the surface were projected onto the same point in the plane; this was a singularity. Isolated points could even be encountered around which three points of S were projected onto the same point in the plane: these were *cusp* singularities. These two were the only local singularities that would survive small perturbations of the sheet.

Thom's elementary catastrophe theory basically extended this classification to higher dimensions, but with a slight difference. In *Structural Stability*, he recognized that the essential characteristics of a smooth function could be analyzed by studying its embedding into a smooth family of functions $F(x,u)$, such that $F(x,0) = f(x)$, which he called an *unfolding* of the function f (where x and u are multidimensional vectors). "The goal of catastrophe theory is to detect properties of a function by studying its unfoldings."[38] An infinite number of unfoldings existed for a given function f. The question was to know if one existed that captured the essential information about all of them. Such an unfolding, when it existed and the number of dimensions of the variable u was minimal, was called *universal*. The fold and the cusp, discussed earlier, were universal unfoldings of $f(x) = x^3$ and x^4, respectively (see figure 3). The tricky part of this program was to find universal unfoldings.

A beautiful, intriguing field of pure mathematics

The relationship between catastrophe theory and mathematics has always been contested. On the one hand, the mathematician John Guckenheimer wrote that *Structural Stability and Morphogenesis* "contains much of interest to mathematicians and has already had a significant impact upon mathematics, but [it] is not a work of mathematics." On the other hand, authors of recent textbooks often feel the need to stress its mathematical nature. One started by emphasizing that "catastrophe theory is a branch of mathematics." Another asserted that this branch had in fact been "discovered" by Whitney and transformed "into a 'cultural' tool" by Thom.[39]

There can be no doubt that Thom's mathematical experience made catastrophe theory possible and shaped his philosophy. As early as 1967, he divided catastrophes into two categories on the basis of his mathematical knowledge: the seven *elementary catastrophes* arising in simple systems and *generalized catastrophes*, which lived in more complex spaces arising with global loss of symmetry.[40] Thom wrote very little about the latter, since the mathematical basis for their classification was lacking. As for the former, they were those sudden discontinuities that occurred in systems whose dynamics were controlled by a gradient (or potential). The classic image "of a ball rolling around a landscape and 'seeking' through the agency of gravitation to settle in some position which, if not the lowest possible, then at least lower than any other nearby" was offered by Tim Poston and Ian Stewart.[41]

One of the most powerful results from singularity theory, and one that made catastrophe theory possible, was a complete classification of the elementary catastrophes that arose in systems described by less than four internal parameters. In this case, Thom conjectured that only seven elementary catastrophes existed: the *fold, cusp, swallowtail, butterfly*, and the three *umbilics*. Later widely known as "Thom's theorem," this conjecture was fully proved by Bernard Malgrange and John N. Mather, who used a heavy arsenal of functional analysis and algebraic topology.[42] Elementary catastrophe theory showed for certain that for gradient dynamical systems with a small number of parameters, abrupt generic changes had to be described locally by one of Thom's elementary catastrophes.

It was Christopher Zeeman's exploitation of Thom's theorem that made the international fame of catastrophe theory and later brought discredit to it. But this barely touched on Thom's own vision for his theory.[43] Too tight a focus on this theorem betrays his philosophy and misses the point of his most important innovations for the practice of modeling, a fact recognized by some catastrophists. "It is not Thom's *theorem*, but Thom's *theory*, that is the important thing: the assemblage of mathematical and physical ideas that lie behind the list of elementary catastrophes and make it work."[44]

Thom emphatically concurred with this view. He granted that advances in topology had made his philosophy possible and that mathematical concerns shaped his theory. Indeed, a larger body of qualitative mathematics would have been quite beneficial for catastrophe theory. But such mathematical tools were just one facet of a general method of scientific inquiry.

Catastrophe theory is not a theory that is part of mathematics. It is a mathematical theory to the extent that it uses mathematical instruments for the interpretation of a certain number of experimental data. It is a hermeneutical theory, or even better, a methodology, more than a theory, aiming at interpreting experimental data and using mathematical instruments whose list is, for that matter, not a priori defined.[45]

The most casual reading of Thom's work reveals that his thought was framed by mathematical language. His emphasis on shapes and qualitative theories can be traced directly back to his work on topology, where measurements are eschewed, and on singularity theory, where global properties can be extracted from the local study of critical points. But Thom did not come up with catastrophe theory until he had experimented with biological theories. These are at least as important as his mathematical practice in explaining catastrophe theory. In fact, it was from his reading of embryology textbooks that he adopted the notion of *attractor*, later to figure prominently in the modeling and experimental practice of chaos.

Toward a theoretical biology?

Overlooking beautiful Lake Como, in the village of Bellagio, Italy, stands Villa Serbelloni owned by the Rockefeller Foundation. There, on 28 August 1966, a select group of computer scientists, mathematicians, physicists, and, of course, biologists (but hardly any molecular biologists!) gathered "to explore the possibility that the time [was] ripe to formulate some skeleton of concepts and methods around which Theoretical Biology [could] grow."[46] There also, René Thom introduced the notion of catastrophe. Far from being the first application of catastrophe theory to another discipline, Thom's theory of morphogenesis, as we shall see, grew out of his foray into embryology, which, at a mathematical level, helped him conceptualize the notion of attractor, and, at a philosophical level, gave Thom an example of a practice that used morphological raw materials.

From pure mathematics to theoretical biology, 1960–1968

In 1963, René Thom joined the faculty of the Institut des hautes études scientifiques (IHÉS), where he would have no teaching obligation and could devote most of his time to research, and where he would slowly move away from mathematics and venture into biology and linguistics.[47] At the IHÉS, he noted, "I had more leisure time, I was less preoccupied by teaching and administrative tasks. My purely mathematical productivity seemed to be declining and I began to be more interested in the periphery, that is, to possible applications." Perhaps he had finally succumbed to a taste for philosophy that he had neglected since his lycée years.[48] For all his success, Thom seemed to have found mathematics hard to practice and somewhat dissatisfying. "If you don't need to work in mathematics for a living you need much courage to do it, because, in spite of all, mathematics is difficult!"[49] However, he did not immediately abandon all concern with pure mathematics: throughout the 1960s he published articles on singularity theory, introducing many concepts picked up by other mathematicians.[50] Ultimately, one might concur with Zeeman: "In a sense Thom was forced to invent catastrophe theory in order to provide himself with a canvas large enough to display the diversity of his interest."[51]

In 1960, while in Strasbourg, Thom had already begun to experiment with caustics—those luminous outlines formed, for example, by sunlight in a cup of coffee. With singularities proving so fruitful in mathematics, he wondered whether they would be just as useful in the study of the physical world. Armed with a few simple instruments, he studied several caustics and their perturbations. The rays reflected by a spherical mirror, for example, formed a luminous curve with a cusp: a singularity. "This cusp has the marvelous property of being stable. If the orientation of the light rays is slightly changed, one sees that the cusp subsists. *This is the physical effect of a theorem of mathematics.*"[52]

Stumbling upon an unexpected behavior in optics, Thom then turned to biology. In 1961, while visiting the Natural History Museum in Bonn, he hit upon a plaster model of the gastrulation of a frog egg. "Looking at the circular groove taking shape and then closing up, I saw . . . the image of a cusp associated to a singularity. This sort of mathematical 'vision' was at the origin of the models I later proposed to embryology."[53] Thom also recalled that around 1962 he was struck by some mathematical models for biology: a proposal by the physicist Max Delbrück in 1949 to account for cell differentiation in terms of transitory perturbations of the cell's chemical environment, and Christopher Zeeman's articles on the topology of the brain, suggesting that topology could be applied to biological phenomena.[54]

In his preface to *Structural Stability and Morphogenesis*, Thom singled out four biologists as his precursors. In addition to D'Arcy Wentworth Thompson's classic *On Growth and Form*, he mentioned two "physiologists": Jakob von Uexküll and Kurt Goldstein.[55] In their works Thom found a way of treating organisms as wholes, a nonreductionist approach to biology that provided mechanisms accounting for the finality of living beings. Above all, he was impressed by the writings of the fourth man he cited: British biologist Conrad Hal Waddington. In 1968, Thom claimed two sources for his theory of morphogenesis: "On the one hand, there are my own researches in differential topology and analysis on the problem called structural stability. . . . On the other hand, there are writings in Embryology, in particular those of C. H. Waddington whose ideas of 'chreod' and 'epigenetic landscape' seem to be precisely adapted to the abstract schema that I met in my theory of structural stability."[56]

This acknowledgment of Thom's—that his catastrophe theory derived also from embryology rather than having been merely applied to it—has rarely been taken seriously by commentators. But it is at the interface with biology that Thom would develop a mathematical picture of competition between attractors in dynamical systems—a picture that would become one of the cornerstones of both catastrophe and chaos theories.

"Wad" and the synthesis of biology

According to Waddington, the main problem of biology was to account for the characteristics that defined living organisms: form and end. "How does development produce entities which have Form, in the sense of integration or wholeness; how does evolution bring into being organisms which have Ends, in the sense of goal-seeking or directiveness?"[57] Organisms retained their shapes in spite of the fact that matter was continuously flowing through them. Development always ended up in the same final state, after having passed through the same stages. These problems of organization were fundamental questions, only to be solved by a synthesis of evolution, embryology, and genetics. Although Waddington believed that genes were the major cause for development, he never denied the influence of the rest of the organism. Thus, he thought that, while part of the answer lay in genetics, the main focus of study should not be the genes themselves but the nature of the causal relationship between the organism and its genes. For this science, he coined the name *epigenetics*.[58]

Being "stuck" with a biological order "in which there [was] an inescapable difference between the *genotype*—what is transmitted, the DNA— and the *phenotype*—what is produced when the genotype is used as instructions," the epigeneticist's task was to come up with mechanisms that could explain the phenotype in terms of the genotype.[59] Epigenetics had two main aspects: changes in cellular composition (cell differentiation), and changes in geometrical form (morphogenesis). Development followed definite pathways, which were resistant to change. The description of these pathways and the genetic influences on them was thus a major task of epigenetics. In 1939 Waddington introduced an intermediary space between the genotype and the phenotype, which he called the *epigenetic landscape*. In a unique visual representation, it combined all the development paths, which were pictured as valleys (figures 4 and 5).[60] The epigenetic landscape had no physical reality, but it helped visualize the various developmental processes.

> Consider a more or less flat, or rather undulating surface, which is tilted so that points representing later states are lower than those representing earlier ones [figure 4]. Then if something, such as a ball, were placed on the surface it would run down towards some final end state at the bottom edge. . . . We can, very diagrammatically, mark along it one position to correspond, say, to one eye, and another to the brain.[61]

The image of the ball rolling down a surface is of course reminiscent of the potential functions of catastrophe theory. Moreover, the valleys formed on the epigenetic landscape had the property of being stable, in the sense that after a small perturbation in its trajectory, the ball tended to go back to the valley. These stable pathways of change, Waddington called *creodes*, and later *chreods*.[62] In his work on *Drosophila* during the 1930s, Waddington had studied the switches that can occur among several developmental paths. If a gene were active at a particular moment in the sequence of events, then the eye had a different tint of red. At the switches an important phenomenon took place. The ball had to choose among several pathways (figure 6). René Thom would see in this a topological change in the set of minima (singularities) of the potential function: a *catastrophe!*

Attractors in dynamical theories of morphogenesis

In his "dynamical theory of morphogenesis," Thom introduced a biochemical model of cellular differentiation. Independently, Waddington and Delbrück had proposed that gradients in the concentrations of some postulated chemical substance might account for the phenomenon.[63] In their schemes, the cell was constantly processing chemical substances so that the different concentrations changed in a complex way—given by coupled, nonlinear equations. In a biological system, a flux equilibrium was eventually reached; that is, concentrations remained stable even though chemical substances always flowed through the cell. Waddington and Delbrück considered that several stable regimes were achievable. The classification of these stable regimes became, in Thom's scheme, the description of the system's morphologies. Hence one of his most innovative ideas: to consider systems, even physical ones, in terms of the different end points they can reach, which he translated as a study of forms in nature. It expressed in a mathematical language adapted to the physical sciences the concept of finality in biology.

These different stable regimes of the system Thom called *attractors*.[64]

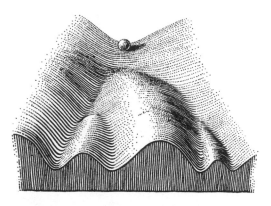

4. "*Part of an Epigenetic Landscape*. The path followed by the ball, as it rolls down towards the spectator, corresponds to the developmental history of a particular part of the egg. There is first an alternative, towards the right or the left. Along the former path, a second alternative is offered; along the path to the left, the main channel continues leftwards, but there is an alternative path which, however, can only be reached over a threshold." Source: Conrad Hal Waddington, *The Strategy of the Genes: A Discussion of Some Aspects of Theoretical Biology* (London: Allen and Unwin, 1957), 29.

5. "*The complex system of interactions underlying the epigenetic landscape.* The pegs in the ground represent genes; the strings leading from them the chemical tendencies which the genes produce. The modelling of the epigenetic landscape, which slopes down from above one's head towards the distance, is controlled by the pull of these numerous guy-ropes which are ultimately anchored to the genes." Source: Waddington, *Strategy of the Genes*, 36.

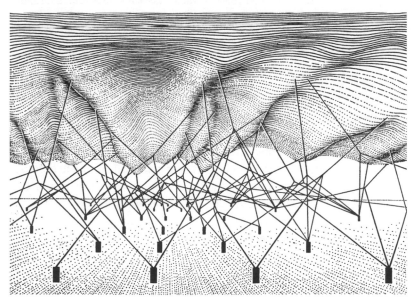

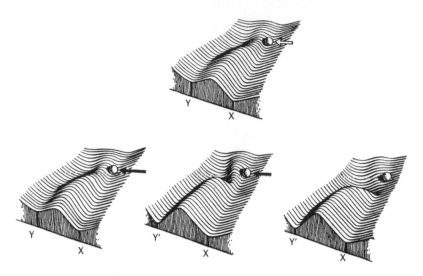

6. "*'Organic selection' (the Baldwin effect) and genetic assimilation.* The diagram above shows part of an epigenetic landscape, with a main valley leading to the adult character X and a side branch leading to Y; the developing tissue does not get into the Y path unless an environmental stimulus (hollow arrow) pushes it over the threshold. The three diagrams below show ways in which the 'acquired character' Y might become incorporated into the genotype. On the left, the original environmental stimulus is replaced by a mutant allele (dark arrow) which happens to turn up; this is 'organic selection.' On the right are two modes of 'genetic assimilation.' In the central one, the threshold protecting the wild type is lowered to some extent, but there is an identifiable major gene which helps push the developing tissues into the Y path. On the right, the genotype as a whole causes the threshold to disappear and there is no identifiable 'switch gene.' Note that in both the genetic assimilation diagrams there has been a 'tuning' of the acquired character, i.e., the Y valley is deepened and its end-point shifted from Y to Y'." Source: Waddington, *Strategy of the Genes*, 167.

They were regions of the configuration space stable under the dynamical equations of the system and such that any configuration close enough to an attractor would approach it asymptotically. The *basin of the attractor* was a region containing the attractor and inside of which any initial condition fell back to it. Of course Thom was aware that to achieve a complete topological description of attractors and basins of a general system would be a difficult task. It was an imaginable one, however, and, in essence, this task became a major focus for research on chaotic systems.

For local systems where, for example, the concentration of chemical substances was given at each point of space and time, attractors could differ from point to point. Thus the domain of space under study—the cell—was divided into several regions associated with different attractors.

These regions were separated by surfaces that Thom called "shock waves." Using Thom's theorem, he could establish that for gradient dynamics, these surfaces could only exhibit a small number of singularities, which were elementary catastrophes. Starting with a local singular situation in a dynamical system, he could say what ulterior catastrophes were contained in the "universal catastrophe space" associated with the singularity. For example, if one started with a local critical cusp situation, the only other catastrophes that could occur later were folds. Of course, all of this was local in a topological sense: some finite time limit existed beyond which anything could happen. There was no way of knowing how large this limit was; it could be as small as one wished but not zero. It could even be impossible to detect; hence Thom was reluctant to accept that catastrophe theory could be submitted to experimental control.

In his theory, Thom saw "a mathematical justification for the idea of 'epigenetic landscape,' suggested 20 years earlier by Waddington."[65] This was not mere gesture: the ideas of conflicting attractors had been described almost word for word by the biologist.

> [1] At each step [of development] there are several genes acting, and the actual development which occurs is the result of a balance between opposing gene-instigated tendencies. [2] At certain stages in the development of an organ, the system is in a more than usually unstable condition, and the slightest disturbances at such times may produce large effects on later events. . . . [3] An organ or tissue is formed by a sequence of changes which can be called the "epigenetic paths." . . . And also each path is "canalized," or protected by threshold reactions so that if the development is mildly disturbed it nevertheless tends to regulate back to the normal end-result.[66]

Although he hardly knew enough mathematics, Waddington claimed that Thom had "shown how such ideas as chreods, the epigenetic landscape, switching points, etc.,—which previously were expressed only in the *unsophisticated language of biology*—can be formulated more adequately."[67] However, catastrophe theory explained biological structures by describing "the basic and universal constraints of stability imposed on epigenetic mechanisms," independently of DNA, and therefore never answered the question that had prompted Waddington to imagine epigenetic landscapes and chreods in the first place, that is, the link between development and genetics. Contentiously, Thom insisted that "only a mathematician, a topologist, could have written [this theory], and the time may be very near when, even in biology, it might be necessary to think."[68]

Revealing contrasts exist between Thom's writings on biology and those of Jacques Monod, the Nobel prize-winning molecular biologist whose work would reach a broad audience. A chapter of Monod's *Chance and Necessity*, published in 1970, was devoted to the problem of spontaneous morphogenesis of living organisms. But Monod's picture was almost totally opposed to Thom's. Indeed, Monod explained his aims as follows:

> In this chapter I wish to show that this process of spontaneous and autonomous *morphogenesis* rests, at bottom, upon the stereospecific recognition properties of proteins; that is *primarily a microscopic process* before manifesting itself in macroscopic structures. . . . But we must hasten to say that this "reduction to the microscopic" of morphogenetic phenomena does not yet constitute a working theory of phenomena. Rather, it simply sets forth the principle in whose terms such a theory would have to be formulated if it were to aspire to anything better than simple phenomenological description.[69]

As opposed to Thom's reduction of morphogenetic processes to a certain mathematical idealism, Monod argued for the "principle" of reducing them to molecular interaction. As Monod's remarks indicate, this was nothing more than a "principle" and certainly not a full theory. But Monod put a great deal of faith in this principle.

> I for my part remain convinced that only the shape-recognizing and stereospecific binding properties of proteins will in the end provide the key to these [morphogenetic] phenomena. . . . In a sense, a very real sense, it is at the level of chemical organization that the secret of life lies, if indeed there is any one such secret.[70]

Emphasizing the molecular and chemical properties of the substratum, the forces acting between organic macromolecules, and quantitative studies, Monod's discourse sounded like a diatribe directed at Thom, or conversely.[71] Just as uncompromising, the mathematician emphasized that no theoretical explanation was conceivable in biology without the aid of mathematics.

> There should not exist any other theorization than mathematical; concepts used in each discipline that are not susceptible of gathering a consensus around their use (let us think, for example, of the concept of information in Biology) should be progressively eliminated after having fulfilled their heuristic function. In this view of science, only the mathematician, who knows how to characterize and generate stable forms in the long term, has the right to use (mathematical) concepts; *only he, at bottom, has the right to be intelligent.*[72]

In this context, one is hardly surprised by the fact that Thom's theory had little impact on biology.[73] However, his forays into embryology provided Thom with crucial intuition about ways to study dynamical systems with finality. In no small sense, his introduction of the concept of the attractor, and the even more important concept of the basin of an attractor, can be seen as stemming from his involvement in biology.

Topology and meaning

Having pointed out the relevance of topological concepts and practices for the modeling of biological phenomena, Thom saw no reason to stop there. Since the early 1970s, his main fields of research, besides philosophy, have been linguistics and semiotics. With his incursion into the human sciences, Thom was bound to confront structuralism. Never himself a structuralist per se, but trained in the mathematical structuralism of Bourbaki, he was attracted by this movement. With some adjustments, his theories could be made to fit into structuralist modes of thought. But because he began to work on linguistics so late, catastrophe theory was only mildly affected by structuralism in practice. Increasingly faced with strong opposition to his ideas about modeling, Thom pondered the epistemological foundations of catastrophe theory. In attempting to articulate the kind of knowledge that the theory produced, he used structuralist resources most obviously.

Catastrophes, man, and language

In his manuscript of *Structural Stability and Morphogenesis*, Thom titles chapter 13 "L'homme." It would be published with substantial additions under the title "From Catastrophes to Archetypes: Thought and Language." The original chapter aimed at extending the techniques and assumptions of catastrophic models of morphogenesis to human thought processes and societies. He actually developed few of the models he suggested. Always a mathematical terrorist, Thom used mathematical notations and language only to express vague correspondences among neurological states, thoughts, and language.

His basic assumption was that there existed a few "functional chreods," later renamed "archetypal chreods," that expressed simple biological actions: to throw a projectile, to capture something, to reproduce, and so forth. These chreods had been internalized in the human brain, whose mental activity (*activité psychique*) he identified with a dynamical system. By analogy with the epigenetic landscape, Thom postulated that this psychological system was divided among basins and attractors. "The sequence of our thoughts and our acts is a sequence of attractors, which succeed each other in 'catastrophes.' "[74]

Language, Thom then claimed, was a translation of these mental attractors. A mental atlas of dynamic chreods existed that was common to all human beings. An idea was a mental attractor. When one wished to formulate a sentence expressing an idea, it was mathematically projected onto a space of admissible sentences, where several attractors competed. One was eventually chosen, and the sentence was uttered. All this was manifestly programmatic and rather vague.

In the Parisian intellectual climate of the late 1960s, Thom could not avoid structuralism, especially since he dealt with language. As early as 1968, he noted that "the problem of meaning has returned to the forefront of philosophical inquiry."[75] Nevertheless, semiotics was first introduced in Thom's work not as a quest in itself, but as a method for biology. Indeed, he considered Saussure's notions of *signified* and *signifier* as congenial to the goals of epigenetics, which were to find the connections between genetics and embryology. "Is not such a discipline which tries to specify the connection between a global dynamic situation [the organism] (the 'signified'), and the local morphology in which it appears [DNA] (the 'signifier'), precisely a 'semiology'?" He portrayed his method for morphogenesis as a problem of semantics. "The decomposition of a morphological process taking place in $\mathbf{R}^m$ can be considered as *a kind of generalized m-dimensional language*; I propose to call it a *'semantic model.'* "[76]

7. Waddington's switching diagram. "The formation of eye colors in Drosophila. The pigment-forming process normally runs down the line through the ca^+ substance, the v^+ substance, and the cn^+ substance, to give wild type pigment. The genes, ca, v and cn interrupt this sequence, so that the process takes an altered course, to give claret, vermilion or cinnabar pigmentation." Source: Conrad Hal Waddington, *Organisers and Genes* (Cambridge: Cambridge University Press, 1940), 77.

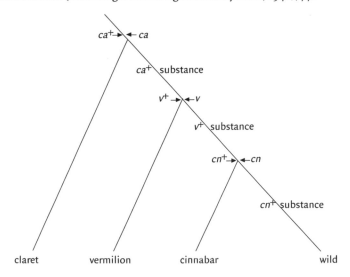

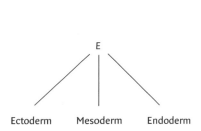
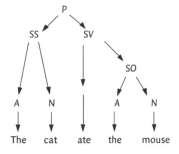

8. Thom's analogy between graphs of sentences and development. Source: Thom, "Structuralism and Biology," in Conrad Hal Waddington (ed.), *Towards a Theoretical Biology* (Edinburgh: Edinburgh University Press, 1972), 4:80.

In 1970, Thom presented a more sophisticated catastrophe-theoretical model of language.[77] His goal was to explain the syntactical structure of atomic sentences (basically, those with one verb), in terms of their meaning. He was struck by the resemblance between the tree-shaped graphs that linguist Louis Tesnière used to analyze the structure of sentences and Waddington's chreods (figures 7 and 8).[78] Indeed, stripped of the out-of-equilibrium position, the epigenetic landscape became a switching diagram, like a tree. In Tesnière's view, verbs were the center of gravity of sentences. They became, in Thom's view, the attractors of mental activities, and words were chreods. He developed a visual representation of the verbs associated with spatiotemporal activities by using sections of elementary catastrophe surfaces. This was, he would say twenty years later, a "geometrization of thought and linguistic activities."[79] The main benefit of such an analysis was to establish a map from signified to signifier, which went against the Saussurean dogma that the relation between signifier and signified is arbitrary. Classifying syntactical structures into sixteen categories, Thom claimed that "the topological type of the interaction determines the syntactical structure of the sentence which describes it."[80] Meaning and structure were no longer independent. Thus, with catastrophe theory and the biological analogy, Thom subverted classical structuralist ideas, which explains why his work would be picked up by philosophers like Michel Serres and Jean-François Lyotard, who explicitly opposed the structuralist project.[81]

Structuralism and biology: explaining forms

Thom confronted structuralism head on in 1972: "Can structuralist developments in anthropological sciences (such as linguistics, ethnology, and so on) have a bearing on the methodology of biology? I believe this is so." By then what he called structuralism was merging with his own method.

> The task of any structuralist theory is: (1) to form a finite lexicon of elementary chreods; (2) to build experimentally the "corpus" of the empirical morphology; . . . (3) to define "conditional chreods," objects of the theory; (4) to describe the internal structure of a conditional (or elementary) chreod by associating a mathematical object to it, whose internal structure is isomorphic to the structure of the chreod.[82]

Catastrophe theory and structuralism reinforced each other. The semantic analogy was common among molecular biologists who were prone to interpret the living order in terms of the DNA code as a static linear string. For Thom, this was wrongheaded because if biology could indeed be seen as a semantic model, it was a dynamical, multidimensional one. The DNA code by itself was a semantic model of dimension one: how could it describe the spatial processes of biology?[83]

In contact with structuralist linguistics, Thom extracted a philosophy of science that he thought would be able to make sense of the knowledge his approach had produced, and not only in the human sciences. Henceforth, Thom distinguished two approaches to scientific knowledge: the *reductionist* and the *structural*.[84] Both approaches aimed at simplifying the description of empirically observed morphologies. But the latter refused to do so by attributing causal effects to factors outside the observed morphology. The only admissible causality was structural.

Thom had obviously modeled his "structural approach" on the linguists' claims to knowledge production. He thought some human sciences had succeeded in building nonreductionist theories, especially formal linguistics and Lévi-Strauss's structural analysis of myths. They held a "paradigmatic value: they show the way in which a purely structural, morphological analysis of empirical data can be engaged." It would indeed be absurd, Thom contended following Lévi-Strauss, to base linguistics on reductionist assumptions. "It would consist in an attempt at explaining the syntactical structure of a sentence of words by an interaction of phonemes of a phonologic character."[85]

Like Swiss psychologist Jean Piaget, Thom saw a serious epistemological problem in structuralism, namely that it could not account for the emergence of its structures because, historically, structuralist linguistics was synchronic, that is, static in time.[86] Thus the hope for the future lay in synthesizing both approaches. Nothing prevented linguists from conceiving time as another dimension of space-time: "we can make a structural theory of the changes of forms, considered as a morphology on the product space of the [conjoined] substrate space and time axis."[87] Indeed, catastrophe theory provided a way of building a dynamic structuralism that would explain the emergence of structure. As he had done with the structuralist mathematics of Bourbaki, Thom used structuralist linguistic practices to undermine the very project of structuralism.

Shapes, logoi, and catastrophes: Thom's philosophy of science

As we have seen, René Thom was among those who loudly contested the success of reductionist science. That science in the twentieth century had been mainly a reductionist enterprise was a commonplace. In their efforts to understand the world—or, more precisely, pursuing the Laplacian dream, to predict its future course—scientists followed Jean Perrin's ideal: "to explain complex visible things with the help of simple invisible things."[88] Thom contended that this approach was far from having lived up to its promises. "The Universe is nothing more than a brew of electrons, protons, [and] photons," he wrote. "How can this brew settle down, on our scale, into a relatively stable and coherent form far from the quantum-mechanistic chaos?"[89]

Thom believed that physicists overreached themselves when they claimed to be able to explain the everyday world. "Realization of the ancient dream of the atomist—to reconstruct the universe and all its properties in one theory of combinations of elementary particles and their interactions—has scarcely been started." Thom adamantly opposed dogmatic reductionism: "this primitive and almost cannibalistic delusion about knowledge, [which demands] that an understanding of something requires first that we dismantle it, like a child who pulls a watch to pieces and spreads out the wheels in order to understand the mechanism."[90] He did recognize reductionism, in principle, as a valid approach to knowledge, but one which was unachievable at the practical level.

"Reality presents itself to us as phenomena and shapes."[91] Thom's program was to make the morphologies of our day-to-day reality the object of a dynamical science of shapes. In a given domain of experience, his modeling practice could be summarized as follows: find the shapes that are usually encountered, establish a list of these shapes according to their topologic character, and find the underlying dynamics that governs their emergence and destruction.[92] Thom took his cue from D'Arcy Thompson, who had recognized the morphological problems arising in the physical sciences. Thompson, however, had confidence that physics was capable of explaining morphologies. "The waves of the sea, the little ripples on the shore, the sweeping curve of the sandy bay between the headlands, the outline of the hill, the shape of the clouds, all these are so many riddles of form, so many problems of morphology, *and all of them the physicist can more or less easily read and adequately solve.*"[93] Listing similar natural shapes, Thom disagreed that traditional physics could do it:

> Many phenomena of common experience, in themselves trivial (often to the point that they escape attention altogether!)—for example, the cracks in an old wall, the shape of a cloud, the path of a falling leaf, or the froth on a pint of beer—are very difficult to formalize, but is it not possible that a mathematical theory launched for such homely phenomena might, in the end, be more profitable for science [than large particle accelerators]?[94]

Catastrophe theory was thus an attempt at formalizing in rigorous mathematical language a dynamics of forms. And in *Structural Stability*, the first seven chapters gave an outline of a general theory of morphology, applicable to all problems of shape.

But it was one thing to focus on forms and quite another to focus on the specific ones that he and Thompson listed. Just as Thom questioned the pertinence to the everyday world of explanations in terms of electrons, he also noticed that science was quite unable to account for "the froth on a pint of beer." The French mathematician Benoît Mandelbrot, inventor of fractals, shared this concern. "Clouds are not spheres, mountains are not cones."[95]

If Mandelbrot saw himself as a new Euclid, Thom thought of himself as picking up a broken line of thought just where Heraclitus had left it. Around 500 BC, Greek philosopher Heraclitus already noticed the difference between knowledge and understanding. "Many people do not understand the sorts of things they encounter! Nor do they recognize them even after they have had experience of them, though they themselves think [they do]."[96] In Heraclitus's fragments, Thom found inspiration. Christening his elementary catastrophes with names like *swallowtail* and *butterfly*, he applied Heraclitus's precept to irregular figures impossible to visualize.[97]

Once a description of natural forms was achieved, the next pressing concern was their stability, especially if one believed that they emerged from a "brew of electrons." Returning to his quarrel with reductionist physics, Thom noticed that "although certain physicists maintain that the order of our world is the inescapable consequence of elementary disorder, they are still far from being able to furnish us with a satisfactory explanation of the stability of common objects and their qualitative properties." In other words, the physicists were not able to understand the morphologies of the world in terms of atoms.

Thom believed that the explanation lay in an ideal mathematical structure. "The stability of a form rests definitively upon a structure of algebraic-geometric character . . . endowed with the property of *structural stability* with respect to the incessant perturbations affecting it. It is this algebraic-geometric entity that I propose, recalling Heraclitus, to call the *logos* of the form."[98] For Heraclitus, the *logos* was the "true discourse according to which everything happens. It was the truth of this world."[99] Thom attributed a logos to each form; it was "a formal structure which insures its unity and stability." One may note here that he was indeed applying Perrin's precept, except that Thom's "simple invisible things" were mathematical structures as opposed to atoms. For all his structuralist talk, Thom's philosophy is well captured by the term "neoreductionism," a word that Giorgio Israel has used to characterize von Neumann's approach.[100]

Thom felt he needed to emphasize that he studied morphology without regard to the substrate. In his manuscript, written in 1966, he had not mentioned this.[101] As a Bourbakist topologist, he believed in the universal relevance of his mathematics. But after having presented his theory to biologists, he underscored its independence from specific material bases. "The essence of our theory, which is a certain knowledge of the properties peculiar to the substrates of the forms, or the nature of the forces at work, may seem difficult to accept, especially on the part of experimenters."[102]

Again, Thom saw himself as heir to D'Arcy Thompson, who, "in some pages of rare insight, compared the form of a jellyfish to that of the diffusion of a drop of ink in water."[103] The only thing that Thompson lacked, Thom added, was a formal foundation in topology, which provided the basis for explaining morphogenesis without relying on material properties. He endorsed a strong idealism: "The hypothesis that Platonic ideas give shape to the universe," he wrote in 1970, "is the most natural and, philosophically, the most economical."[104]

There was a drawback to this all-encompassing vision. Based on topology, Thom's method was not suited to quantitative analysis and measurement. It had to remain qualitative. Traditionally, this was a serious problem: "probably [one of scientists'] most deeply held values concerns predictions; . . . quantitative predictions are preferable to qualitative ones."[105] But Thom saw the qualitative aspect of catastrophe theory in a positive light. To understand the world one had to rid oneself of "the intolerant view of dogmatic quantitative science." To expose this common prejudice, he loved to recall Rutherford's dictum: "Qualitative is nothing but poor quantitative!" But "what condemns these speculative [qualitative] theories in our eyes," Thom wrote, "is not their qualitative character but the relentlessly naive form of, and the lack of precision in, the ideas they use." Now, he claimed, everything had changed since one could "present qualitative results in a rigorous way, thanks to recent progress in topology and differential analysis, for we know how to define a form."[106] Catastrophe theory was the rigorous way to think about quality.

Intelligibility of the world was the benefit and the ultimate goal of Thom's approach. He defended Descartes against Newton: "Descartes, with his vortices, his hooked atoms, and the like, explained everything and calculated nothing; Newton, with the inverse square law of gravitation, calculated everything and explained nothing." Again and again, Thom opposed explanation to prediction, intelligibility to control, understanding to action: "just as precise knowledge of a pathology often makes us antici-pate, powerlessly, the sickness and death of a dear one, it is not impossible that an increased understanding will make us foresee the development of a catastrophe, a catastrophe whose theory will make us know the very reason of our powerlessness."[107] Ultimately, he believed that a theory would be totally intelligible when the theory itself could decide on its own validity: "a theory of meaning whose nature is such that the act of knowing itself is a consequence of the theory."[108] While Thom never claimed that catastrophe theory could live up to this feat, he nevertheless thought that it made the world more intelligible.

He often insisted that catastrophe theory was not a proper scientific theory. It was a language, a method. Nowhere was this more evident than when he confronted the delicate question of experimental control. He always admitted that an experiment that would falsify or, for that matter, confirm his theories was in principle impossible because catastrophe the-ory was inherently qualitative. It might eventually provide the basis for elaborating a quantitative model susceptible of experimental control, but in general, the necessary mathematics did not yet exist. And even if it were possible to analyze mathematically the dynamical processes that insured the stability of a form, "this analysis is often arbitrary; it often leads to several models between which we can only choose for reasons of economy or mathematical elegance."[109]

But, once again, according to Thom, the drawback was not fatal. He saw at least two reasons to justify scientists' interest in the theory. First, catastrophe theory questioned the traditional "qualitative carving out of reality . . . into the big disciplines: Physics, Chemistry, Biology."[110] It would integrate this taxonomy of experience into "an abstract general theory, rather than blindly accept[ing] it as an irreducible fact of reality." Second, catastrophe theory could replace the "lucky guess" of previous model con-struction in science. "The ultimate aim of science is not to amass un-differentiated empirical data," he wrote, "but to organise this data in a more or less formalised structure, which subsumes and explains it."[111] On the path toward a "general theory of models," catastrophe theory showed the way of the future.

Conclusion

With catastrophe theory, René Thom believed that he was breaking away from centuries of reductionist thinking. He developed models for biology, linguistics, and semiotics displaying his vision of a structural science. He introduced a new modeling practice and tried to codify its epistemological rules. Based on his mathematical experience, catastrophe theory used topology as a resource for grasping a world of qualities and shapes. Embryology suggested to him a new starting point for theory, namely the end of a dynamic process: its morphology. Thom never argued for the intrinsic superiority of his method, but rather for its greater capacity to explain the world as it is perceived. Catastrophe theory provided "schemes of intelligibility. And this seems quite valuable to me."[112]

In developing catastrophe theory, Thom introduced important mathematical concepts and attempted to extend them beyond their rigorous limits. In doing so, his speculations were often rejected by mathematical communities. His insistence on denying the possibility of experimentation was met with suspicion by practicing biologists. Finally, it was the nongenericity of structural stability for nongradient systems that discredited the general ambitions of catastrophe theory. As for elementary catastrophe theory applied to the physical sciences, it did not seem to explain anything that was not already known.

Thom's program, however, was richer than just concepts, models, theorems, and theories. His modeling practice presented some appealing aspects, some explanatory strategies that would be taken up by "chaologists." In 1971, using Thom's concept of attractors and his geometric vision of dynamical systems, David Ruelle and Floris Takens conjectured that the attractor usually posited for turbulence was not structurally stable, and thus introduced the notion of *strange attractors* at the roots of chaos theory.[113] But, contrary to Thom's model, theirs was successfully submitted to the verdict of experiments, in the laboratory and on the computer. This would be a decisive difference.

Notes

For their help and comments I am indebted to the participants in the Princeton workshop, to the graduate students of Princeton University's program in the history of science, and to the members of my doctoral committee, M. Norton Wise, Amy Dahan Dalmedico, and Michael S. Mahoney. I thank the Institut des hautes études scientifiques of Bures-sur-Yvette and its director Jean-Pierre Bourguignon for generous access to their archives. I am grateful to René Thom for having commented on an earlier version and responding to my questions. This work was supported by the Social Science and Humanities Research Council of Canada and the John C. Slater Fellowship of the American Philosophical Society.

1   René Thom, "La science malgré tout . . . ," *Encyclopædia Universalis*, vol. 17: *Organum* (1975), 5, 7. All translations are mine except when I quote from a published translation. For a French view of the crises facing science in the 1970s, see Alain Jaubert and Jean-Marc Lévy-Leblond (eds.), *(Auto)critique de la science* (Paris: Seuil, 1975).

2   Thom, "La science malgré tout . . . ," 10; my emphasis. Thom mentions the "prime vocation" of science on p. 5.

3   Only accounts written by scientists exist concerning the history of catastrophe theory. See Alexander Woodcock and Monte Davis, *Catastrophe Theory* (New York: E. P. Dutton, 1978), which is also a good nontechnical introduction to the subject; Ivar Ekeland, *Le Calcul, l'imprévu: Les figures du temps de Képler à Thom* (Paris: Seuil, 1984), translated as *Mathematics and the Unexpected* (Chicago: University of Chicago Press, 1988); Tito Tonietti, *Catastrofi: Una controversia scientifica* (Bari: Dedalo, 1983); and Vladimir I. Arnol'd *Catastrophe Theory*, 3d rev. and expanded ed., trans. G. S. Wasserman, based on a translation by R. K. Thomas (Berlin: Springer, 1992); see also John Guckenheimer, "The Catastrophe Controversy," *Mathematical Intelligencer* 1 (1978): 15–20; and Alain Boutot, "Catastrophe Theory and Its Critics," *Synthese* 96 (1993): 167–200.

4   Domenico P. L. Castrigiano and Sandra A. Hayes, *Catastrophe Theory* (Reading: Addison-Wesley, 1993), xii; see also Thom's assessment of this in "René Thom répond à Lévy-Leblond sur la théorie des catastrophes," *Critique* 33(361/362) (1977): 681.

5   René Thom, *Prédire n'est pas expliquer*, interview by Émile Noël (Paris: Eshel, 1991), 47; my emphasis.

6   David Aubin, "Chaos et déterminisme," in Dominique Lecourt (ed.), *Dictionnaire d'histoire et de philosophie des sciences* (Paris: Presses universitaires de France, 2000), 166–168.

7   The notion that most domains of science have emphasized shapes and forms in their development during the last decades has been shown in a recent special issue of *La Recherche*, no. 305 (January 1998), devoted to the "origin of forms."

8   René Thom, "La linguistique, discipline morphologique exemplaire," *Critique* 30 (1974): 245.

9   On the link between mathematics and the social sciences in France at an earlier period and the manner in which catastrophe theory intervened in this context, see David Aubin, "The Withering Immortality of Nicolas Bourbaki: A Cultural Connector at the Confluence of Mathematics, Structuralism, and the Oulipo in France," *Science in Context* 10(2) (1997): 297–342.

10 For a historiographical review of chaos, see David Aubin and Amy Dahan Dalmedico, "Writing the History of Dynamical Systems and Chaos: *Longue Durée* and Revolution, Discipline and Culture," *Historia Mathematica* 29 (2002): 273–339. The standard account of chaos as scientific revolution is James Gleick, *Chaos: Making a New Science* (New York: Viking, 1987). The philosopher Stephen Kellert speaks of "dynamic understanding" in his book *In the Wake of Chaos: Unpredictable Order in Dynamical Systems* (Chicago: University of Chicago Press, 1993).

11 See David Aubin, "A Cultural History of Catastrophes and Chaos: Around the Institut des Hautes Études Scientifiques, France" (Ph.D. diss., Princeton University, 1998); and "From Catastrophe to Chaos: The Modeling Practices of Applied Topologists," in U. Bottazzini and A. Dahan Dalmedico (eds.), *Changing Images in Mathematics: From the French Revolution to the New Millennium* (London: Routledge, 2001), 255–279.

12 René Thom, *Stabilité structurelle et morphogénèse* (Reading, Mass.: W. A. Benjamin, 1972; Paris: InterÉditions, 1977); *Structural Stability and Morphogenesis*, trans. David H. Fowler (Reading, Mass.: Benjamin, 1975); hereafter cited as *SSM*.

13 René Thom, "Une théorie dynamique de la morphogénèse," in Conrad Hal Waddington

(ed.), *Towards a Theoretical Biology*, vol. 1 (Edinburgh: Edinburgh University Press, 1968); and "Topologie et signification," in *L'Âge de la science* 4 (1968). Both of these articles are reprinted in René Thom, *Modèles mathématiques de la morphogénèse: Recueil de textes sur la théorie des catastrophe et ses applications* (Paris: Union générale d'éditions, 1974; Paris: C. Bourgois 1980); *Mathematical Models of Morphogenesis*, trans. W. M. Brookes and D. Rand (Chichester: Ellis Horwood, 1983), 1–38, 166–191; hereafter cited as *MMM*.

14 See, e.g., René Thom, "Towards a Revival of Natural Philosophy," in W. Güttinger and H. Eikemeier (eds.), *Structural Stability in Physics* (Berlin: Springer, 1979), 5–11. See also Jean Largeault, "René Thom et la philosophie de la nature," *Critique* 36 (1980): 1055–1060.

15 Thom, "Une théorie dynamique," *MMM*, 15.

16 René Thom, "Exposé introductive," in Jean Petitot (ed.), *Logos et Théorie des catastrophes: À partir de l'oeuvre de René Thom*, Actes du colloque international de Cerisy-la-Salle, 1982 (Geneva: Patiño, 1988), 24, where he also wrote that "sociologically" (*sociologiquement*), he was a mathematician. Thom recounts his memories in two published interviews: *Paraboles et catastrophes: Entretiens sur les mathématiques, la sciences et la philosophie*, interview by G. Giorello and S. Morini (Paris: Flammarion, 1983) and *Prédire n'est pas expliquer*. See also "Problèmes rencontrés dans mon parcours mathématique: un bilan," *Publications mathématiques de l'IHÉS* 70 (1989): 199–214, and his interview in Marion Schmidt (ed.), *Hommes de sciences: 28 portraits* (Paris: Hermann, 1990), 228–234.

17 Thom, "Problèmes rencontrés," 200.

18 Liliane Beaulieu, "Bourbaki. Une histoire du groupe de mathématiciens français et de ses travaux (1934–1944)," thèse de l'université de Montréal (1989), 1; see also Aubin, "The Withering Immortality."

19 David Hilbert and S. Cohn-Vossen, *Geometry and the Imagination*, trans. P. Nemenyi (1932; New York: Chelsea, 1952), iii; their emphasis.

20 Interviewer's comment, to which Thom agrees, in *Paraboles et catastrophes*, 24.

21 Thom, *Paraboles et catastrophes*, 23; see also André Haefliger, "Un aperçu de l'œuvre de Thom en topologie différentielle (jusqu'en 1957)," *Publications mathématiques de l'IHÉS* 68 (1988): 15.

22 Jean Dieudonné, *Panorama des mathématiques pures: Le choix bourbachique* (Paris: Gauthier-Villars, 1977), 14; see also Thom's paper "Quelques propriétés globales des variétés différentiables," *Commentarii Mathematici Helvetici* 28 (1954): 17–86.

23 Heinz Hopf, "The Work of R. Thom," in *Proceedings of the International Congress of Mathematicians*, Edinburgh, August 1958 (Cambridge: Cambridge University Press, 1960), lxiii–lxiv.

24 Tim Poston, quoted by Woodcock and Davis, *Catastrophe Theory*, 16.

25 René Thom, "Mathématiques modernes et mathématique de toujours," in Robert Jaulin (ed.), *Pourquoi la mathématique?* (Paris: Union générale d'éditions, 1974), 49; and in *Entretiens avec "Le Monde,"* interview by Jean Mandelbaum (Paris: La Découverte, 1984), 3:52, 80; see also René Thom, "Modern Mathematics: An Educational or Philosophical Error?" *American Scientist* 51 (1971): 697; repr. in *New Directions in the Philosophy of Mathematics* (Boston: Birkhäuser, 1986), 67–78; originally published in French in *L'Âge de la science* 3(3) (1970): 225–236; repr. in *Pourquoi la mathématique?*, 57–88.

26 Thom, "Quelques propriétés globales"; see also "Sous-variétés et classes d'homologie des variétés différentiables," *Séminaire Bourbaki* 5 (February 1953), exposé #78; and "Variétés différentiables cobordantes," *Comptes-rendus de l'Académie des Sciences* 236 (1954): 1733–1735.

27 Thom, "Exposé introductif," 27; my emphasis.

28 René Thom, "La vie et l'œuvre de Hassler Whitney," *Comptes-rendus de l'Académie des sciences—La vie des Sciences* 7 (1990): 473–476.

29 Paul Montel, "Sur les méthodes récentes pour l'étude des singularités des fonctions analytiques," *Bulletin des sciences mathématiques* 2d ser., 56 (1932): 219.

30 Thom, "Exposé introductif," 26.

31 René Thom, "Les singularités des applications différentiables," *Annales de l'Institut Fourier de Grenoble* 6 (1955–56): 87; see also "Les singularités des applications différentiables," *Séminaire Bourbaki* 8 (May 1956), exposé #134; and see Bernard Teissier, "Travaux de Thom sur les singularités," *Publications mathématiques de l'IHÉS* 68 (1988): 19–25; Haefliger, "Un aperçu," 16.

32 René Thom, "Mémoire de la théorie des catastrophes," in R. Thom, M. Porte, and D. Bennequin (eds.), *La genèse de formes*. I thank R. Thom and M. Porte for providing me a copy of this text.

33 Thom, "Exposé introductif," 31. On Lefschetz, see Amy Dahan Dalmedico, "La renaissance des systèmes dynamiques aux États-Unis après la deuxième guerre mondiale: l'action de Solomon Lefschetz," *Rendiconti dei circolo matematico di Palermo*, ser. II, Supplemento, 34 (1994): 133–166. On the Russian school, see Simon Diner, "Les voies du chaos déterministe dans l'école russe," in Amy Dahan Dalmedico, Karine Chemla, and Jean-Luc Chabert (eds.), *Chaos et déterminisme* (Paris: Seuil, 1992), 331–370; see also Aubin and Dahan Dalmedico, "Writing the History of Dynamical Systems" and references therein.

34 Aubin, "From Catastrophe to Chaos." About Smale, see Stephen H. Smale, *The Mathematics of Time: Essays on Dynamical Systems, Economic Processes, and Related Topics* (New York: Springer, 1980); Morris W. Hirsch, Jerrold E. Marsden, and Michael Shub (eds.), *From Topology to Computation: Proceedings of the Smalefest* (New York: Springer, 1993); and Steve Batterson, *Stephen Smale: The Mathematician Who Broke the Dimension Barrier* (Providence, R.I.: American Mathematical Society, 2000).

35 Marston Morse, "The Calculus of Variation in the Large," *Collected Papers* (Singapore: World Scientific, 1987), 1:423; *The Calculus of Variation in the Large* (New York: American Mathematical Society, 1934); and the famous textbook by John Milnor, *Morse Theory* (Princeton, N.J.: Princeton University Press, 1963).

36 See H. Whitney, "Singularities of Mappings in Euclidean Spaces," *Symposium internacional de topología algebraica* (Mexico City: Universidad Nacional Automa de México and UNESCO, 1958), 285–301.

37 H. Whitney, "On Singularities of Mappings of Euclidean Spaces. I. Mappings of the plane into the plane," *Annals of Mathematics* 62 (1955): 374–410; repr. in *Collected Papers* (Berlin: Birkhäuser, 1992), 370–406.

38 Castrigiano and Hayes, *Catastrophe Theory*. See *SSM*, 29–34; and *MMM*, 59–77.

39 John Guckenheimer, review of *SSM*, *Bulletin of the American Mathematical Society* 79 (1973): 878–890; A. Majthay, *Foundations of Catastrophe Theory* (Boston: Pitman, 1985), 1; Michel Demazure, *Catastrophes et bifurcations* (Paris: Ellipse, 1989), 167. The title of this section is a quote from Castrigiano and Hayes, *Catastrophe Theory*, xii.

40 Thom, "Une théorie dynamique," and *SSM*. For the date, see Thom, "Problèmes rencontrés," 203.

41 Tim Poston and Ian Stewart, *Catastrophe Theory and Its Applications* (London: Pitman, 1978), 2.

42 Two recent books are essentially dedicated to a pedagogical reproduction of this proof: Demazure, *Catastrophes et bifurcations*, and Castrigiano and Hayes, *Catastrophe Theory*. Poston and Stewart present an intermediate-level explanation of the notions that articulate this theorem; see their chapter 7 in Castrigiano and Hayes, *Catastrophe Theory*, 99–122.

43   E. C. Zeeman's most famous articles on catastrophe theory were gathered in his *Catastrophe Theory: Selected Papers, 1972–1977* (Reading, Mass.: Addison-Wesley, 1977). See in particular the Thom-Zeeman debate, 615–650.

44   Poston and Stewart, *Catastrophe Theory*, 7.

45   Thom, *Paraboles et catastrophes*, 98; see also René Thom, "Le statut épistémologique de la théorie des catastrophes," *Morphologie et imaginaire*, *Circé*, 8/9 (1978): 7–24; repr. in *Apologie du logos* (Paris: Hachette, 1990), 395–410.

46   Waddington, preface to *Towards a Theoretical Biology*, 1.

47   On this institute, see David Aubin, "Un pacte singulier entre mathématiques et industrie: L'enfance chaotique de l'Institute des hautes études scientifiques," *La Recherche*, no. 313 (1998): 98–103; and A. Jackson, "The IHÉS at Forty," *Notices of the American Mathematical Society* 46(3) (1999): 329–337.

48   Thom, *Prédire n'est pas expliquer*, 27. For his early philosophical interest, see ibid., 14; and Jacques Nimier (ed.), *Entretiens avec des mathématiciens (L'heuristique mathématique)* (Villeurbanne: Institut de Recherche en Enseignement des Mathématiques, 1989), 96–97.

49   Thom, "Exposé introductif," 27. He also said, "I never mistook myself for a mathematician"; see *Paraboles et catastrophes*, 29.

50   René Thom, "La stabilité topologique des applications polynomiales," *L'Enseignement mathématique*, 2nd ser., 8 (1962): 24–33; and "Ensembles et morphismes stratifiés," *Bulletin of the American Mathematical Society* 75 (1969): 240–284.

51   Zeeman, *Catastrophe Theory: Selected Papers*, 373.

52   Thom, *Prédire n'est pas expliquer*, 27; my emphasis. About the catastrophe theory approach of caustics, see Michael V. Berry, "Les jeux de lumières dans l'eau," *La Recherche* 9 (1978): 760–768.

53   Thom, *Paraboles et catastrophes*, 45. Gastrulation is the process by which the first internal layer of cells is formed in an animal embryo.

54   Thom, "Exposé introductif," 30; see also Max Delbrück's comment in *Unités biologiques douées de continuité génétique* (Paris: CNRS, 1949), 33–34, trans. in *MMM*, 29–31. E. C. Zeeman, "The Topology of the Brain and Visual Perception," in M. K. Fort (ed.), *Topology of 3-Manifolds and Related Topics: Proceedings of the University of Georgia Institute 1960–61* (Englewood Cliffs: Prentice-Hall, 1962), 240–256; and "Topology of the Brain," *Mathematics and Computer Science in Biology and Medicine* (London: Her Majesty's Stationery Office, 1965), 277–292.

55   *SSM*, xxiii. About von Uexküll and Goldstein, see Ann Harrington, *Reenchanted Science: Holism in German Culture from Wilhelm II to Hitler* (Princeton, N.J.: Princeton University Press, 1996).

56   Thom, "Une théorie dynamique," 152; see also *MMM*, 14.

57   Conrad Hal Waddington, *The Strategy of the Genes: A Discussion of Some Aspects of Theoretical Biology* (London: Allen and Unwin, 1957), 4, 9. On Waddington, see Alan Robertson, "Conrad Hal Waddington," *Biographical Memoirs of Fellows of the Royal Society* 23 (1977): 575–622; and Donna J. Haraway, *Crystals, Fabrics, and Fields: Metaphors of Organicism in Twentieth-Century Developmental Biology* (New Haven: Yale University Press, 1976).

58   Conrad Hal Waddington, "The Basic Ideas of Biology," in Waddington (ed.), *Towards a Theoretical Biology*, 1:9. See also Waddington, *Organisers and Genes* (Cambridge: Cambridge University Press, 1940) and *Principles of Embryology* (London: Allen and Unwin, 1956).

59   Conrad Hal Waddington, "The Theory of Evolution Today," in Arthur Koestler and J. R. Smythies (eds.), *Beyond Reductionism: New Perspectives in the Life Sciences* (London: Hutchinson, 1969), 363.

60   Scott F. Gilbert has examined the source of this idea in his "Epigenetic Landscaping:

Waddington's Use of Cell Fate Bifurcation Diagrams," *Biology and Philosophy* 6 (1991): 135–154. The epigenetic landscape first appeared in *An Introduction to Modern Genetics* (New York: MacMillan, 1939) and was treated extensively in *Organisers and Genes* and *Strategy of the Genes*.

61 Waddington, *Strategy of the Genes*, 29.

62 From the "Greek roots χrh, it is necessary, and οδοξ, a route or path." Waddington, *Strategy of the Genes*, 32.

63 See "Correspondence Between Waddington and Thom," in Waddington (ed.), *Towards a Theoretical Biology*, 1:166–179.

64 It is unclear whether Smale or Thom first introduced this concept. According to Robert Williams, "each says the other invented it"; see *From Topology to Computation*, 183. For their first definitions, see *SSM*, 39, and Smale, *Mathematics of Time*, 20. At least once, however, Thom claimed responsibility for the term, while borrowing Smale's definition: "Problèmes rencontrés," 203–204.

65 Thom, "Une théorie dynamique," 158; a translation appears in *MMM*, 19.

66 Waddington, *Organisers and Genes*, quoted in Robertson, "Conrad Hal Waddington," 593.

67 C. H. Waddington, *SSM*, xxi; my emphasis. See also his "Theory of Evolution," 367.

68 René Thom, "A Global Dynamical Scheme for Vertebrate Embryology," *Lectures on Mathematics in the Life Sciences* 5 (*Some Mathematical Questions in Biology IV: Proceedings of the Sixth Symposium on Mathematical Biology*) (Providence, R.I.: American Mathematical Society, 1973), 44.

69 Jacques Monod, *Chance and Necessity* (New York: Knopf, 1971), 81, 88; my emphasis.

70 Ibid., 89, 95.

71 Note, however, that neither Thom nor Monod mentioned the work of the other in their writings. Indeed, Monod wished to counter vague approaches based on "general systems theory"; see ibid., 80.

72 René Thom, "D'un modèle de la science à une science des modèles," *Synthese* (1975): 359–374; my emphasis.

73 Françoise Gail, "De la résistance des biologistes à la théorie des catastrophes," in Jean Petitot (ed.), *Logos et théorie des catastrophes* (Geneva: Patiño, 1988), 269–279. One may note the more ambivalent position defended by François Jacob in "Le modèle linguistique en biologie," *Critique* 30(322) (1974): 197–205; Henri Atlan, *Entre le cristal et la fumée: Essai sur l'organisation du vivant* (Paris: Seuil, 1979), 219–229; and Michael A. B. Deakin, "The Impact of Catastrophe Theory on the Philosophy of Science," *Nature and System* 2 (1980): 177–288.

74 Thom, manuscript for *SSM*, sect. 13.3.C., Fine Library, Princeton University.

75 Thom, "Topologie et signification," *L'Âge de la science* 4 (1968); repr. in *MMM*, 166–191.

76 Thom, "Topologie et signification," in *MMM*, 169; "Topological Models in Biology," in Waddington (ed.), *Towards a Theoretical Biology* (Edinburgh: Edinburgh University Press, 1970), 3:89–116, 103; repr. from *Topology* 8 (1969): 313–335.

77 René Thom, "Topologie et linguistique," in André Heafliger and R. Narasimhan (eds.), *Essays on Topology and Related Topics (Dedicated to G. de Rham)* (Berlin: Springer, 1970), 148–177; repr. in *MMM*, 192–213.

78 Louis Tesnière, *Élements de syntaxe structurale* (Paris: Klincksieck, 1965).

79 René Thom, *Semiophysics: A Sketch*, trans. Vendla Meyer (Redwood City: Addison-Wesley, 1966), viii.

80 Thom, "Topologie et linguistique," in *MMM*, 197. See a figure of sixteen archetypal types in *SSM*, 307.

81 See Aubin, "Withering Immortality"; Michel Scrres, *Hermès V. Le Passage du Nord-Ouest*

(Paris: Seuil, 1980), 99; and Jean-François Lyotard, The Postmodern Condition: A Report on Knowledge, trans. G. Bennington and B. Massumi (Minneapolis: University of Minnesota Press, 1984), 60.

82 René Thom, "Structuralism and Biology," in Waddington (ed.), Towards a Theoretical Biology (Edinburgh: Edinburgh University Press, 1972), 4:68, 70; this article also appeared in the first French edition of MMM (1974) but was absent from later editions.

83 For a contemporary attempt at articulating a multidimensional structuralism, see a book by one of Thom's followers, Paul Scheurer, Révolutions de la science et permanence du réel (Paris: Presses Universitaires France, 1979).

84 Thom calls this second approach "l'approche structurale." We must note a difference between French qualifiers: structurel (as in "stabilité structurelle," simply the translation of an English phrase) refers to actual structures, while structural refers to structures as syntax, which can be realized in several instances of actual structures; see Jean-Marie Auzias, Clefs pour le structuralisme (Paris: Seghers, 1967), 18.

85 Thom, "La science malgré tout . . . ," 6; "La linguistique," 239.

86 Concerning Piaget, see Aubin, "Withering Immortality."

87 Thom, "La linguistique," 240.

88 "Expliquer du visible compliqué par de l'invisible simple," introduction to J. Perrin, Les Atomes (Paris: Félix Alcan, 1913).

89 Thom, "Topologie et signification," in MMM, 174.

90 Thom, SSM, 159.

91 Thom, "Généralités sur les morphologies: la description," in French edition of MMM (1974), 9, but absent from later editions.

92 Note that there is nothing absolute about the relation between the study of forms and nonreductionism; see, for example, Norma E. Emerton, The Scientific Reinterpretation of Form (Ithaca: Cornell University Press, 1984), a historical study that focuses on the struggle to find molecular accounts of crystal shapes.

93 D'Arcy Wentworth Thompson, On Growth and Form (1916), abridged ed. by John Tyler Bonner (Cambridge: Cambridge University Press, 1961), 10; my emphasis. Note that Thom used this quotation as an epigraph to his introduction in SSM, 1, but dropped the last part where Thompson confidently asserts the success of physics.

94 SSM, 9. The allusion to accelerators was added for the second French edition.

95 Benoît B. Mandelbrot, "Towards a Second Stage of Indeterminism in Science," Interdisciplinary Science Review 12 (1987): 117.

96 Heraclitus, Fragments, trans. T. M. Robinson (Toronto: University of Toronto Press, 1987), fragment 17.

97 "Whatsoever things are objects of sight, hearing, and experience, these things I hold in higher esteem." Heraclitus, Fragments, fragment 55.

98 Thom, "Topologie et signification," in MMM, 174–175.

99 M. Conche, in Heraclitus, Fragments (Paris: Presses Universitaires de France, 1986), 65.

100 Giorgio Israel, La Mathématisation du réel (Paris: Seuil, 1996), 198.

101 Thom, manuscript for SSM, 13–14, Fine Library, Princeton University; cf. SSM, 8–10.

102 Thom, "Une théorie dynamique," 153; MMM, 14.

103 Thom, SSM, 9; Thompson, On Growth and Form, 72–73.

104 Thom, "Modern Mathematics," 697.

105 Thomas S. Kuhn, "Postscript–1969," The Structure of Scientific Revolutions, 2d ed. (Chicago: The University of Chicago Press, 1970), 185.

106 Thom, SSM, 159. For Rutherford's quote, see p. 4, for example.

107 Thom, "La science malgré tout . . . ," 9–10.

108 Thom, *SSM*, 5; and "Topologie et signification," in *MMM*, 170; italics in the original text.

109 Thom, "Une théorie dynamique," in *MMM*, 21.

110 Thom, "Le découpage qualitatif de la réalité . . . en grandes disciplines: Physique, Chimie, Biologie," in *SSM* (1972), 323. The English translation of *SSM*, 322, misses Thom's point here.

111 Thom, *SSM*, 322; and "Une théorie dynamique," *MMM*, 22.

112 Thom, *Prédire n'est pas expliquer*, 45–46.

113 David Ruelle and Floris Takens, "On the Nature of Turbulence," *Communications in Mathematical Physics* 20 (1971): 167–192; and their "Note" in ibid. 23 (1971): 343–344.

Coping with complexity in technology

4 From Boeing to Berkeley: civil engineers, the cold war, and the origins of finite element analysis

Ann Johnson

Structural engineering is the art of molding materials we do not really understand into shapes we cannot really analyze, so as to withstand forces we cannot really assess, in such a way that the public does not really suspect.
—E. H. Brown, *Structural Analysis*

E. H. Brown's statement rings particularly true when thinking about the problem of analyzing airframes, the complicated skeletal structures of airplanes. In the early years of aircraft production, plane designers often approached structural design problems as a series of trial-and-error experiments. However, during the cold war, the United States military began to funnel more money into aircraft development, buoyed by the success of air power in World War II. At the same time, the jet engine and an increasing array of cockpit electronics began to drive the cost of planes higher and higher. Trial-and-error engineering was becoming prohibitively expensive. Mathematical modeling of the airframe's structural behavior was an interesting alternative to trial and error, but existing mathematical tools were not well adapted to designing airframes. Aircraft designs featuring sheet metal skins over tubular ribs were difficult to idealize, and the increasing sophistication of testing instruments led to an increasing divergence between experimentally measured values for structural coefficients and those obtained through numerical methods. The influx of military funding to the aircraft industry coupled with structural engineers' desire to solve mathematical puzzles in airframe design led them to develop a considerable array of new mathematical tools. By the 1970s the practice of structural engineering had changed fundamentally.

Research and development and the culture of the cold war

While changes in physics communities attributable to the cold war have been well-documented,[1] less historical work has been done on the changing roles of academic engineers.[2] One common claim in the history of cold war science is that disciplinary distinctions between the sciences, as well as between the sciences and engineering, were blurring.[3] However, not all sciences or engineering fields were affected by the cold war in the same way. As the status of certain areas of research in the physical sciences rose because of their ability to attract military funding, a similar redistribution of status occurred with engineering. Still, neither scientists nor engineers sat by idly as their disciplines were reordered; instead, many people worked to forge a place for their work in the new cold war order.

For structural engineers, attracting federal funding often posed a peculiar problem because their academic appointments were often in civil engineering departments. Civil engineering has historically developed as engineering knowledge not applicable to military concerns, focusing not on fortresses or weapons but on bridges, earthworks, harbors, roads, canals, and so forth. While structural engineers had skills necessary for military projects, those from civil engineering departments had to create a place for themselves and their discipline in a technological culture that potentially excluded them.

This military and technological culture of the 1950s and 1960s was a defining aspect of what Dwight D. Eisenhower called the military-industrial complex. Eisenhower coined the phrase in his farewell address delivered on 17 January 1961. In this speech he defined the aspects of the cold war most important for the changes in technology. Eisenhower declared, "A vital element in keeping the peace is our military establishment. Our arms must be mighty, ready for instant action, so that no potential aggressor may be tempted to risk his own destruction."[4] The job of producing these ever-ready, mighty arms fell to the defense industries, an economic sector in rapid growth since the bombing of Pearl Harbor, and the focus of the cold war.[5] When Eisenhower referred to the military-industrial complex (MIC), he underscored its necessity saying, "We can no longer risk emergency improvisation of national defense. We have been compelled to create a permanent armaments industry of vast proportions. Added to this, three-and-a-half million men and women are directly engaged in the defense establishment. We annually spend more on military security than the net income of all United States corporations."[6] But in the same speech he also warned against allocating too much influence and control to the MIC, cautioning that "we must never let the weight of this combination endanger our liberties or our democratic process."[7] Given the importance of the military-industrial complex, Eisenhower feared that public policy would become captive to a scientific and technological elite.[8]

It is this tension between the utility of the relationship between the military and industry and the fear that the system will expand its influence beyond the control by democratic means that characterizes the MIC.[9] In the 1950s, before the taint of Vietnam, most scientists and engineers were far more concerned with the utility of the military-industrial relationship than its ethics. Individuals from numerous disciplines were trying to carve out their niches and establish places for their research, their colleagues, and their students in this new system. This is also the environment in which the structural engineers were working to establish themselves.

World War II and the aircraft industry

World War II marks a watershed in three areas critical for the development of finite element analysis (FEA). During the war, scientific work became an issue of national security to an unprecedented degree. The success of the air war, especially in the Pacific, elevated both the strategic importance and the economic power of the aircraft industry and laid the groundwork for an independent air force after the war. The industrial remapping of the country began as the West Coast gained a new military and economic importance.[10] After the war, the migration of labor to the West Coast and the military importance of the industries that had been established there before the war—namely, aircraft and shipping—proved to be compelling resources for universities trying to establish ties with both industry and the government to secure status and funding, not to mention their genuine patriotism. A new set of relationships was forged in the years following the war, a direct outgrowth of the war's successes. No academic department could afford to ignore these changes. For many departments, as well as for individuals, the question was not whether to get on board with the MIC, but how. This was especially true for departments that had not done research for the war effort. Academics looked to wartime physics, to chemistry, and to radio and electrical engineering for examples.

The legacy of World War II helped shape the postwar MIC in many important ways, through conflict as much as through funding. In 1947 the new independence of the Air Force opened up several internal divisions within the MIC. The most important fight occurred first between the Army and the Air Force. Although both service branches continued doing aeronautical research, their modes of production and styles of contracting were strikingly different. The Army wanted to continue with the traditional arsenal research model, where Army engineers and scientists performed the research and development and only brought in industry when a prototype had been designed.[11] Then corporations bid for the production of that unit. The Army's aeronautical research center was Redstone Arsenal in Huntsville, Alabama, and was led in 1950 by Wernher von Braun, former head of the German V-2 program at Peenemunde.

On the other hand, the Air Force set up two departments to monitor out-sourced contracts. Military managers, such as General Bernard Schriever, allocated resources to various projects and chose which companies received which contracts for research needs.[12] In *The Rise of the Gunbelt*, Ann Markusen, Peter Hall, Scott Campbell, and Sabina Deitrick explain this strategy in political and economic terms: "By eschewing any in-house capacity, the Air Force commanded a big well-heeled constituency of scientists, organized labor, and industry. Its impeccable free-enterprise viewpoint removed much of the stigma from the surge in public spending that the arms revolution entailed."[13] Although the Army, Navy, and Marines still contracted independently with aircraft producers, research contracts for the Air Force became the prize for both the aircraft industry and universities, particularly those located in the west. Over time the Air Force provided the most lucrative contracts and gave researchers the greatest freedom in development.[14] Gaining research contracts was both more glamorous and more profitable than earning production contracts.

Another schism, this time within the Air Force, shaped the aerospace industry in its infancy. By the early 1950s, the wartime Air Force regime was increasingly at odds with supporters of a missile-oriented Air Force. The "flying generals," headed by General Curtis LeMay, were committed to strategic bombing, which meant that they focused their research priorities on increasingly faster and longer-range manned bombers and advanced avionics, while the "silo-sitters," best exemplified by Schriever, supported projects like intercontinental ballistic missiles (ICBMs) and radar-triggered defensive cruise missiles.[15] The political struggle within the Air Force between these two regimes challenged the aircraft producers, who did not want to offend or alienate either group. By 1954 the ICBM supporters had achieved some level of dominance, and aircraft companies began transforming themselves into aerospace companies. Some companies even dropped most of their aircraft production for a while, as they made themselves integral parts of the missile industry. Others, such as Boeing, avoided making such a strong commitment to missiles. Those that ignored altogether the transition from aircraft to aerospace tended to disappear as independent entities. A company as large as Boeing could focus on both, albeit more strongly on aircraft, and Boeing, in particular, was able to take advantage of the rise of commercial airlines to accommodate its relative weakness in missile designing.[16]

The business culture of the aerospace industries

A new industrial system of cooperating competitors invited new business practices and with it a new culture of speculation. Research was elevated from a necessary, but extremely expensive, component of business to the aspect of business that most effectively attracted funding. Productivity was rarely an issue in the MIC because units were not produced in enough quantity to warrant concerns about production capacity. Cost-plus contracting forced some companies to rethink profit-making and taxation schemes so as to maximize their profits without looking too profitable. Although aerospace companies' contracts could be renegotiated post-facto in the face of illegitimate or "excessive" profits, research was still the cash cow of these businesses. Speculative, futuristic technologies were particularly attractive to the MIC. Markusen and colleagues write,

> Key individuals helped build institutions, both in the private and public sectors that were dedicated to undertaking high-risk research and to selling to the government. In turn, they reacted on one another synergistically to create a technical climate based on the suspension of disbelief, that the impossible could be made possible. There was an atmosphere of optimism, of invention, free of the old industrial and business traditions left behind in the East.[17]

It was this atmosphere which was particularly appealing to the scientists and engineers working in these industries and which attracted the participation of academic scientists and engineers as consultants. Vannevar Bush, director of the Office of Scientific Research and Development during the war, voiced his vision of this atmosphere in his 1945 tract, *Science, the Endless Frontier*: "The pioneer spirit is still vigorous within this nation. Science offers a largely unexplored hinterland for the pioneer who has the tools for his task. The rewards of such exploration both for the Nation and the individual are great."[18] Identifying science as the frontier was only half the battle. Universities also needed pioneers to stake claims in companies and government organizations for their colleagues and their students. Senator J. William Fulbright was referring to these arrangements when he changed Eisenhower's term military-industrial complex to military-industrial-academic complex (MIAC). And like Eisenhower's, Fulbright's attitude toward this nexus was cautionary, even chastising the academy for its apparently materialistic overtures to industry and the military.[19]

Civil engineers build ties to the MIAC

Physics departments in West Coast universities had little difficulty making ties to these industries—companies sought the scientists out. But challenges existed for other academic departments that had skills and information useful to the MIAC, but that might not be the most obvious sources for research-based information. Faculty in these departments, whether they were the life sciences or various engineering disciplines, had to market their abilities to the Pentagon as well as to industry. For many disciplines, the key strategy was to get one individual into a company to open doors for his colleagues or to bring cutting edge methods from industry back to the academy for further investigation. For academic civil engineers building ties to the aerospace industry, Ray W. Clough acted as such a pioneer.

Ray Clough, born in 1920, completed his BS degree in civil engineering at the University of Washington in 1942 and immediately joined the war effort. As an engineering graduate, Clough secured a position in the Army Air Force's aviation engineers, for whom he worked until 1946. This early connection to aviation would serve him well in developing not only his career but his discipline's place in the MIAC. After completing his doctorate in structural engineering at MIT, Clough returned west to become an assistant professor in the civil engineering department at the University of California, Berkeley, in 1949. He arrived at Berkeley with three important resources: his personal ties to Seattle, experience in aviation, and an insider's knowledge of MIT, a university where the connection between industry, the military, and research had already been forged and exploited.[20] At MIT, Clough had obtained firsthand knowledge and experience of a model for interaction between university and industrial research.

In 1952, Clough sought a summer consulting position and joined the Boeing Summer Faculty Program. In a 1990 account Clough described his participation: "This was a program in which young engineering professors were hired from all over the country to work on various special research projects. I was attracted to the program because it offered an opportunity to work in the field of structural dynamics and I was particularly fortunate to work directly under the head of the Structural Dynamics Unit, Mr. M. J. Turner."[21] Clough had found a place to work for the summer; it was close to his boyhood home in Seattle and offered work on an exciting new set of problems for a prestigious company in the MIAC. Furthermore, the position he took with Boeing placed few constraints on his research—he was a "hired gun," solving particularly intractable problems in structural dynamics with the considerable intellectual as well as financial resources of Boeing at his disposal. In working directly under the head of the Structural Dynamics Unit, he would have little contact with the bureaucracy of a large MIAC business. It was an ideal position.

For Clough, as a member of a civil engineering department, spending the summer at an aerospace company brought him into contact with more advanced methods of analysis. Until the end of World War II, the structural theories used in aircraft engineering were the same as those used in the structural analysis of buildings and bridges. However, Clough writes that after the war the practices and tools of structural analysts in different subdisciplines began to diverge:

> In the years immediately following World War II, the aircraft structural engineering profession began to move ahead of the civil engineers in the theory of structural analysis due to the pressures resulting from the increasing complexity of airplane configurations and the compelling need to eliminate excess weight. An important advance initiated almost exclusively by the aeronautical engineers was the introduction of matrix notation in formulating the analysis. A factor contributing to this step was that the major airplane design companies had access to the best computers available at that time.[22]

Building ties between civil engineering and aerospace engineering could help structural engineers keep up to speed and prevent them from appearing outmoded, since new, higher standards of investigation were increasingly set by the aerospace-based structural engineers.

Aeronautical engineers moved to the forefront of structural engineering as a result of the development of jet planes, particularly given their success in the Korean War. While the jet engine had been invented before World War II, its incorporation into American aircraft depended on the development of an airframe that would both endure the structural stresses created by the jet engine and provide an aerodynamic shape that would optimize its performance. The jet engine and the swept-back wing are two prime examples of what Edward Constant calls coevolutionary technologies in his book *Origins of the Turbojet Revolution*.[23] Neither the new airframe nor the jet engine could succeed unless the other component was successfully developed. Therefore, structural engineers in aerospace industries encountered new sets of problems related to the necessary changes in airframe geometry and loading brought about by the jet engine. Boeing's xb-47 jet fighter placed Boeing firmly at the cutting edge of jet airframe design.

Yet the problems Clough was hired to address were not new in the same sense that the H-bomb presented new problems to physicists. Unlike other research projects spawned by the cold war, structural dynamics was a problem of long standing. Designers of railroad bridges in the nineteenth century had run up against intractable structural dynamics problems.[24] Closer to home, the collapse of the Tacoma Narrows Bridge in 1940, just four months after its completion, provided a bold and graphic symbol of a structural dynamics problem.[25] For bridge builders, overbuilding could often eliminate issues of structural dynamics. Even though the results were occasionally aesthetically unattractive, overbuilding was usually a viable and conservative solution. For aircraft designers striving for lighter and thinner airframes, overbuilding was not acceptable, and conservative solutions were generally undesirable. With that option closed to them, aerospace engineers instead sought to increase the accuracy of their theoretical predictions: Just how accurately could one predict structural behavior? Jet flight and thin, light airframes required an increasing level of accuracy in predicting this behavior.

Although experimental testing often provided a high level of accuracy in determining values for stress and strain, practical tests became increasingly expensive and time consuming. Not only was the machinery in jet engines more expensive; the airframes were more expensive to build. The new materials of the cold war—ceramics, plastics, electronics—further increased production costs. If failed designs could be caught before the prototype phase, the costs to develop new craft could be slashed. More accurate predictions would mean fewer disastrous tests. Clough explained the situation in a 1956 paper, written with his Boeing colleagues:

> Elaborate models are expensive, they take a long time to build, and tend to become obsolete because of design changes; for these reasons it is considered essential that a continuing research effort should be applied to the development of analytical methods. It is to be expected that modern developments in high-speed digital computing machines will make possible a more fundamental approach to the problems of structural analysis; we shall expect to base our model on a more realistic and detailed conceptual model of the real structure than has been used in the past.[26]

While it was obvious to associate the accuracy of prediction with the accuracy of modeling, engineers had to (and still have to) idealize structures to a significant extent to make them soluble at all.

The time required to develop new aircraft and weapons also played a significant role during the cold war. Whereas World War II forced companies to step up production, the cold war presented a different challenge: it forced companies to produce new designs more quickly. When faced with what seemed to be an imminent war at some unknown time in the future, engineers struggled to cut the total time elapsed from concept to usable prototype. A design still on someone's drawing board would not help the war effort, nor would one insufficiently tested and liable to fail at thirty-five thousand feet. More accurate predictions would cut design time and help eliminate unnecessary tests. A war with the Soviet Union would occur on a huge scale, but the cold war always remained in the very immediate future. These factors gave a unique intensity to the scientific and engineering research tasks, and this is what Eisenhower was referring to in his farewell address when he talked of being "ready for instant action."

When Clough arrived at Boeing in 1952, he knew that the primary task before him was to find a way to predict structural behavior. Since experimental testing provided knowledge of the behavior that would occur, he did not need to discover and explain the behavior. He knew that the wing would flex; his task was to predict the degree to which it would occur in various wing geometries. Clough needed a method to calculate the wing's stiffness, that is, its flexibility influence coefficients. That information would allow engineers at Boeing to predict the amount of wingtip flutter, which would negatively affect the stability of the plane. Experimental values had been found in the wind tunnel and laboratory for one wing shape. Clough's problem was one of modeling: How could he model the wing so that the numerical method would duplicate the wind tunnel findings?[27]

He first reviewed the work done by aeronautical researchers and applied mathematicians on the use of matrix transformations in structural analysis.[28] With these techniques he was able to compare the results of existing methods with the measured values, but using the flexibility coefficients he had derived, the calculations of flutter exceeded the measured results by an unacceptable 13 to 65 percent.[29] Clough believed that the problem lay in accurately modeling the wing's skin. A more accurate model would provide more accurate results, he reasoned. But engineers had no way of modeling the skin of the wing; it was not a beam, column, truss member, or even a shell.[30] As a result, until this time the wing had been modeled as a skeletal form, virtually ignoring the structural contributions of the skin. Clough thought the skin was adding stiffness to the frame in a way that the model did not take into account. He thought of modeling the steel panels as a patchwork of discrete triangular shapes to find out how much stiffness each section was contributing to the wing, but no mathematical procedure existed for this kind of discrete element modeling. By the end of the summer program in 1952, he and Turner had decided to pursue the problem, but he admitted that "little progress had been made toward the proper representation of the wing skin in the mathematical modeling of low aspect ratio wing structures."[31]

Returning to Berkeley in the fall of 1952, Clough continued to work on the problem of modeling the aircraft wing, including the skin. He started to look for resources in other disciplines for new mathematical approaches to the skin elements; at Berkeley this was where the predominance of resources lay. He had no access to a computer at this time. Using the method of direct stiffness calculation—then a common approach in structural analysis—Clough took the displacements of joints as the unknowns and solved simultaneous equations of equilibrium. Since every stable structural system must satisfy two sets of conditions—the forces have to be in equilibrium, and the deformations must be compatible between members—simultaneous equations are extremely common in structural analysis problems. In structures with a high degree of redundancy—airframes, for example—this method had advantages over taking the forces acting in structural members as the unknowns. Clough decided to use direct stiffness methods and to try to fit the modeling problem within a tried and true method of calculating displacements.[32]

Meanwhile Turner continued to work on the problem at Boeing. Clough credits him with conceiving a better way to model the skin of the wing. Turner proposed that the deformations of the skins be approximated by combining strain fields within the skin.[33] This way he was able to take into account the variation in strain over the entire wing. Although the wing would deform as a physical whole, the patterns of strain within the wing could help to calculate how much stiffness the skin was providing to the skeleton. The wing's skin was visualized as a geometrical shape with a mesh of triangles laid over it. The triangular subdivisions of the skin were arbitrary but discrete. Strain of three different types could be calculated for each triangular space: normal strain in the x and y directions, and shear strain. Calculating the total deformation of the skin allowed designers to calculate the amount of stiffness it was adding to the wing's frame, as well as indicating particular areas of high strain in the skin. Although the imaginary elements were discrete, if continuity held at the corner intersections between elements, called nodes, then continuity could be assumed to hold between the sides of the elements, giving the model structural integrity.

Clough tested Turner's breakthrough during the summer of 1953, when he returned to the Boeing Summer Faculty Program. He combined Turner's insight with his work on the use of the direct stiffness method.[34] The engineers found that the smaller the triangular elements were, the greater their agreement with the experimental data. Since the mesh was of arbitrary size, they could fine tune it to achieve the desired level of accuracy, taking into account that finer meshes entailed larger matrices. However, matrices quickly became too large to solve by hand. Even the method's earliest stages required serious number-crunching resources. Clough and Turner's method would depend on electronic computation from the beginning. Although the computer was in its infancy, Boeing's computer provided Clough and Turner with notions of how they might solve large numbers of simultaneous equations.

Working with H. C. Martin, a professor of aeronautical engineering at the University of Washington, and L. J. Topp, a Boeing engineer, Clough and Turner prepared a paper detailing the success of their method. Turner presented their findings for the aeroelasticity session of the 1954 meeting of the Institute for Aeronautical Sciences, and the four Boeing collaborators subsequently published the work in that organization's journal in the fall of 1956. This paper presented the findings of the 1953 Summer Faculty Program and the authors' hopes for future developments. Nevertheless, the paper was "concerned exclusively with methods of theoretical analysis"[35] and did not at that time present a workable method for engineers to use. The general theory had been laid out, but to transform it into a set of mathematical tools would require a great deal of development. It could not change engineering practice—the stated intent of the authors—until those tools were available. They would depend on fast calculations.

In the 1954 presentation and the 1956 paper, Clough, Turner, Martin, and Topp presented a new way to calculate stiffness influence coefficients for shell-type structures. Their readers knew the important role these coefficients played in doing dynamical analyses. The paper begins with a short overview of existing methods of analysis and points out their shortcomings, especially in analyzing complex, multilayered structures, such as the skinned airplane wing. They claim the method presented in the paper "avoids drastic modification of the geometry of the structure or artificial constraints of the elastic elements." To achieve this, they rely on "modern developments in high-speed digital computing."[36]

In the paper, they summarize the method in six steps:[37]

1. Idealize the structure as an assembly of discrete shapes connected to each other at node points (figure 1).
2. Determine a stiffness matrix for each structural type in the model. This matrix is a constant for each individual type of element. There would be different general stiffness matrices for different structural types, for example, beam, truss, plate, or shell elements.[38]
3. Displace a single element, while holding all others fixed, using the applied loads. This provides the reaction forces at the nodes. These nodal forces will affect adjacent elements. Reactions and forces fill in one column of the matrix. Through this technique, all the deflections can be calculated, and a complete stiffness matrix is determined. This matrix will be $3n \times 3n$, where n is the number of nodes.
4. Where nodes have zero displacement, columns and rows can be eliminated from the matrix. These areas represent the support conditions of the system. At a support, the reactions must cancel and deflection is not physically possible.
5. Known external forces can be checked against the aggregate nodal displacements.
6. Internal forces can be derived from the nodal displacements.

Clough and his colleagues point out that operations 3 through 6 are purely mathematical transformations, derived solely from the method and the

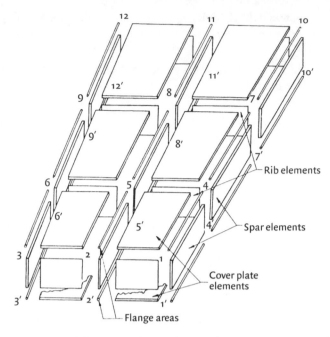

1. Wing structure breakdown. Numbers refer to nodes. Source: H. C. Martin, R. W. Clough, M. J. Turner, and L. J. Topp, "Stiffness and Deflection Analysis of Complex Structures," *Journal of the Aeronautical Sciences* 23 (1956): 810.

given model. These operations were to be executed by the computer or by "non–engineering-trained personnel."[39] The authors give details of how to form the stiffness matrix for a number of different structural types, and the rest of the paper is devoted to three examples: a simple truss, a flat plate, and a box beam. Although they discuss how to idealize the airplane wing, the wing is not presented as an example. The choice not to include this problem clearly indicated the theoretical nature of the paper.

Clough did not return to Boeing after 1953, but remained involved in the development of matrix methods and stayed in contact with Turner. A sabbatical leave in 1956 in Trondheim, Norway, allowed Clough time to study the work being done in Britain by John Argyris.[40] Argyris was working on expressing traditional force and displacement methods in matrix form, so that engineers could use these concepts when analyzing a variety of structural assemblies comprising different types of structural members, as was common in aeronautical engineering. Working with traditional methods in matrix form would facilitate the use of these methods on the computer.[41] According to the paper published by Turner, Clough, Martin, and Topp, using the computer was paramount: "it is our object to outline the development of a method that is well-adapted to the use of high-speed digital computing machinery."[42] Clough's contact with the computer at Boeing laid a critical foundation in furthering the method's development because neither Berkeley's civil engineering department nor Trondheim had computer access. After his work at Boeing, Clough thought of many ways to use the computer in existing methods of structural analysis. Structurally speaking, he was beginning to see the world as a numerical model that could be applied to the computer, and he searched for new ways to use the computer in structural engineering.

Role of the computer

While anyone familiar with ENIAC (electronic numerical integrator and computer) knows the importance of military funding in its development, it is still striking to see how many aerospace companies were customers for the first commercially produced computers. The business connection between aerospace—or defense more generally—and the computer industry was made early. During the Korean War, Thomas J. Watson, Sr., chairman of IBM, personally offered the services of IBM to Harry Truman for anything useful to the war effort. This offer led IBM to help in designing and producing a bombsight for the B-52, and it forged a connection between IBM and Boeing, the B-52's contractor. In 1979, Cuthbert Hurd, the first director of IBM's applied science department explained another outcome of this collaboration with the Pentagon:

> These visits [at the Pentagon] verified my view that government agencies had problems whose solutions required large amounts of processing and calculations. Some of these problems were being solved on large-scale analog machines, but I concluded that most could be performed better on a general-purpose computer. I explained my reasons and conclusions to Birkenstock [special assistant to Tom Watson, Jr., executive vice-president of IBM] as we traveled around together, and he was convinced that IBM should build a computer for the defense industry. During this same period, Palmer [of the Poughkeepsie engineering department] and I visited several defense laboratories and contractors, and he too was convinced of the need for a high-performance general-purpose computer, which was to be binary.[43]

Two years later, Hurd and his colleagues had developed the IBM 701—the "Defense Calculator." IBM announced the 701 in May 1952, available for lease at fifteen thousand dollars a month.[44] They produced eighteen machines, plus the prototype, which remained in IBM's New York headquarters. Of those eighteen, eleven were let to aerospace companies, including

Boeing.[45] The 701 was the computer Clough had encountered at Boeing. Given the rental cost and difficulty of programming the machine, it seems doubtful that many structural engineers had access to the 701. However, working in proximity with the 701 certainly gave a number of structural engineers notions of what formulations could be adapted to computer usage in the future.

The vast majority of 701s were delivered to West Coast aerospace companies in 1953, and almost immediately companies began working together and with their IBM representatives to get their machines up and running. The companies also began to develop an informal program library. IBM was pleased to support and sponsor this collaboration and dubbed it SHARE.[46] This was particularly important for the 701 because the work of programming fell to the user. The machine arrived with literature that described the machine's basic instruction set and how the machine processed data but with no programs. Sharing programs benefited all participants in the face of this enormous task. Since aerospace companies required similar mathematical subroutines for many different projects, the industry was a particularly fertile ground for program-sharing.[47]

With the introduction of the 701, the era of big, commercially produced computers had begun.[48] IBM followed the success of the 701 with the 650, the first truly mass-produced machine. It was produced in far greater quantity (250) than any other computer had been. But because the 650 was aimed at the business community and was less powerful than the 701, scientists and engineers had to wait until the 704. The 704 benefited from the development of SAGE, the air defense and air traffic monitoring system produced by IBM and MIT's Lincoln Laboratory. The FORTRAN language was co-developed with the 704 to facilitate scientific programming. FORTRAN broadened the user group for computers and eliminated many of the 701's programming difficulties.[49] Together, the 704 and FORTRAN proved to be a system highly compatible with endeavors in structural engineering.

Marrying the IBM 701 to FEA

When Clough returned from his sabbatical, he was pleased to find that the College of Engineering at Berkeley had procured an IBM 701.[50] He immediately began developing a matrix algebra program, which would be the first ingredient of a structural analysis program. In 1958, he presented a paper on his program at the ASCE Conference on Electronic Computation in Kansas City.[51] The organization of this conference indicated that civil engineers were becoming familiar with the aeronautical engineers' matrix methods. Furthermore, the rapid growth of the computer industry provided an increasing number of engineers with access to computing time. Berkeley emerged as an important center for structural engineers to work on computer applications in a civil engineering department. Clough himself ensured that civil engineering at Berkeley remained at the leading edge of computer use for structural analysis, an area of investigation that continued to be dominated by aerospace designers. Still, matrix methods were slowly becoming something that all structural engineers needed to know about, including those outside of the MIAC.

Clough, as a professor at Berkeley, was in a good position to proselytize for matrix methods. In 1958 he began teaching the methods that he and Turner had worked on earlier.[52] Graduate students in particular were encouraged to use the matrix methods and especially to program the computer to perform subroutines. Edward L. Wilson, one of Clough's graduate students, immediately began writing FORTRAN programs for the new IBM 704. Wilson's project was to automate the mesh formation of structural elements.[53] Clough had used the computer only to solve the matrix equations; setting up the coordinates of the mesh was even more time-consuming in many cases. Wilson's work would result in what he called a structural analysis program (SAP). In addition to making the method more efficient to use, Wilson, with Clough's aid and support, was also broadening its fields of application. Wilson modeled gravity dams instead of airframes (figure 2). Although gravity dams resist completely different forces, SAP proved flexible enough for modeling them and many other new uses.

Some time in the late 1950s Clough and his students began to refer to their matrix method as the "finite element method," although this term did not catch on outside of Berkeley until the mid-1960s. With hindsight, the 1956 paper has been credited with introducing the two techniques critical to FEA—the means of assembling the mesh and the means of approximating the partial differential equations through matrix transformations—but the name "finite element method" did not appear publicly until Clough's 1960 paper at the Second ASCE Conference on Electronic Computation in Pittsburgh.[54] In Clough's account of formulating the finite element method, he claimed that selecting a name became one of the principal problems of the 1960 paper. His goal in writing that paper was to introduce new quadrilateral elements and their stiffness matrix formations to engineers unfamiliar with the aeronautical industry's applications. The name chosen for public consumption needed to indicate both the element assembly procedure and the process of evaluating the stiffness of the elements by solving the direct stiffness matrix equations. Clough referred to the pieces of the mesh as "finite elements" to tie them to the finite difference method and to show that the equations would be solved for discrete

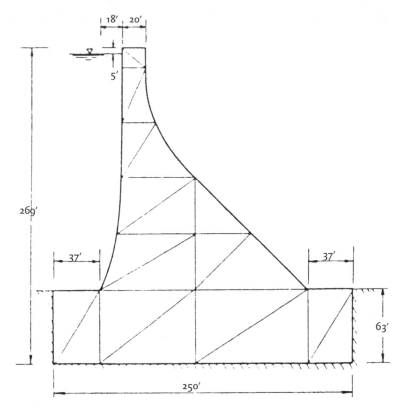

2. First finite element analysis of a gravity dam. Source: Edward L. Wilson, "Automation of the Finite Element Method—A Personal Historical Overview," *Finite Elements in Analysis and Design* 13 (1993): 92.

regions.[55] The 1960 paper, "The Finite Element Method in Plane Stress Analysis," also detailed a way to calculate stress, as opposed to strain, making the method a much more flexible tool for civil engineers.

Although in 1960 the method still required more computing power than was likely to be available to most civil engineers, civil engineering had become established as a fruitful research discipline in the MIAC. As more structural engineering students were introduced to FEA, uses for the method in new areas proliferated. In addition to using FEA for an increasing number of structural dynamics problems, including automobile design and earthquake engineering, the technique was profitably applied to fluid mechanics and heat transfer problems—areas which had provided intractable engineering problems for decades. Problems of nonlinear behavior became particular targets for FEA-wielding engineers. In addition during the 1960s, a mathematical theory of finite elements was slowly developed.[56] When a rigorous, independent, mathematical theory had been established,

FEA had matured as a subdiscipline in its own right. Like most useful tools, it was not contained within a single discipline, and the more the use of the method spread, the more innovative its applications became.

FEA and complexity

FEA was invented because the old methods of analysis were insufficient for the work that engineers in the defense industry were performing. As observed previously, experimental testing became too expensive to do indiscriminately, and overbuilding was impractical in light of aircraft performance requirements. The accuracy desired for airframe design was greater than previous methods could provide. Many engineers argued that the airframe structure was simply more complex than structures they had previously analyzed and this new level of complexity drove their investigation into matrix methods. But in what way were the stresses and strains themselves *more* complex than those found in a suspension bridge? In what way did a skinned wing represent a *more* complex structure than a reinforced concrete shell? Certainly the dynamic loads created by a jet engine were no more complex than those created by a locomotive. In fact, the situations contained comparable complexity. The problem arose in the need to model that complexity. Achieving greater accuracy in modeling required creating ever more complex models to approximate more closely the real structure. Thus, the difference in the level of complexity between an airframe and a bridge lay in the constraints on the solution. The simplest solution, overbuilding, was not open to the airframe designers.

Recalling E. H. Brown's humorous proclamation in *Structural Analysis* about the difficulty of structural engineering, engineers thought FEA enabled them to understand the structure in a more *fundamental* way, as Clough and his colleagues claimed in the 1956 paper. FEA offered solutions that were more detailed, more precise, and more accurate. It simply provided more information about what stresses and strains existed, their locations, and their magnitude. The arbitrarily determined elements were smaller and therefore offered a more precise basis for modeling the aggregate behavior of the steel plate. In this sense, its creators saw that it had something in common with reductionist programs in the sciences—getting down to the smallest relevant unit of investigation. Certainly engineers likened themselves to physicists searching for the key pieces that composed the world. Finite elements seemed analogous to subatomic particles.

In many ways this was a rather idealistic view of FEA. It did not explain *why* structures failed; it was simply a more accurate way to pinpoint where the failure occurred. Engineers often had to spend long hours trying to understand the FEA results themselves—and understanding them meant knowing the underlying physics.[57] Engineers had to explain strange results, either by finding them inaccurate relative to experimental results or by conceiving the structural problem in a new way. J. T. Oden, a professor of engineering mechanics and one of the foremost FEA theorists, referred to it (and computational mechanics generally) as "a good science. . . . It uncovers more questions than it answers."[58] FEA was not going to change the dynamic Brown wrote of; it was designed to better predict behaviors that were still not always understood. Engineers needed more accurate models, but a better understanding of the structure's complexity would not necessarily yield better predictions.

More importantly, engineers needed more flexible tools that allowed a selectable degree of accuracy. An analysis performed early in the development process might not require a high level of accuracy; instead, performing several quick analyses might be a greater priority.[59] Later analyses require an increasing level of accuracy until a prototype is selected. While the structure may not be any more complex than one a century old, the method of analysis is. All engineering analysis uses some idealization of the actual structure, and FEA provides a way of selecting the level of idealization by the fineness of the elements chosen. The engineer gains control over the level of accuracy. While this is not at all a new phenomenon, placing the choice in the modeling phase so methodically requires the user to know up front the level of accuracy needed. FEA provides this flexibility, since the user can choose to make the mesh finer in places of known high stresses and coarser in areas where less information is needed.

FEA, information, and the concerns of its inventors

The knowledge needed to make choices about the mesh has raised concerns within the structural engineering community and, in particular, among the founders of the finite element method. Clough writes, "I have noticed an even more troubling tendency among many engineers: to accept as gospel the stress values produced by a finite element computer program—without realizing the limitations of the approximation procedure being applied."[60] What makes computerized finite element analysis extraordinarily useful as a tool is also what gives Clough his concern—it distances the theoretical and mathematical complexity of the method from the user. The result is a paradox: a complex and flexible method exists to analyze complex structures with greater accuracy, but it presents an intuitive interface of increasing simplicity.

Furthermore, finite element practitioners have raised questions about expertise in the computer applications. When creating a program, expertise resides in the programmer. But once a program is sold, does the expertise of the programmer then reside in the program? Clough's admonition about accepting the computer output as gospel indicates a concern about users believing that the program was expert. In addition, the computer output could not improve on the mathematical model used to set up the program. Without ongoing research into the output's accuracy, the expertise of the program could not be evaluated.[61] The judgment of the programmer, whether it be good or flawed, always resided in the output of values for each node in each model. The accuracy of the results depended on the foundational mathematical models. The computer's algorithms ensured the precision of the output; once the user selected the parameters, the results were always the same. Importantly, the structure itself did not necessarily follow the algorithms. No matter how complex, the model was still an idealization.

The acceptance of FEA reflected major changes in the aesthetics of engineering theory as well. Engineering theories had been judged in part on their elegance, which meant simplicity, formalism, and primacy of ideal geometric forms. Karl Culmann's graphical statics method of the 1850s harkened to nearly Platonic ideals of geometric shapes. Geometry had always been the underlying fundamental mathematics of structural engineering. But Clough's, and later others', references to the fundamental nature of FEA questioned the old standard of elegance and fundamentality. Instead, FEA's value derived from its flexibility and power. Users admitted that FEA's power was obtained at the cost of mathematical elegance. FEA was referred to as "a crude engineering tool . . . a brute force technique for the solution of problems, undignified by the mathematical elegance and formalism proper to decent analyses, relying instead on the prodigious computational feat of solving up to 100 simultaneous equations."[62]

In addition, it was an iterative approximation technique, what physicists might have referred to, in a derisive tone, as "curve-fitting." When results did not fit, algorithms for FEA models were changed. The standards for good mathematical practice in engineering changed in light of FEA's utility. The method was referred to as more fundamental, but its fundamentality lay in its accuracy rather than in more abstract attributes drawn from physical or mathematical conceptions. To some extent, the role of the MIAC in the development of FEA gave this work prestige derived from its usefulness. It created a new standard of usefulness that overrode previous standards of mathematical elegance. If expertise had been redefined as accurately matching predictions to experimental results, then an important parallel development lay in utility getting the upper hand over elegance.

FEA has gradually but nevertheless fundamentally changed the practices of structural engineers, particularly in the design phase. Traditionally, design and analysis were two interactive but sequentially defined activities for the engineer. First, design engineers created plans to meet technical specifications, then more mathematically-minded engineers, or even specialists in partial differential equations, were released on a problem to see if it was feasible. If not, the design reverted to the design engineers for another go-round. This loop continued until both designers and analysts were satisfied with a version of the project, at which point it was up to the testing and evaluating engineers to make a prototype and check it for another level of feasibility.[63] FEA blurred the distinctions between design, analysis, and testing, which cut costs and time; in other words, it made the practice of invention more rational and efficient. As recently as 1992, Colin Macilwain of the British magazine, *The Engineer*, explained industry's enthusiasm for this development:

> Instead of being used by skilled analysts to test component designs, finite element analysis is becoming a routine means for the design team to check its work as it proceeds. . . . The main thing is to get as much up-front engineering analysis done as possible and to get a lot in early, instead of waiting to test the component in service. . . . The main issue is that engineering analysis takes place too late in the design and development process. We need to pull it right back, to the early design stage.[64]

Although the move to interactive design and analysis is still in process today, this basic change in engineering practice began with Clough's FEA and Wilson's SAP.

It is an oversimplification to think of design as a visual process and analysis as a mathematical one, and FEA blurs this distinction even further. Early in the 1960s programmers wanted to create graphical interfaces for finite element modeling. At first, a mesh was drawn by a draftsman, who would further define that mesh using a coordinate system. The coordinates were required as input to the computer. Later, the drudgery of creating the mesh became the task of the computer. The number activities between the design and the numerical output of structural values was diminishing as the computer performed more and more of the work. Wilson and other programmers envisioned a process of automating the finite element method that would require the user to input only a visual model from a computer-aided design program, and, assuming that the user chose the correct parameters, the computer would do the rest. Analysis would thereby become a part of design. Rather than relegating the analysis to a department of equation solvers, necessary when partial differential equations were involved, the airframe designers could tell the computer how they wanted it to construct a mathematical model of the structure. Structurally unacceptable designs could be eliminated immediately. Thus practitioners wanted the finite element methods to unite design and analysis so as to create a virtual testing lab in the computer. Their goal was a more organic procedure for creating the prototype, all in the name of producing new designs in less time.

Conclusion: the key events of 1968

In the fall of 1968, Brigadier General R. A. Gilbert, director of laboratories for the U.S. Air Force Systems Command, welcomed 250 participants to the Second Conference on Matrix Methods in Structural Mechanics at Wright-Patterson Air Force Base in Dayton, Ohio. In his keynote address, Brigadier General Gilbert began:

> To search, to find, to evaluate, to learn, to teach, to apply and to develop are the chosen goals of the community of scientists, engineers, and educators represented here today. It is fitting therefore that this second conference on matrix methods in structural mechanics should again be sponsored by the Air Force Institute of Technology, an institution dedicated to higher research, and by the Air Force Flight Dynamics Laboratory, an organization charged with the responsibility for advancing the technology of flight vehicle structures, working with other government agencies, universities, and industrial organizations.[65]

This opening address reinforces the MIAC's crucial role in developing these new methods of structural analysis. The MIAC created an environment as well as an economic climate that fostered particular attitudes toward scientific and engineering research. In the MIAC environment, renewed emphasis was placed on the investigation of problems—heretofore intractable—which could benefit from new approaches, both theoretical and instrumental. Such was the case with FEA. The cold war provided new incentives for increased accuracy in predicting the behavior of the complex structures in airframes. Just as aircraft had been a key to victory over the Japanese, so jets and ICBMs would provide victory over the Soviets. Yet studies of cold war engineering generally focus on physicists and the "Buck Rogers" engineers, those designing missiles and H-bombs to fit in ICBM nose cones, not those working to solve the more mundane but nevertheless critical problems of airframe design and analysis.[66]

In the same year as the conference, Clough's student Edward L. Wilson, who was subsequently hired in the civil engineering department at Berkeley, began work on his structural analysis program, which immediately and intentionally became known as SAP. Wilson later explained, "The slang name was selected to remind the user that this computer program, like all computer programs, lacks intelligence."[67] For two reasons Wilson felt compelled to remind users that a program only does what one tells it to do. First, the hope had existed for a decade that users of the program could eventually forgo the underlying mathematics, and Wilson and most other software developers worried about this approach. Secondly, general computer use was in its infancy in the 1960s. Although many engineering students would have been familiar with FORTRAN by the late 1960s, using a packaged piece of software which required little user programming was a relatively new phenomenon. Wilson's program was important in the history of structural engineering because it marked the culmination of over a decade of work on the finite element method of structural analysis, and with SAP the civil engineering department of Berkeley achieved its goal of making the first commercially available, computerized, structural engineering program. The practice of structural engineering would change as a result of SAP and subsequent structures programs. But in the fifteen years it took to develop FEA, from the inspiration of an insoluble problem in structural dynamics to a computer program that might be mistaken for an intelligent structural designer, we find an odyssey leading through the unbridled military and industrial culture of the cold war.

Notes

I thank M. Norton Wise, Michael S. Mahoney, Stuart W. Leslie, David C. Brock, Louis Bucciarelli, George Smith, and Michael Latham for their input, as well as the members of the 1993 Princeton University graduate seminar on Postmodern Science and participants in the Workshop on Growing Explanations.

1 See, for example, Stuart W. Leslie, *The Cold War and American Science* (New York: Columbia University Press, 1993); Peter Galison and Bruce Hevly, *Big Science: The Growth of Large Scale Research* (Stanford, Calif.: Stanford University Press, 1992); Paul Forman, "Behind Quantum Electronics: National Security as Basis for Physical Research in the United States, 1940–1960" *Historical Studies in the Physical and Biological Sciences* (hereafter HSPBS), 18 (1987): 149–229; Allan Needell, "Preparing for the Space Age: University-based Research, 1946–57," HSPBS 18 (1987): 89–109; Daniel J. Kevles, "The Cold War and Hot Physics," HSPBS 20 (1990): 239–64; David Kaiser, "Cold War Requisitions, Scientific Manpower, and the Production of American Physicists after World War II," HSPBS 33 (2002): 131–59.

2 The one important exception being Bruce Seely's research, best represented by "Research, Engineering and Science in American Engineering Colleges," *Technology and Culture* 34 (1993): 346–387.

3 Peter Galison, *Image and Logic: A Material Culture of Microphysics* (Chicago: University of Chicago Press, 1997), especially chapter 4.

4 Dwight D. Eisenhower, "Liberty Is at Stake," in Herbert I. Schiller and Joseph D. Phillips (eds.), *Super-state: Readings in the Military Industrial Complex* (Urbana: University of Illinois Press, 1970), 31.

5 For an overview of the economics of the defense industries during the cold war period, see Jacques S. Gausler, *The Defense Industry* (Cambridge: MIT Press, 1980).

6   Eisenhower, "Liberty Is at Stake," 31.

7   Ibid., 32.

8   Ibid.

9   Literature of the military-industrial complex dealing with the development of new technologies includes Schiller and Phillips (eds.), *Super-state*; Carroll Pursell, *The Military Industrial Complex* (New York: Harper and Row, 1972); Steven Rosen (ed.), *Testing the Theory of the Military-Industrial Complex* (Lexington, Mass.: D. C. Heath, 1973); Paul Koistinen, *The Military-Industrial Complex: A Historical Perspective* (New York: Praeger, 1979); Merritt Roe Smith (ed.), *Military Enterprise and Technological Change: Perspectives the American Experience* (Cambridge: MIT Press, 1985); John Tirman, *The Militarization of High Technology* (Cambridge, Mass.: Ballinger Publishing, 1984); Leslie, *Cold War and American Science*; Paul Edwards, *The Closed World: Computers and the Politics of Discourse in Cold War America* (Cambridge: MIT Press, 1996).

10   For accounts of the industrial shift to the West Coast, especially at the expense of traditional manufacturing areas such as the Great Lakes region, see Gregory Hooks, *Forging the Military Industrial Complex: America's Battle of the Potomac* (Urbana: University of Illinois Press, 1991); Ann Markusen, Peter Hall, Scott Campbell and Sabina Deitrick, *The Rise of the Gunbelt: The Military Remapping of Industrial America* (Oxford: Oxford University Press, 1991); and Roger W. Lotchkin, *Fortress California, 1910–1961* (Oxford: Oxford University Press, 1992).

11   Markusen et al., *Rise of the Gunbelt*, 31.

12   These are the Air Research and Development Command (ARDC) in Baltimore and the Air Materiel Command (AMC) in Dayton. Schriever headed the Western Development Division of the ARDC in Inglewood, California. See ibid., 95–96.

13   Ibid., 32.

14   Until NASA began contracting, that is. Once NASA became a research and development contractor, its contracts garnered the most prestige. It is interesting to note that although NASA had planned to work with the arsenal style of research and development (R&D), they quickly converted to the Air Force outsourcing style.

15   Markusen et al., *Rise of the Gunbelt*, 30. Also see R. Futrell's study of the Air Force, *Ideas, Concepts, Doctrine: A History of Basic Thinking in the United States Air Force, 1907–1964* (Maxwell Air Force Base, Ala.: Air Force University, 1974).

16   In fact, Boeing diverted funds into airliner research to avoid taxation, and this allowed Boeing to establish itself in the commercial airline market using federal R&D funds. By the time a tax court found Boeing guilty of this charge in 1962, the company's ties to commercial airlines had been made. See Hooks, *Forging the Military Industrial Complex*, 247. This is not to say that Boeing had no interest in missile production, as Boeing was the original contractor of the Minuteman series. Until the 1970s, however, Boeing remained the most successful military as well as commercial aircraft developer in the industry.

17   Markusen et al., *Rise of the Gunbelt*, 100.

18   Vannevar Bush, *Science—The Endless Frontier: A Report to the President* (Washington: Government Printing Office, 1945), vi.

19   J. William Fulbright, "The War and Its Effects: The Military-Industrial-Academic Complex," in Schiller and Phillips (eds.) *Super-state*, 173–178.

20   Interestingly, Clough also had a master's of science in meteorology from Cal Tech and would have had some knowledge of how connections between industry and academia operated there as well. For an excellent account of forging the connections between industry and MIT, see Larry Owens, "Straight Thinking: Vannevar Bush and the Culture

of American Engineering," (Ph.D. diss., Princeton University, 1987); and Leslie, *Cold War and American Science.*

21 Ray W. Clough, "Original Formulation of the Finite Element Method," *Finite Elements in Analysis and Design* 7 (1990): 91.

22 Ibid., 91.

23 Edward W. Constant II, *Origins of the Turbojet Revolution* (Baltimore, Md.: Johns Hopkins University Press, 1980), 14.

24 David Billington, *Tower and the Bridge: The New Art of Structural Engineering* (Princeton, N.J.: Princeton University Press, 1983); see chapter 3 for an account of Brunel and Stephenson's encounters with the structural dynamics of the locomotive and the bridge.

25 As engineering students at the nearby University of Washington when the bridge collapsed, Clough and his fellow students must have known about its structural dynamics problems. I do not know whether the experience played any part in his decision to investigate structural dynamics as a career, but that would not be surprising.

26 H. C. Martin, R. W. Clough, M. J. Turner, and L. J. Topp, "Stiffness and Deflection Analysis of Complex Structures," *Journal of the Aeronautical Sciences* 23 (1956): 805.

27 Wind-tunnel testing had its theoretical difficulties, too. Wind-tunnel test results could rarely be taken at face value because of scale effects—that is, deformation does not increase linearly with load or structure size, problems related to structural dynamics. Therefore, a set of experimental results in aerodynamic research often has numerical methods already tagged onto it to get the experimental results to match the actual flight conditions. Knowledge of these scale effect problems predate World War I and was developed by Prandtl at Göttingen. See Hashimoto Takehito, "Theory, Experiment and Design Practice: The Formation of Aeronautical Research, 1909–1930" (Ph.D. diss., Johns Hopkins University, 1990); James R. Hansen, *Engineer in Charge: A History of the Langley Aeronautical Laboratory, 1917–1958* (Washington: NASA, 1987); Alex Roland, *Model Research: The National Advisory Council on Aeronautics, 1915–1958* (Washington: NASA, 1985).

28 Clough was particularly interested in Falkenheimer, "Systematic Calculation of the Elastic Characteristics of Hyperstatic Systems," *La Recherché Aéronautique* 17 (1950) and 23 (1951); and B. Langfors, "Analysis of Elastic Structures by Matrix Transformation," *Journal of Aeronautical Sciences* 19(7) (1952): 450–60.

29 Clough, "Original Formulation," 92.

30 It is important to note that different types of structures react differently to different loads. There are important differences between a beam member and a truss member in a structural analysis. Whereas the analysis of the beam usually hinges on an analysis of its bending deformation, trusses carry load by distributing the tension and compression forces into individual structural members, eliminating internal bending forces to a great extent. While a truss may deform as a unit, its internal members are not designed to bend. Different structural elements are conceived of to model only the internal forces that appear in them. This is, of course, an idealization. In addition, connections must be idealized by their degrees of freedom or the degree to which they transfer loads and moments.

31 Clough, "Original Formulation," 93.

32 Martin et al., "Stiffness and Deflection," 807.

33 Clough, "Original Formulation," 93.

34 While Clough was employed at Boeing, a competing approach appeared in an article by Samuel Levy, "Structural Analysis of Influence Coefficients for Delta Wings," *Journal of Aeronautical Sciences* 20 (1953): 449. In the early 1950s, many articles appeared regarding influence coefficients on various wing shapes; see, e.g., J. Archer and C. Samson, "Struc-

tural Idealization for Digital-Computer Analysis," *Proceedings of the Second Conference on Electronic Computation,* 8–9 September 1960 (New York: ASCE, 1960), 283–325.

35 Martin et al., "Stiffness and Deflection," 805.

36 Ibid., 806.

37 Ibid., 810.

38 See note 30.

39 Martin et al., "Stiffness and Deflection," 810.

40 Argyris's most important work in this period was "Energy Theorems and Structural Analysis," *Aircraft Engineering* 26 (1954): 10–11; ibid. 27 (1955): 2–5.

41 C. P. Hunt worked on the computer applications with Argyris; see C. P. Hunt, "The Electronic Digital Computer in Aircraft Structural Analysis," *Aircraft Engineering* 28 (1956): 3–5.

42 Martin et al., "Stiffness and Deflection," 805.

43 C. C. Hurd, "Early IBM Computers: Edited Testimony," *Annals of the History of Computing* 3 (1981): 167.

44 Project engineers might make up to five hundred dollars a month, by comparison.

45 Hurd, "Early IBM Computers," 169.

46 W. F. McClelland and D. W. Pendery, "701 Installation in the West," *Annals of the History of Computing* 5 (1983): 167–170; see also C. J. Bashe, L. R. Johnson, J. H. Palmer, E. W. Pugh, IBM's Early Computers (Cambridge: MIT Press, 1986), 348–349.

47 Randall E. Porter, "First Encounter with the 701: Boeing Airplane Company," *Annals of the History of Computing* 5 (1983): 202–204.

48 F. Gruenberger, "A Short History of Digital Computing in Southern California," *Annals of the History of Computing* 2 (1980): 249.

49 See Bashe et al., IBM's Early Computers, chapters 5 and 9.

50 Since Berkeley's Engineering College is not on the original list of leasers for the 701, the machine must have been let second hand.

51 R. W. Clough, "Structural Analysis by Means of a Matrix Algebra Program," *Proceedings of the ASCE Conference on Electronic Computation* (New York: ASCE, 1959), 109–32.

52 For more examples of the impact military research had on the university science and engineering curriculum, see also Leslie, *Cold War and American Science.*

53 Edward L. Wilson, "Automation of the Finite Element Method—A Personal Historical Overview, *Finite Elements in Analysis and Design* 13 (1993): 92.

54 J. T. Oden, "Historical Comments on Finite Elements," in Stephen G. Nash (ed.), *A History of Scientific Computing* (New York: ACM Press, 1990), 155.

55 Clough, "Original Formulation," 99.

56 Oden, "Historical Comments," 156–157.

57 Wilson, "Automation of the Finite Element Method," 103.

58 J. T. Oden and K. J. Bathe, "A Commentary on Computational Mechanics," *Applied Mechanics Reviews* 31 (1978): 1053.

59 According to Wilson, "Finite element programs must be fast in order that the design engineer can conduct a large number of 'what if' studies on a large number of alternate structural designs"; see Wilson, "Automation of the Finite Element Method," 103.

60 Clough, "Original Formulation," 101. Wilson shares this concern; see "Automation of the Finite Element Method," 103.

61 Clough, "Original Formulation," 101. "The results of a finite element analysis cannot be better than the data and the judgment used in formulating the mathematical model, regardless of the refinement of the computer program that performs the analysis. The main purpose of that word of caution was to emphasize the continuing need for experi-

mental observations of structural behavior; it is only with such experimental evidence that computer analysis procedures can be validated, and there is no question that the need for such validation is as great now as it was 10 years ago."

62 S. Kelsey, "Finite Element Methods in Civil Engineering," in R. H. Gallagher and Y. Yamada (eds.), *Recent Advances in Matrix Methods of Structural Analysis* (Huntsville: University of Alabama Press, 1971), 775.

63 For an excellent description of and commentary on the practices of design engineers, see Walter Vincenti, *What Engineers Know and How They Know It* (Baltimore, Md.: Johns Hopkins University Press, 1990).

64 Colin Macilwain, "Analysis Finds a Place in the Toolbox," *The Engineer* (23 July 1992), 29.

65 R. A. Gilbert, "Keynote Address," *Proceedings of the Second Conference on Matrix Methods in Structural Mechanics*, 15–17 October 1968 (Dayton, Ohio: Air Force Institute of Technology, 1969), 3.

66 "Buck Rogers engineering" is a term used by Markusen et al. in *Rise of the Gunbelt*; see especially chapter 5. Buck Rogers engineers designed missiles and rockets rather than the more traditional aircraft.

67 Wilson, "Automation of the Finite Element Method," 98.

5 Fuzzyfying the world: social practices of showing the properties of fuzzy logic

Claude Rosental

The boat of uncertainty reasoning is being rebuilt at sea. Plank by plank fuzzy theory is beginning to gradually shape its design. Today only a few fuzzy planks have been laid. But a hundred years from now, a thousand years from now, the boat of uncertainty reasoning may little resemble the boat of today. Notions and measures of overlap A $^\wedge$ A^c and underlap A U A^c will have smoothed its rudder. Amassed fuzzy applications, hardware, and products will have broadened its sails. And no one on the boat will believe that there was a time when a concept as simple, as intuitive, as expressive as a fuzzy set met with such impassioned denial.—Bart Kosko, "Fuzziness vs. Probability"

Can one identify a recent and general trend in scientific research which would lead academics to value more complex, diverse, and indefinite explanations? This essay will investigate one aspect of this question by examining the way in which advocates of the logical theory called "fuzzy logic" have put forward the need for an adequate formalism to capture aspects of qualitative reasoning, reasoning under uncertainty, or reasoning under incomplete information.

Some proponents of fuzzy logic, whom I have met and interviewed over a five year period, perceive the development of fuzzy logic as part of a wider transformation in a whole emerging field—the field of handling imperfect knowledge. For them, fuzzy logic is one of a set of competing or complementary approaches to probability theory, which has for a long time monopolized this field. Such proponents generally include within this set approaches for the handling of uncertainty, vagueness, imprecision, incompleteness, and partial inconsistency, and theories like nonmonotonic logics, modal logics, Bayesian and non-Bayesian probability theories, belief function theory, fuzzy sets, possibility theory, theory of evidence, and belief networks.[1] In this view, fuzzy logic would contribute to addressing a growing number of questions in handling imperfect knowledge that could not even be raised in probability theory. Furthermore, the development of fuzzy logic would be a significant historical event for successfully bridging the field of logic and the field of uncertainty reasoning.

But what is "fuzzy logic" in the first place? The juxtaposition of the two terms *fuzzy* and *logic* is by itself immediately problematic. How, at all, did the virtues of fuzziness come to be celebrated in logic?

These questions cannot be quickly answered. The proponents of fuzzy logic themselves offer various definitions and histories of their objects, which are both changing and competing. It will be useful at this stage to analyze some of them, as it allows one to address the central issue of this essay and to portray at the same time the actual state of the historiography of fuzzy logic.[2]

Growing numbers of issues and resources in fuzzy logic

Among the diverse definitions of fuzzy logic, many insist on the fact that while this logic allows one to consider an infinite number of degrees of truth, binary logic offers only two degrees of truth (zero and one, corresponding to "false" and "true"). The histories of fuzzy logic put forward by its proponents are often histories of ideas, which, though very diverse, converge to build a history of the invention of a theory. Such histories attribute fuzzy logic's invention to Lotfi Zadeh by virtue of an original article published in 1965.[3] Some narratives argue that the fuzzy logic developed out of the need to devise a formalism which could capture some aspects of qualitative human reasoning so that this ability could be incorporated into computers. This type of presentation can be found for example in the following passage, where the author introduces Zadeh's individual motivations.

> The theory of Fuzzy Logic was introduced to the world by Professor Lotfi A. Zadeh of the University of California at Berkeley. Professor Zadeh observed that conventional computer logic is incapable of manipulating data representing subjective or vague human ideas, such as "an attractive person" or "pretty hot." Computer logic previously envisioned reality only in such simple terms, as on or off, yes or no, and black or white. Fuzzy Logic was designed to allow computers to determine valid distinctions among data with shades of gray, working similarly in essence to the processes which occur in human reasoning. Accordingly, Fuzzy technologies are designed to incorporate Fuzzy theories into modern control and data processing, to create more user-friendly systems and products.[4]

Other historical narratives portray "classical binary logic" as having encountered a growing number of problems and limitations which were highlighted by classical logical paradoxes. These narratives present the emergence of fuzzy logic as a solution to these problems and limitations, and their corresponding histories of fuzzy logic start far earlier.

Hence, Bart Kosko, a visible figure in fuzzy logic (and the author of my epigraph), relates the development of fuzzy logic to Lukasiewicz's multi-valued logics and to Russell's philosophy of vagueness. According to him, the rise of fuzzy logic constitutes a response to the "failure of classical logic" and to the failure of Aristotle's law of the excluded middle in the face of the Greek paradoxes that Russell highlighted and developed. According to Kosko, Max Black, a philosopher of quantum mechanics, drew the first fuzzy set diagrams, by applying Lukasiewicz's multivalued logic developed in the 1920s. Black called these "vague sets," in accordance with Russell's work, and it was not until thirty years later that Zadeh applied Lukasiewicz's logic to each member of a set and created a complete algebra for fuzzy sets.[5]

Some proponents of fuzzy logic go still further back and start the history of this logic in Greek antiquity. A tutorial on fuzzy logic, written by James Brule, clearly illustrates this type of narrative. In this text, the author draws a genealogy of ideas that links Heraclitus and Plato to Zadeh, going through Hegel, Marx, Engels, and Knuth.[6] According to Brule, when Parmenides proposed the first version of the law of the excluded middle, which states that every proposition must be either true or false, there were strong and immediate objections, such as the one from Heraclitus who proposed that things could be simultaneously true and not true. In Brule's narrative, it was Plato who laid the foundation for what became fuzzy logic by indicating that there was a third region (beyond true and false) where these opposites "tumbled about." It was in this spirit that Zadeh designed fuzzy logic as an infinite-valued logic, following Hegel, Marx, Engels, and more recently the work of Lukasiewicz and Knuth on three-valued logic.

But not all histories of fuzzy logic consist of histories of ideas. Some narratives try to situate the invention of fuzzy logic in the frame of a factual history.[7] They not only involve researchers other than Zadeh, but also other resources such as technological devices or products, laboratories, journals, articles, associations, and symposiums. The introduction of human and nonhuman actors shows the rise of fuzzy logic as the product of a collective action[8] and *incorporation*, and leads to its *excorporation* from Zadeh's mind. The following extended quotation will illustrate this point:

> The year 1990 witnessed the 25th anniversary of the invention of Fuzzy theory. It has undergone numerous transformations since its inception with a variety of Fuzzy Logic applications emerging in many industrial areas. Dividing these past years into different stages, the

early 1970s are the "theoretical study" stage, the period from the late 1970s to early 1980s the stage of "developing applications for control," and that from late 1980s to the present the stage of "expanding practical applications." Here are the major events in the history of Fuzzy Logic:

1965: Professor L. A. Zadeh of the University of California at Berkeley introduces "Fuzzy sets" theory.

1968: Zadeh presents "Fuzzy algorithm."

1972: Japan Fuzzy Systems Research Foundation founded (later becoming the Japan Office of the International Fuzzy Systems Association [IFSA]).

1973: Zadeh introduces a methodology for describing systems using language that incorporates fuzziness.

1974: Dr. Mamdani of the University of London, UK, succeeds with an experimental Fuzzy control for a steam engine.

1980: F. L. Smidth & Co. A/S, Denmark, implements Fuzzy theory in cement kiln control (the world's first practical implementation of Fuzzy theory).

1983: Fuji Electric Co., Ltd., implements Fuzzy theory in the control of chemical injection for water purification plants (Japan's first).

1984: International Fuzzy Systems Association (IFSA) founded.

1985: 1st IFSA International Conference.

1987: 2nd IFSA International Conference. (Exhibit of OMRON's Fuzzy controller, a joint development with Assistant Professor Yamakawa.) Fuzzy Logic–controlled subway system starts operation in Sendai, Japan.

1988: International Workshop on applications of Fuzzy Logic–based systems (with eight Fuzzy models on display).

1989: The Laboratory for International Fuzzy Engineering Research (LIFE) established as a joint affair between the Japanese Government, academic institutes and private concerns. Japan Society for Fuzzy Theory and Systems founded. . . .

1987 marked the start of Japan's so-called "Fuzzy boom," reaching a peak in 1990. A wide variety of new consumer products since then have included the word "Fuzzy" on their labels and have been advertised as offering the ultimate in convenience.[9]

Such a presentation suggests that grasping the rise of fuzzy logic requires one to take into account the growing number of heterogeneous resources, actions, and networks which have been progressively involved since the mid-1960s. A thorough examination of the existing literature on fuzzy logic certainly confirms this view. The progressive involvement of fuzzy logic within various activities appears to correspond to the building of very wide networks linking science, technology, and society.[10] A description of the development of fuzzy logic should then start with the previous list of elements and extend it to other resources including at least the following: thousands of researchers and engineers dispatched throughout the world (in particular in the United States, Western Europe, Japan, and China) covering a multiplicity of disciplines and research fields (logic, philosophy, mathematics, control systems, computer science, linguistics, psychology, social sciences, physics, biology, medicine, decision analysis, pattern recognition, information processing, artificial intelligence, etc.); various industrial and distribution firms developing and promoting computer or household electrical fuzzy-named products; a considerable textual production which includes books, expert reports, and newspaper articles.[11]

Taking into account the considerable number of these resources, giving here a full history of fuzzy logic is out of the question. On the other hand, it is feasible to focus on a few important elements involved in its development in the first half of the 1990s. During that period I had the opportunity to conduct interviews and make observations related to the case. But what methodological approach should be adopted to build such a history?

Mediations of fuzzy logic and social practices of showing

As we have seen, even the most refined history of ideas would still be insufficient to describe the development of fuzzy logic. Neither can its rise be captured through the routine study of a discipline, for a simple reason: the objects of fuzzy logic cannot be simply allocated to one or even to several disciplines. One possible start then would be to focus on what most proponents consider to be their primary object—the specific formalism of fuzzy logic—and to look at how a mobilization and networking of heterogeneous resources might have grown from it. This involves a very close examination of the actors' practices and especially those which are linked to their textual production.[12]

Essential for this purpose is a material approach to abstraction. Indeed, the production of abstraction is itself often mistakenly viewed as an abstract process and therefore unreachable for an empirical study. In particular, the material resources involved in writing or reading activity (a scriptovisual setting), which involves eyes and hands as well as visual and writing devices, are easily forgotten.[13] Jack Goody, for example, has analyzed the material activity of writing as a way of acting on communities of people. Instead of viewing writing as an act of representation synonymous with bald visualization, Goody shows how it can transfigure things, people, and their everyday activity.[14]

Hence, materializing logical activity is intended here to be a first and essential step in showing all the mediations that give rise to logical objects. The term *mediation* is imported here from the field of the sociology of art and especially the sociology of music, which studies the media. This social history of music seems to encounter the same problems with the so-called immaterial and unmediated objects as does the social history of mathematics and logic. So the term *mediation* rather than *intermediary* is used here as a generic analytical tool to highlight and characterize all the resources which might be considered as in-betweens (material texts or instruments, for example), and to show them as proper beings.[15] In this essay, written symbols or material devices such as computers, which are generally presented by the actors as embodying fuzzy logic, will be considered as mediations of this logic.[16]

Furthermore, the five years of observations and interviews that I have conducted among logicians in the United States and in France to develop a sociology of logic have led me to focus on what would seem to be one of their most intensive, resource-consuming, and high-stakes activities: "showing." It might seem surprising to say "showing" and not "proving," given that these two terms are part of a dichotomy which has been stabilized by philosophy and especially philosophy of science. By looking at replication in the so-called experimental sciences, social studies of science have opened a large black box and raised fundamental questions about obtaining evidence in science.[17] By comparison, reopening the question of proving in the so-called deductive sciences might be now of equal, if not greater, interest. I intend here to provide some elements for understanding how this question might be usefully formulated in the frame of a social history of practices of showing. It is unfortunate that the French translation of "to prove," *démontrer*, is not used in English, as it would help make this project more explicit. *Démontrer* is indeed formed with the verb *montrer*, literally "to show." The project is, in this sense, an attempt to depict the activity of *demonstration* as involving practices of *de-monstration*, or of *monstration*.[18]

We can begin looking at practices of proving as ways of showing, as opposed, in particular, to mere technologies of conviction, by examining the case of fuzzy logic. Looking at activities of showing will reveal how heterogeneous resources might be mobilized and networked around a formalism such as fuzzy logic.[19] This approach will in fact organize which of fuzzy logic's mediations I choose that helped to develop it in the first half of the 1990s.

As mentioned earlier, it is important to begin with the material study of fuzzy logic's formalism through the analysis of its proponents' textual production. I will therefore analyze a passage of an article written by Bart Kosko, the importance of which will appear a posteriori.

Tacit showing practices and graphic resources

In an article titled "Fuzziness versus Probability," Kosko tries to show that fuzzy logic can solve a classical logical paradox formulated originally by Russell—the barber's paradox. The demonstration provided is reproduced in the following passage (I have numbered some statements):

> Russell's barber is a bewhiskered man who lives in a town and shaves a man if and only if he does not shave himself. So who shaves the barber? If he shaves himself, then by definition he does not. But if he does not shave himself, then by definition he does.
>
> [1] So he does and he does not—contradiction ("paradox"). Gaines observed that this paradoxical circumstance can be numerically interpreted as follows.
> [2] Let S be the proposition that the barber shaves himself and not-S that he does not. Then since S implies not-S and not-S implies S, the two propositions are logically equivalent:
> [3] S = not-S.
> [4] Equivalent propositions have the same truth values:
> [5] t(S) = t(not-S)
> [6] = 1 − t(S)
> Solving for t(S) gives the point of the truth interval . . . :
> [7] t(S) = 1/2. . . .
> . . . In bivalent logic both statements S and not-S must have truth value zero or unity. The fuzzy resolution of the paradox only uses the fact that the truth values are equal. . . . The midpoint value 1/2 emerges from the structure of the problem.[20]

Many of Kosko's contemporaries consider Russell's barber paradox as among those that marked the end of logicism at the beginning of the twentieth century.[21] Kosko mobilizes this emblematic paradox to show that binary logic, as opposed to fuzzy logic, leads to contradiction. Later in his text, he indeed insists on the fact that *paradox* is a euphemism used to avoid admitting the contradictions of classical logic.

To reach this goal, Kosko shows that having truth values between 0 and 1, in other words having statements not entirely true or false, as in fuzzy logic, allows one to solve the barber's paradox. What resources and practices do then correspond to the term *solving*?

The first step consists of formulating a little scenario, a prose statement to introduce a specific manipulation of symbols. In more Wittgensteinian terms, graphic operations to be visualized on paper are introduced and justified by exhibiting a form of life that naturalizes them.[22] Indeed, the little prose scenario on the barber stages a form of life which, followed by relevant translations, naturalizes the use of a truth value between 0 and 1. Kosko expresses it in his own way when he writes that the truth value 1/2 "emerges from the structure of the problem."

Kosko then moves on to translations. He links a series of statements, shown as equivalent or as having the same meaning. He uses detours of formulation (in other words, periphrases or circumlocutions) to prepare the progressive display of terms classically recognized as logical symbols (like not) associated with specific graphic signs. This whole series of exhibitions, implying the display of a scenario, translations, the production of detours of formulation and of graphic signs consecrated as logical symbols, proceeds, in fact, from well-stabilized tacit practices.[23] This procedure is characteristic of a widespread ritualized exercise known as solving a logical paradox.

Let us follow Kosko's demonstration step by step. Kosko substitutes for statement [1] the periphrasis [2]. He displays this periphrasis [2] as equivalent to [1], that is to say, as likely to serve as a visual substitute for the former written sentence (or to use a more precise term, for the former inscription).[24] This detour of formulation allows him to graphically prepare a symbolization or formalization in [3]. By symbolizing or formalizing, I mean introducing written symbols. Formalizing, in fact, can be described as a material operation which consists of introducing graphic signs. Next, Kosko once again substitutes for statement [3] a translation using consecrated graphic representations [5]. For this purpose, he introduces a sentence corresponding to a usual hypothesis [4] (another widespread logical practice). Through this production of inscriptions, Kosko translates the initial scenario into an equation. Then, by implicitly imposing a manipulation (which is common to many fuzzy logicians), "t(not S) = 1 − t(S)," he obtains a simple algebraic equation [6]. He finally conducts a calculation, again a very common tacit practice, to obtain the final result [7].

Hence, the visual display of fuzzy logic's ability to solve the barber's paradox proceeds first by mobilizing tacit practices governing the production of inscriptions, whether in detail or with respect to the whole process. But it also proceeds from graphic resources which impose, one might say, their own logic. Material traces organize the skilled reader's reading act.[25] They punctuate it. The text of the proof in this sense interacts with its readers in a specific way. This process does not merely rely on immaterial conventionalism.[26] It escapes from the descriptive tools of a social relativist who would dismiss the inscriptions as arbitrary or as entirely determined by social relations involving only humans. The proving or showing action, far from being itself abstract, simply localized in the actors' heads, or exclusively delocalized among "contextual" practices of "scientific communities," takes place first in the writing or reading of a text from a determined workplace. This is not to say that the proof has itself an absolute power to drive all readers from the first step to the last and to convince them of its validity. As we shall see later, some researchers deny the validity of the proof. The point here, at the same time trivial and important, is that the showing action is distributed right up to the material use of inscriptions.[27]

Before examining further mediations which, apart from the text just analyzed, put forward fuzzy logic's capacity to solve the barber's paradox, I return briefly to the emblematic dimension of solving this paradox. One might think that we have so far just encountered the establishment of a purely formal property. To grasp all the dimensions of the showing activity that takes place in Kosko's text, however, as well as the way fuzziness comes to be celebrated as a virtue in logic, we need to analyze more passages in the article.

Following his demonstration, Kosko displays historical, cultural, and realistic properties of fuzzy logic connected with the singular history of the barber's paradox. Still through a writing device, he mixes various registers to state that fuzzy logic is the only realistic description of the world (i.e., statements about the world are neither entirely true or entirely false). The barber's paradox, he claims, unequivocally shows the contradictions of binary logic and of the law of the excluded middle. Adopting a rationalist stance, he infers that adopting binary logic and the law of the excluded middle is itself irrational. This irrationality can only be explained for him through a cultural and historical approach. The historian Kosko then states that the "arbitrary insistence on bivalence" is "simply a cultural preference, a reflection of an educational predilection that goes back at least to Aristotle." Fuzzy logic marks a turning point between the end of western binary logic and the rise of eastern thinking and civilization.[28]

Hence, demonstrating fuzzy logic's formal property of being able to solve the barber's paradox constitutes one part of a larger demonstration mixing heterogeneous registers that intends to reveal fuzzy logic's realistic, historical, and cultural properties. Kosko articulates this celebration of fuzziness versus binarism in logic using the process of solving the barber's paradox and the viewpoint of a historian of science and civilization. Social studies of experimental sciences have shown how some scientists try to rebuild the world from their laboratories.[29] Similarly, one can see here how the graphic display of a formal property can serve as a crucial mediation, not only to show the need to express numerous aspects of fuzziness in logic, but also to rebuild "universal" categories of representation and the world history of civilizations.

In fact, writing the history of logic, or even of civilizations, might be seen as a proper means for the fuzzy logic's proponents to show the value of fuzziness in logic and of the work done in fuzzy logic so far. In other words, the histories of fuzzy logic should be seen in retrospect as authentic demonstrations, involving specific showing practices. In particular, the histories that show the rise of fuzzy logic from 1965 to the 1990s might be seen, in part, as promoting the role of contemporary researchers by insisting on an *ex nihilo* rise of fuzzy logic through their singular actions. On the other hand, some narratives that place the origins of fuzzy logic further back in time correspond to an explicit quest for respectability.

Mediations of visibility and countervisibility

Complementary to the graphic display and historical narrative operating within Kosko's simple text are a variety of other mediations that carry his interpretation of the properties of fuzzy logic to a wide audience. They include a fairly technical book that Kosko published in the early 1990s and that attained a high volume of sales titled *Neural Networks and Fuzzy Systems*.[30] It was widely used as a textbook in computer science and electrical engineering departments in universities throughout the United States. Many people I have interviewed say that far more copies of this book were sold in the United Sates than any other book on fuzzy logic. Two similar vehicles of visibility were another popular book by Kosko titled *Fuzzy Thinking* and a book titled *Fuzzy Logic* co-authored by two journalists, Dan McNeill and Paul Freiberger, which formulated similar claims.[31] A whole series of other mediations, involving Kosko in various ways, can be added to this list: articles, course lectures at the University of Southern California where Kosko was teaching, talks, tutorials, seminars, signature sessions in libraries, interviews given on the radio or reproduced in newspaper articles, campaigns conducted on electronic bulletin boards and in specialized journals to promote his books, and so on. These are of course standard forms of promotion in academic life and are not specific to logical properties.

Faced with this onslaught of mediations promoting Kosko's view of the properties of fuzzy logic, and with the growing visibility of its corresponding definition and portrait, other proponents of fuzzy logic tried to form coalitions to put forward visible alternative definitions. They celebrated the virtues of fuzziness in logic quite differently. In fact, the term *fuzzy logic* should be considered a "collective statement," to borrow an expression from the historian Alain Boureau,[32] for it was shared by academics with competing definitions. The shared use of the term linked different objects and the actions of various researchers in complex ways. Though each actor used and appropriated the terminology in a singular way, it played a role in their general development.

It will be worthwhile to look at a few mediations in the fight for the "performative definition of fuzzy logic," which expression insists that the definitions given of fuzzy logic, in the act of statement itself, participated in constructing it. During an interview conducted at the beginning of 1994, a major proponent of fuzzy logic (who will remain anonymous here) stated that Quine had definitively shown that the barber's paradox could not be solved with a multivalued logic.[33] He thought Kosko's translations contained a trick used to solve the paradox because fuzzy logic did not have the appropriate properties. And the historical and cultural properties that Kosko attributed to fuzzy logic were for him just extravagant. He thought in particular that Kosko's criticisms of the Aristotelian tradition were completely irrelevant because mulivalued logics had been developed in this tradition. The fight was once again on the field of the history of logic.

In addition to the mediation of the oral denunciation expressed during this interview to convey a different definition of fuzzy logic, the interviewee carried out other actions. He had published in a journal a virulent critique of a popular book purveying Koskoian claims. He also contributed to the writing of a textbook on fuzzy logic which was meant to provide an alternative to Kosko's. In its first chapter, he proposed an exercise to show that Kosko's solution of the barber's paradox was wrong. This exercise proceeded from a graphic deconstruction and a new exhibition. It consisted in locating and contesting one of the translations involved in Kosko's demonstration and in substituting at this very location another series of translations leading once again to the display of a paradox.[34] One sees here that Kosko's textual demonstration could not by itself lead this researcher from the first step to the last and convince him of its validity. Quite important differences were involved in their practices of doing logic.

Consider a final example of mediations involved in this fight over definitions. The organizers of an important conference on fuzzy logic thought that Kosko's claims were erroneous, giving a false picture of fuzzy logic and threatening its credibility as they became more and more visible. They therefore tried to minimize Kosko's visibility by assigning him a theme for his talk which was not central to the definition of fuzzy logic. Backstage they were also expressing their fears to other academics, encouraging them to pass the word around and to denounce publicly Kosko's mistakes during his talks. They insisted on the fact that Kosko's false claims were threatening the whole research activity in this domain, especially in the context of a cold war with alternative methods in artificial intelligence, resulting from stiff competition for the same sources of funding. And indeed, an actor who has been described as a proponent of classical artificial intelligence later published an article to show some inconsistencies of fuzzy logic, starting from Kosko's presentation. This article was awarded a prize at the American Association for Artificial Intelligence conference in 1993 and led to important debates on fuzzy logic and on his own proof as well.[35] This was only one of many such critiques and actions.

Why should the question of the funding of artificial intelligence intervene here? One might also wonder about Kosko's claim in my epigraph that "amassed fuzzy applications, hardware, and products" should "broaden fuzzy logic's sails" within a few centuries. In fact, the development and promotion of fuzzy logic was largely achieved at the beginning of the 1990s by embodying its properties in technological devices. Computer systems and control systems for a variety of mechanisms were used by scholars and engineers to contrast fuzzy logic with alternatives like Boolean logic. In this way they were investing in various industrial domains.[36] In what follows, I will focus on a privileged set of such exhibitions: the demos (abbreviated from demonstration) of technological devices.

Demos

In the first half of the 1990s, the exhibition of fuzzy logic's properties in technological devices was an important form of mediation mobilized to obtain industrial contracts. It allowed some proponents to obtain autonomous networks of short- and middle-term funding to subsidize their research. It allowed them in particular to bypass any appeal to hostile peers who were developing other approaches and involved them in a hard-nosed competition for funding. Some major researchers in fuzzy logic believed the sale of products under the emblem of fuzzy logic would ensure in the long-run that all ontological forms of this logic were disseminated. Equivalently, displaying neurofuzzy technological devices, under the emblems of both neural networks and fuzzy logic, was a new showing strategy developed at the beginning of the 1990s. It proceeded at a certain level from a sociotechnological alliance against what was called classical artificial intelligence and its funding.

In this strategy, the demo played an important role as a form of showing. It generally occurred in a closed setting in front of a limited audience. A demonstrator spoke for a running device, linking its operation to general properties. The properties were exhibited to the witnesses in an elevation move. *Elevation move* seems the most adequate expression to describe the way these properties were introduced, as they were shown as emanating from, but also detachable from, the materiality and the specificity of the device.[37]

Demos were used in a wide variety of situations. A fuzzy logician could use a demo to try to enroll engineers during a tutorial. An engineer who had developed fuzzy logic products could use a demo at a stand at an international conference on fuzzy logic to try to convince a manager of a firm to buy a fuzzy system. Or a demo could be mobilized in the wider consumer market. In such cases, a product such as a camera, a particular brand, and the emblem *fuzzy logic* would generally be promoted simultaneously. A demo could also be used in front of academic peers, and this form of demonstration is becoming more and more common.

For some proponents of fuzzy logic, bringing demos into play occurred during a long process of learning the most relevant forms of showing in various situations. One French scholar described to me how he had progressively realized the importance of demos in seducing industrial managers who were insensitive to academic publications. He had decided to hire an engineer to build devices for demos. It took them many months to hone both the working devices and the repertoire (or stabilized narrative) to comment on their operation. Conversely, another researcher told me of an experience that he thought characteristic of French technocrats oriented toward mathematics. As he could not convince a research manager of an important French industrial group of the usefulness of fuzzy logic by displaying the principle of working devices, he spoke of a formal result on fuzzy logic. In this case, the formal result itself, rather than a demo, provoked the enthusiasm of his interlocutor.

But let us come back to the demo process itself. To repeat, a demo generally involves exhibiting a technological device in action in front of an individual or a small group of people. The demonstrator attributes properties of fuzzy logic to its working, making himself a representative of the system (sometimes a sales representative). This point can be illustrated through a brief account of a demo, taking place at a meeting of a fuzzy logic association. A researcher in fuzzy logic was exhibiting properties of fuzzy logic in front of a gathering of French academics, engineers, and industrial research managers. "In particular, he [the demonstrator] illustrated the efficiency and the simplicity of this type of control in the case of the inverted pendulum (balanced on its base), in dynamic and even randomly modified circumstances: a glass of water filled during the experiment, and even a live mouse moving on the plate at the top of the pendulum, did not cause the pendulum to lose its balance, even though the rules of control were not changed."[38]

The purpose of such a demo is clear: convincing an audience that some properties of fuzzy logic are real, thereby creating witnesses who can testify to their reality.[39] The demonstrator, making himself a representative of the device in action, exhibits properties such as "efficiency" and "simplicity" and attributes them to fuzzy logic. To underline these properties, a whole scene is set.[40] The demonstrator puts here on stage the extreme and spectacular circumstances of the working of the device. But by not mentioning the role of any other components in the working of the system (mechanical subdevices for example), the demonstrator tacitly ascribes its attributes exclusively to fuzzy logic. The specificity of the whole system disappears. What appears instead in this elevation move are the general properties of fuzzy logic, which are extricated from the materiality of the device and thereby shown as exportable to other devices.[41] On this last point, some engineers stated that controlling an inverted pendulum was emblematic of controlling a space rocket during takeoff. It was important to control the rocket's oscillations as it could break in half if they became too intense. To this extent, a research manager working in the aerospace industry could literally see the demo mastering a technological problem involving an investment of billions of dollars.

In an ultimate phase, this particular demo was relayed through a written account sent by mail to a larger audience. Similarly, through the mediation of this essay, it is now relayed to more readers. The scene of exhibition is continually being extended.

It is important to note that demos can be mobilized in diverse situations. For example, a professor of control at MIT exhibited the properties of fuzzy logic to his frequent visitors through laboratory tours which consisted of a series of demos. These tours were organized in the greatest detail and were strongly stabilized. Several devices in action were shown in a row. The tours were especially intended for industrial investors who subsidized the laboratory's research or from whom the professor hoped for new funding. Alternatively, he often made himself into a sales representative, traveling throughout the United States and Japan looking for new investors. He took with him videos of devices in action. The videos were staged following a carefully elaborated scenario. This showing strategy obviated the need to carry cumbersome and fragile mechanisms. It also allowed him to avoid a considerable investment in time and the risk of failure involved in randomly replicating real-time demos.

Some conferences intended for a mixed audience of academics and industrialists combined talks and demos, as in a conference in Paris in May 1992. A quota of participants had been fixed in advance explicitly to prevent the groups forming around demonstration devices from being too large. A written account of this event described it in the following terms:

> This event was not only interesting for the talks given by the experts, . . . but also for computer demonstrations of software based on fuzzy logic, which were located in an adjacent room. Demonstrations

focused on car braking, numerical control, data merging, decision help, financial or medical applications, etc. Many various domains where the participants could *de visu* acknowledge the performances of fuzzy logic. Movie projections and displays of equipment crowned it all. This potential remains to be put into practice, and it therefore remains to arouse the interest of some industrialists and governmental authorities.[42]

This account, written by a member of an association whose statutory goal is to promote fuzzy logic in France, shows succinctly how staging demos can be part of a broad strategy to arouse the interest of various actors. In effect, the conference exhibited experts as representatives of fuzzy logic. It simultaneously exhibited its embodied properties through various mediations: actual material systems, videotaped demos, computer simulations. The production of witnesses who would see in situ the properties of fuzzy logic proceeds here from an explicit enterprise. The conference was intended to arouse an interest in fuzzy logic in a large audience by addressing in a differentiated way the specific concerns of industrialists and engineers engaged in a wide variety of activities (banking, medicine, car industry, etc.). The very diversity of these exemplifications helped at the same time to show to all participants the potentially universal implementation of fuzzy logic.

We can now see how demos could participate in linking more and more actors by exhibiting more and more properties of fuzzy logic in front of various audiences. Demos can be seen as mediations that contribute to solidifying associations between science, technology, and society.

Of course, as in the case of the exhibition of fuzzy logic's formal properties, some actors challenged the reality of the embodied properties. In the context of competition between proponents of different approaches to control, exhibitions of properties supposed to be embodied in devices were sometimes deconstructed through countermediations of a different kind. An author in a French control journal, for example, attempted to do this textually. In a short article, a control scientist reporting on a laboratory tour in Japan described the backstage of the laboratory in an effort to show that a famous camcorder's operation, exhibited as embodying the properties of fuzzy logic, was in fact due to other mediations.[43] Once again, the display of fuzzy logic properties should be understood within a framework that includes contradictory demonstrations.[44]

Conclusion

The preceding analysis has presented a few important mediations involved in promoting fuzzy logic in the early 1990s, whether in the graphic display of formal properties, in the mobilization of vehicles of visibility, or in demos intended to exhibit properties in devices. It shows that, far from being constituted as any single transcendent entity, fuzzy logic followed multiple trajectories of incorporation and excorporation in different ontological forms. It was distributed in and constituted by all of these differentiated forms of mediation: devices, articles, graphic symbols, and so on. And the variety of forms, their emblems and their human representatives, helped to constitute in turn the large associations which were built around them, linking science, technology, and society. Thus the exhibition of every ontological form of fuzzy logic further disseminated and extended it as a real object. This ostentatious activity was an essential element in its rise.

I hope to have shown the usefulness of a sociological approach to forms of showing for the social history of such an apparently abstract thing as logic. This methodology helps us to understand how, through the mobilization and networking of heterogeneous resources, such a large number of scientific issues could recently have been addressed in terms of fuzziness. The same methodology helps us to grasp some of the modalities through which the virtues of fuzziness could be celebrated in logic and representations of the world could be partly fuzzyfied.

In his article "Fuzziness vs. Probability" Bart Kosko gives his vision of how the world would have been different today if fuzziness had been developed, taught, and applied before probability theory. He mentions in particular how the existence of real-valued dice might have then helped to "support the economy of Las Vegas." It remains to be seen, however, whether Las Vegas gamblers will contribute in their own way to the celebration of fuzziness.

Notes

For his comments and help at all levels, I would like to express my most grateful thanks to Norton Wise. For their comments, I also would like to thank Michelle Lamont and all the participants in the workshop on "Growing Explanations" who attended the session corresponding to this essay. John Richardson and Ruth McNally have generously commented on both earlier and later versions. I would like to dedicate this essay to the memory of John Richardson.

1   D. Dubois, H. Prade, P. Smets, "Partial Truth Is Not Uncertainty: Fuzzy Logic versus Possibilistic Logic," *IEEE Expert* (August 1994): 15–19; P. Smets, "Varieties of Ignorance and the Need for Well-founded Theory," *Information Sciences* 57–58 (1991): 135–144.

2   That the existing histories of fuzzy logic are essentially written by their proponents is certainly not a unique state of affairs. Most past and current histories of mathematics have been written and rewritten by mathematicians or former mathematicians. On this topic, and on the very important stakes that the writing of the history of mathematics often holds for mathematicians, see C. Goldstein, Un *théorème de Fermat et ses lecteurs* (Saint Denis: Presses Universitaires de Vincennes, 1995).

3 L. A. Zadeh, "Fuzzy Sets," *Information and Control* 8 (1965): 338–353.

4 Archives of the Internet electronic bulletinboard "comp.ai.fuzzy," article no. 124. Electronic address: ⟨ftp.cs.cmu.edu; user/ai/pubs/news/comp.ai.fuzzy⟩. One could note that the presentation of Zadeh's motivations are sometimes reframed within inspirational models or little stories such as the following (see comp.ai.fuzzy, article no. 124):

> Here is a little story about how Fuzzy Logic was invented. One day, Dr. Zadeh got into a long argument with a friend about who was more beautiful, his wife or his friend's. Each thought his own wife was more beautiful than the other's wife. There is, of course, no objective way to measure beauty. The concept of 'beautiful' greatly differs among people. Although they continued the argument for a long time, they could not arrive at a satisfactory conclusion. This argument triggered Dr. Zadeh's desire to express concepts with such fuzzy boundaries numerically, and he thereby devised Fuzzy sets. Thus goes the legend.

5 B. Kosko and S. Isaka, "La logique floue," *Pour la Science* 191 (September 1993): 62–68; M. Black, "Vagueness: An Exercise in Logical Analysis," *Philosophy of Science* 4(4) (1937): 427–455.

6 Archives of comp.ai.fuzzy, article no. 40.

7 For an analysis of the many ways to build the history of an invention, see H. Mialet, "Le sujet de l'invention" (Ph.D. diss., Université de Paris I Sorbonne, 1994).

8 B. Gaines and L. Kohout, "The Fuzzy Decade: a Bibliography of Fuzzy Systems and Closely Related Topics," *International Journal of Man-Machine Studies* 9 (1977): 1–68. This article presents the development of fuzzy logic as the product of a collective action involving many actors by offering a large bibliography on fuzzy logic and related work up to the mid-1970s.

9 Archives of comp.ai.fuzzy, article no. 124.

10 The case of fuzzy logic seems to confirm the claim that formalism has a unique ability to concentrate associations, thereby linking a great range of activities (the linkage itself being a source of novelty for the corresponding activities); see B. Latour, *Science in Action: How to Follow Scientists and Engineers through Society* (Cambridge: Harvard University Press, 1987).

11 See the archives of comp.ai.fuzzy, article nos. 124, 1555, 2072; *Synthèse Technologies de l'Information*, vol. 1 (Paris: Aditech, 1989–90); *La lettre du club logique floue* (Paris: Association Ecrin, September 1991).

12 Lakatos remarks that "unfortunately, even the best historians of logic tend to pay exclusive attention to the changes in logical theory without noticing their roots in changes in logical practice"; see I. Lakatos, *Proofs and Refutations: The Logic of Mathematical Discovery* (Cambridge: Cambridge University Press, 1976), 81. However, for an approach comparable to the one developed here, see E. Brian, *La mesure de l'Etat—Administrateurs et géomètres au XVIIIème siècle* (Paris: Albin Michel, 1994). Brian describes how practices of mathematical analysis in France by the end of the eighteenth century proceeded from practices of decomposition, which simultaneously organized procedures of integration, demonstrative compositions, methods of administrative counting, and *tableaux* of *genre* of knowledge.

Examining the actor's practices of textual production has the advantage of avoiding the lapse into sociologism and denunciation. Indeed, by placing the locus of knowledge not only in human-human interactions, but also in their manipulation of symbols, it recognizes what the researchers themselves consider to be their objects. Competing and often nuanced views are held in social studies of science around this issue. See, for example, Harry Collins's view that "the locus of knowledge is not the word or symbol but the community of expert practitioners," *Changing Order: Replication and Induction in Scientific Practice* (London: Sage, 1985), 159.

13 See the methodologies proposed for the material study of abstraction in B. Latour, "Sur la pratique des théoriciens," in J. M. Barbier (ed.), *Savoirs théoriques et savoirs d'action* (Paris: Presses Universitaires de France, 1996), 131–146; and E. Brian, "Le livre des sciences est-il écrit dans la langue des historiens?," in B. Lepetit (ed.), *Les Formes de l'expériences—Une autre histoire sociale* (Paris: Albin Michel, 1995), 85–98. See also H. R. Alker, "Standard Logic versus Dialectical Logic: Which Is Better for Scientific Political Discourse?," paper presented at the Twelfth World Congress of the International Political Science Association, Rio de Janeiro, 8–15 August 1982: "there is a lot to be said for treating and teaching the Frege-Russell program of axiomatic, anti-metaphysical formal, extensional logic as an extraordinarily dialectical, anti-Hegelian exercise, in some ways parallel to Marx and Lenin's efforts to recast Hegelian thought in a materialist vein" (29).

14 J. Goody, *The Domestication of the Savage Mind* (Cambridge: Cambridge University Press, 1977).

15 A. Hennion, *La Passion musicale* (Paris: Métailié, 1993). See also the irreductionist prospect developed in B. Latour, *The Pasteurization of France* (Cambridge: Harvard University Press, 1988), 153–236.

16 See a similar concern, which could be related to what I call a sociology of mediation, in A. Warwick, "The Laboratory of Theory or What's Exact about the Exact Sciences?," in M. N. Wise (ed.), *The Values of Precision* (Princeton, N.J.: Princeton University Press, 1995), 311–351.

17 Collins greatly contributed to developing this important research theme. See Collins, *Changing Order*.

18 The term *monster* has a similar origin. A *monstre* (monster) in old French is a man that one shows by pointing at with one's finger. In a sense, I would like to extend Lakatos's analytical frame of description, when he deals with counterexamples in mathematics as monstrous examples, or "monster-barring." See Lakatos, *Proofs and Refutations*. Another project for developing Lakatos's frame of analysis (for a sociology of mathematics) may be found in D. Bloor, "Polyhedra and the Abominations of Leviticus," in M. Douglas, *Essays in the Sociology of Perception* (London: Routledge, 1982), 131–218.

19 The mathematician G. H. Hardy adopted a denunciatory stance, and not a methodological position, to express a similar view when he wrote, "There is strictly speaking no such thing as mathematical proof; we can, in the last analysis, do nothing but point." See G. H. Hardy, "Mathematical Proof," *Mind* 38 (1928): 18.

20 Kosko, "Fuzziness vs. Probability," *International Journal of General Systems* 17 (1990): 219. See also B. Gaines, "Precise Past, Fuzzy Future," *International Journal of Man-Machines Studies* 19 (1983): 117–134.

21 See J. Largeault, *La logique* (Paris: Presses Universitaires de France, 1993), 3–24. Logicism generally refers to Frege-Russell's attempt to build the foundations for a unitary and universal logic, underlining the whole of reality. It is opposed to formalist or algebraic constructions of logic, which involve scattered models with distinct domains and relations. For a discussion of the relative importance of the barber's paradox and questions of similarity and difference, see R. M. Sainsbury, *Paradoxes* (Cambridge: Cambridge University Press, 1988). For another approach to the barber's and the liar's paradoxes, see J. Barwise and J. Etchemendy, *The Liar: An Essay on Truth and Circularity* (Oxford: Oxford University Press, 1987).

22 Wittgenstein suggests, for example, that taking the square of minus one became an acceptable operation once it was associated with a form of life that naturalized it. This

form of life was itself a geometrical manipulation: "Similarly, by surrounding $\sqrt{-1}$ by talk about vectors, it sounds quite natural to talk of a thing whose square is -1. That which at first seemed out of the question, if you surround it by the right kind of intermediate cases, becomes the most natural thing possible." See L. Wittgenstein, *Wittgenstein's Lectures on the Foundations of the Mathematics* (New York: Harvester, 1976), 226.

23 The investigations I have conducted on the work of logicians show that producing detours of formulation is a widespread and essential tool in the formalization process. The sedimentation of such practices can be traced in the teaching of logic in the United States back to the end of World War II. See C. Rosental, "Ethnography of the Teaching of Logic," paper presented at a workshop on historical epistemology, Institute for the History of Science, University of Toronto, 1993.

24 *Inscription* has been widely used in social studies of science to describe an important and specific practice in scientific activity, the writing (or inscribing) of results on paper, using various graphic representations. See Latour, *Science in Action*.

25 See Goody, *Domestication of the Savage Mind*, on graphic reason, and Jacques Bouveresse's reading of Wittgenstein in *Le mythe de l'intériorité, expérience, signification et langage privé chez Wittgenstein* (Paris: Editions de Minuit, 1976), which should itself be read in a minimalist material way: "The fact that it [a statement] is composed of elements in such and such a quantity, linked with each other in such and such a way, gives them an internal potentiality of meaning, that is to say, [it gives them the potentiality] to be assigned to the representation of a certain kind of situation" (97, my translation). See also E. Livingston, *The Ethnomethodological Foundations of Mathematics* (London: Routledge, 1985).

26 See also A. Pickering and A. Stephanides's attempt to escape from conventionalism by distinguishing free and forced moves in the mathematician Hamilton's work, in "Constructing Quaternions: On the Analysis of Conceptual Practice," in A. Pickering (ed.), *Science as Practice and Culture* (Chicago: University of Chicago Press, 1992), 139–167. For a different sociology of mathematics developed around the case of Hamilton, see D. Bloor, "Hamilton and Peacock on the Essence of Algebra," in H. Mehrtens, H. Bos, and I. Schneider (eds.), *Social History of 19th Century Mathematics* (Boston: Birkhauser, 1981), 202–232.

27 Compare the work on distributed cognition, linked with the mobilization of formalisms, in E. Hutchins, *Cognition in the Wild* (Cambridge: MIT Press, 1995).

28 Kosko, "Fuzziness vs. Probability," 219. To deny the validity of the law of the excluded middle has a strong historical significance. What is often described as logical absolutism has argued that it is one of the strongest foundations of logic. E. Husserl displayed this law throughout his text as one of the few principles which warrant the ideal foundation of logic; see *Recherches logiques*, vol. 1 (1913; Paris: Presses Universitaires de France, 1969). Specifically, Husserl exhibited it as a rampart to any kind of relativism in his fight with the post-Brentanian school.

29 See B. Latour, "Give Me a Laboratory and I Will Raise the World," in K. D. Knorr-Cetina and M. Mulkay (eds.), *Science Observed: Perspectives on the Social Studies of Science* (London: Sage, 1983), 141–170.

30 B. Kosko, *Neural Networks and Fuzzy Systems* (Englewood Cliffs, N.J.: Prentice Hall, 1992).

31 B. Kosko, *Fuzzy Thinking: The New Science of Fuzzy Logic* (New York: Hyperion, 1993), and D. McNeill, and P. Freiberger, *Fuzzy Logic* (New York: Simon and Schuster, 1993).

32 See A. Boureau, "Propositions pour une histoire restreinte des mentalités," *Annales Economies, Sociétés, Civilisations* 6 (1989): 1491–1504. See also the use of the notion of collective statement to give an account of the building and the role of the statement "scientific policy" in France after World War II, in F. Jacq, "Pratiques scientifiques,

formes d'organisation et représentations politiques de la science dans la France de l'après-guerre: la 'politique de la science' comme énoncé collectif (1944–1962)," (Ph.D. diss., École des Mines, 1996).

33 W. V. O. Quine, *The Ways of Paradox and Other Essays* (1966; Cambridge: Harvard University Press, 1976).

34 These processes of visual transformation and displacement are to be distinguished from what Kuhn calls the changes of worldview. Indeed, they do not put into play paradigm modifications. From this prospect, potential modifications of the vision of the world for relatively large groups of actors should be seen as proceeding from chains of mediations which can be traced down even to an interactionist level. Besides, the analysis of the diversity of mediations cannot be reduced to general claims of cognitive psychology, in particular of Gestalt psychology, which constitutes the starting point of the Kuhnian analysis. See T. S. Kuhn, *The Structure of Scientific Revolutions* (Chicago: University of Chicago Press, 1970), 111–135.

35 C. Elkan, "The Paradoxical Success of Fuzzy Logic," *Proceedings of AAAI 1993* (Cambridge: MIT Press, 1993), 698–703.

36 It is interesting to compare the development of fuzzy logic with the development of mathematical analysis through the investment of calculating machines in a whole series of activities in England by the end of the nineteenth century. See Warwick, "Laboratory of Theory."

37 Compare the notion of monstrance, corresponding in particular to the act of showing the host as a part of Christ's body, in an elevation move. What is shown (the apparition of Christ's body) is again emanating from and detached from the material device (the host). See also L. Daston, "Marvelous Facts and Miraculous Evidence in Early Modern Europe," *Critical Inquiry* 18 (1991): 93–124; E. Claverie, "Voir apparaître, regarder voir," *Raisons Pratiques* 2 (1991): 1–19.

38 Account written on 15 May 1991 of a meeting which occurred on 19 April 1991 in Paris (my translation); see *Club Crin Logique floue* (Paris: Association Ecrin, May 1991), 3.

39 The notion of "witnessing" is developed in S. Shapin and S. Schaffer, *Leviathan and the Air-pump* (Princeton, N.J.: Princeton University Press, 1985).

40 The theatrical metaphors of E. Goffman, *The Presentation of Self in Everyday Life* (1959; Harmondsworth: Penguin, 1971), are remarkably relevant to describing the production of these logical demonstrations.

41 By comparing this situation with the sense of monstrance previously mentioned, and the apparition it involves, the term *apparition* would also be relevant to describe the product of the process.

42 *La lettre des clubs Crin*, no. 8 (Paris: Association Ecrin, October 1992), 7 (my translation). "Visually" could be a translation of "de visu."

43 See J. Lévine, "Où sont passés les contrôleurs flous nippons?," *La Lettre de l'Automatique* 12 (Septembre 1991): 2–4. For the study of other cases of travelers whose "accurate sense of observation" made them blind to some facts shown, see S. Schaffer, "Self Evidence," *Critical Inquiry* 18 (Winter 1992): 327–362, and M. Ashmore, "The Theatre of the Blind: Starring a Promethean Prankster, a Phoney Phenomenon, a Prism, a Pocket, and a Piece of Wood," *Social Studies of Science* 23 (1993): 67–106.

44 For a case where engineers' presentation of formal properties in a computer device (a microprocessor) leads to controversy, see D. Mackenzie, "Negotiating Arithmetic, Constructing Proof: The Sociology of Mathematics and Information Technology," *Social Studies of Science* 23 (1993): 37–65.

PART II   The organism, the self, and (artificial) life

Self-organization

6 Marrying the premodern to the postmodern: computers and organisms after World War II

Evelyn Fox Keller

Historically, . . . there could be no mechanical explanation of life functions until men had constructed automata: the very word suggests both the miraculous quality of the object and its appearance of being a self-contained mechanism.
—Canguilhem, *A Vital Rationalist*

Despairing of the possibility of a fully physicochemical (or mechanistic) account of vital phenomena, Claude Bernard felt obliged to conclude in 1878 that "some invisible guide seems to direct [the phenomenon] along the path it follows, leading it to the place it occupies."[1] But fewer than one hundred years later, biology found its homunculus, and it was, after all, a molecule. Quoting Bernard in 1970, the molecular biologist François Jacob was able to claim that "there is not a word that needs to be changed in these lines today: they contain nothing which modern biology cannot endorse. However, when heredity is described as a coded programme in a sequence of chemical radicals, the paradox disappears."[2] Jacob is no mathematician, but when turning to history he adopts the most rudimentary of mathematical principles: Given two points, he draws a straight line between them, giving us a marvelously linear narrative that starts with Bernard, develops through genetics and early molecular biology, and leads— inexorably, as it were—to the resolution of nineteenth-century biology's paramount dilemma. In his account, Jacob recasts Bernard's "invisible guide," the agent responsible for the apparent purposiveness of biological organization, as the genetic "program" encoded in a string of nucleotide bases. In the DNA's sequence, one finds the long sought source of living order, the self within a self. No telos here, only the appearance of telos (or teleonomy, as it is now renamed); no purpose, only, as Monod put it, "chance and necessity."[3]

Yet history lends itself to linear narrative only retrospectively—or perspectively, if you will. Shift the vantage point and other narratives come into view. In this essay, I want to focus on a rather different (and far less linear) biological history than Jacob's now canonical version—a narrative that is rendered distinctive only in retrospect, evinced first by the advent of the computer in the mid-twentieth century (and with it, of new meanings of mechanism), and later, lent particular significance by late-twentieth-century convergences between the computer and biological science. This other narrative begins not with Bernard's invisible guide, but with the notions of organization and self-organization that had been formulated almost a century earlier to counter both mechanistic and design accounts of life, and which, as such, were built into the very definition of *biology*. I begin, therefore, with a brief overview of the declining fate of these notions between 1790 and 1940 (as it were, their prehistory), before turning to their relegitimation by the advocates and architects of the new machine.

The problem of "organization": 1790 to 1940

What is an organism? What is the special property or feature that distinguishes a living system from a collection of inanimate matter? This was the question that first defined biology as a separate and distinctive science at the beginning of the nineteenth century. And by its phrasing (i.e., implicit in the root meaning of the word *organism*), it specified at least the form of what would count as an answer. For what led to the common grouping of plants and animals in the first place—that is, what makes "the two genres of organized beings" (as Buffon referred to them) *organisms*—was a new focus on just that feature: on their conspicuous property of being organized, and of being organized in a particular way. As Jacob observes, by the end of the eighteenth century it was "by its organization [that] the living could be distinguished from the non-living. Organization assembled the parts of the organism into a whole, enabled it to cope with the demands of life and imposed forms throughout the living world."[4] Only by that special arrangement and interaction of parts that brings the wellsprings of form and behavior of an organism *inside* itself could one distinguish an organism from its Greek root, *organon*, or tool. A tool, of necessity, requires a tool user, whereas an organism is a system of organs (or tools) that behaves as if it had a mind of its own; in other words, it governs itself.

Indeed, the two words, *organism* and *organization* acquired their contemporary usage more or less contemporaneously. Immanuel Kant, in 1790, gave one of the first modern definitions of an organism—not as a definition per se, but rather as a principle or maxim which, he wrote, "serves to define what is meant as an organism"—namely:

an organized natural product is one in which every part is reciprocally both end and means. In such a product nothing is in vain, without an end, or to be ascribed to a blind mechanism of nature.[5]

Organisms, he wrote, are the beings that

first afford objective reality to the conception of an *end* that is an end of *nature* and not a practical end. They supply natural science with the basis for a teleology . . . that would otherwise be absolutely unjustifiable to introduce into that science—seeing that we are quite unable to perceive *a priori* the possibility of such a kind of causality.[6]

Elaborating on this kind of causality, he writes:

In such a natural product as this every part is thought as *owing* its presence to the agency of all the remaining parts, and also as existing *for the sake of the others* and of the whole, that is as an instrument, or organ. . . . [T]he part must be an organ *producing* the other parts— each, consequently, reciprocally producing the others. . . . Only under these conditions and upon these terms can such a product be an *organized* and *self-organized being*, and, as such, be called a *physical end.*[7]

Indeed, it is here that the term "self-organized" first makes its appearance in relation to living beings. It is invoked—and underscored—to denote Kant's explicit opposition to argument by design. No external force, no divine architect, is responsible for the organization of nature, only the internal dynamics of the being itself.

The beginnings of biology thus prescribed not only the subject and primary question of the new science, but also the form of answer to be sought. To say what an organism is would be to describe and delineate the particular character of the organization that defined its inner purposiveness, that gave it a mind of its own, that enabled it to organize itself. What is an organism? It is a bounded body capable not only of self-regulation and self-steering, but also, and perhaps most importantly, of self-formation and self-generation. An organism is a body which, by virtue of its *peculiar and particular organization*, is made into an autonomous and self-generating "self." The obvious task for biology was to understand the character of this special kind of organization or self-organization. At the close of the eighteenth century and the dawn of the nineteenth, it was evident—to Kant, as to his contemporaries—that neither blind chance nor mere mechanism, and certainly no machine that was then available, could suffice. "Strictly speaking," Kant writes, "the organization of nature has nothing analogous to any causality known to us."[8] Necessarily, the science of such a mechanism would have to be a new kind of science, one which, given the technology of his age, Kant not surprisingly assumed to be irreducible to physics and chemistry.

Tim Lenoir has persuasively argued that, for the first half of the nineteenth century, such Kantian notions of *teleomechanism* (Lenoir's term) provided German biologists with a "fertile source for the advancement of biological science on a number of different fronts."[9] But by the second half of that century, the authority of the teleomechanists (e.g., Blumenbach, Kielmeyer, Reil, Müller, von Baer, Bergmann, and Leuckart) gave way to that of a younger generation of biologists who espoused the sufficiency of more conventional mechanism (most notably, Ludwig, Helmholtz, and DuBois-Reymond), and who read the invocation of any form of teleology as inherently theological and hence as beyond the realm of science. By the end of the century, Kant's founding notion of "self-organization" had all but disappeared from biological research, except in embryology.

Lenoir attributes the decline of this tradition in part to generational conflict and in part to the increasing specialization and competition prompted by changes in the structure of professional advancement in German universities. But the arrival of a new kind of machine, the steam engine, and with it a new kind of physics called thermodynamics was at least equally important. By the second half of the century, the first law of thermodynamics succeeded in providing a stunningly persuasive account of (at least some) vital functions—most notably, of metabolism and respiration. The production of heat, once thought to be the essence of animal life, was effectively shown to be no different in animals than in machines.[10] As Helmholtz wrote, "The animal body therefore does not differ from the steam-engine as regards the manner in which it obtains heat and force, but [only] in the manner in which the force gained is to be made use of."[11] If animal heat, once the essence not only of ardor but of life itself, is itself mechanical, if the sustenance of animal life accords with the first law of thermodynamics, then any notion of life's special qualities might seem to have been rendered superfluous.

But organisms do more than convert fuel into sustenance; even lowly organisms are not merely workers but lovers and creators. Above all, they must form themselves anew in every generation. From the simplest beginnings, they must construct themselves in the forms appropriate to their kind and, once formed, craft the parts needed to repeat the cycle. In short, they must procreate. And not even the steam engine could do that. Moreover, while metabolic processes obeyed the first law of thermodynamics well enough, it remained impossible to reconcile the second law with the manifest increase in order evidenced in embryogenesis. In the emergence of a fully formed individual from a fertilized egg remained as conspicuous a demonstration as ever of the seeming purposiveness of living matter, leading even so assiduous a student of the role of physicochemical processes in physiology as Claude Bernard to conclude, "What characterizes the living machine is not the nature of its physicochemical properties, complex though they may be, but the creation of that machine, which develops before our eyes under conditions peculiar to itself and in accordance with a definite idea, which expresses the nature of the living thing and the essence of life itself."[12] But in embryology itself, a different tradition was invoked to deal with the dilemma of apparent purposiveness—epigenetic rather than preformationist, indebted more to Kant, say, than to Haller. Indeed, at the close of the nineteenth century, embryology was one of the few areas of biology in which notions of teleomechanism still survived.

One might even say that the need to give precise meaning to older notions of organization, epigenesis, and regulation constituted one of the dominant impulses of *Entwicklungsmechanik*. Hans Driesch and Wilhelm Roux championed different strategies for doing this, with Driesch focusing on the problem of regulation and Roux on that of determination. In particular, Driesch sought to resurrect Aristotle's concept of "entelechy" as a way both of making sense of his own experimental findings and of characterizing the essential purposiveness of the life force. For Driesch, the principle of entelechy captured the difference between the "whole-making causality" (*Ganzheitkausalität*) evident in embryological development and "merely mechanical causality"; entelechy was the essence of the self-organizing capacity of biological organisms. But Driesch's influence was short-lived. His views were contested within embryology itself (especially by Roux). More importantly, however, by the early decades of the twentieth century, his concerns and even the entire tradition of experimental embryology came under direct challenge by the rising discipline of genetics.

The story of the disarticulation of genetics from embryology has been frequently told and does not need repeating here. In the immediate aftermath of his conversion, T. H. Morgan seemed content to grant embryology a separate but equal status, but by the early thirties he and his colleagues had the confidence to take on the older discipline, recasting its preoccupations with organization and regulation as problems of "gene action."[13] By 1932, Morgan was ready to dismiss the entire tradition of Entwicklungsmechanik, faulting it for running "after false gods that landed it finally in a maze of metaphysical subtleties."[14] Especially, he faulted Driesch's notion of entelechy and with it other, more recent attempts to define a "neomechanist" organicism (e.g., Needham, Woodger, Watson):

> The entelechy . . . was postulated as a principle, guiding the development toward a directed end—something beyond and independent of

the chemical and physical properties of the materials of the egg; something that without affecting the energy changes directed or regulated such changes, much as human intelligence might control the running or construction of a machine. The acceptance of such a principle would seem to make it hardly worth while to use the experimental method to study development. . . . In fact, the more recent doctrine of 'the organism as a whole' is not very different from the doctrine of entelechy.[15]

The growth of experimental genetics—its accumulation of solid experimental results—offered a healthy contrast. Even without tangible successes in embryology, an extraordinary confidence held sway in the community of American geneticists, bolstered in part by their success in correlating many phenotypic traits with particulate genes, in part by the availability of a large agricultural market for these successes, in part by the heady expectation that their work would lead to the breeding of new and better races of men, and in part by the efficacy of their discourse of gene action in focusing attention away from the questions they could not answer. The question of the spatial and temporal organization so conspicuously manifested in embryogenesis was one of these.

Less than two decades later, Morgan and his school could claim a stunning vindication when DNA was identified as the genetic material and, in 1953, when Watson and Crick identified a mechanism for self-replication in the double helical structure of that molecule. In the 1930s, geneticists had to contend with constant challenges from embryologists over the inadequacy of genetics for dealing with their concerns, but by the 1950s few embryologists were left to argue, at least in the United States, and one might even say that the entire question of spatial and temporal organization had effectively disappeared from view.

World War II and the computer

The years between Morgan and Watson and Crick were the years in which the revolution of molecular biology took shape; they were also the years in which a revolution of far more cataclysmic proportions erupted, one that left in its wake a world irrevocably transformed. After World War II, no aspect of human existence could be recognized in prewar terms, not even in science. Historians of science have devoted much attention to the ways in which the war transformed the status, the structure, even the content of the physical sciences, but they have paid rather less attention to its impact on the biological sciences. Yet even there, shock waves were felt. The influx of physicists into molecular biology has of course been well noted. But equally important, the disciplinary status of embryology, already substantially weakened by its long contest with genetics, was reduced to an all time low—in part because of its strong association with European (and especially with German) biology. Driesch's "vitalism" and his postulate of an entelechy were seen as standing for the forces of darkness, irrational and antiscientific. To a new generation of molecular geneticists, even Spemann's notion of an "organizer," in which Spemann hoped to find the causal locus of development, carried the taint of an antiscientific metaphysics. Indeed, so discredited had this subject become in the United States in the aftermath of World War II that later, when conditions seemed propitious for its revival, it was deemed tactical to rename it "developmental biology." But far and away the most consequential impact of the war on scientific life—even for the life sciences—resulted from the development of the computer.

This newest of machines, a machine that was to transform the very meaning of mechanism, had itself been developed to cope with the vast increase in complexity of life outside the biological laboratory that had come with the war. The *computer*, as we now understand the term, came into being precisely because those people (mostly women) who had by tradition been responsible for data processing and computation—the original meaning of the word *computer*—could no longer handle the masses of data required to coordinate wartime military operations. *Cybernetics* originated in the self-steering and goal-seeking gunnery designed to meet the need for more effective control, and information theory arose out of the need to maximize the efficiency of communication.

"Out of the wickedness of war," as Warren Weaver put it, arose not only a new machine, but a new vision of science: a science, contra Descartes, based on principles of feedback and circular causality and aimed at the analysis and mechanical implementation of exactly the kind of purposive "organized complexity" so vividly exemplified by biological organisms. Wiener called this new science "cybernetics," and his vision quickly attracted the interest of a motley crew of mathematicians, physicists, and engineers.

I have argued elsewhere that these young Turks (or "cyberscientists") were inhabiting a different world from molecular biologists.[16] To the latter, life still seemed simple. Insulated equally from problems of military coordination and from the "organized complexity" of multicellular organisms, they retained the confidence that life could be managed by "gene action," by a unidirectional chain of command issuing from the instructions of a central office. Not surprisingly, the efforts of these two disciplines pointed in different directions.

Though happy to borrow the metaphors of "information" and "program" from the new computer science,[17] molecular biologists remained committed to the strategy that had so spectacularly proven itself in physics (for many, their home discipline): streamlining their new subject and reducing it to the simplest possible units. This new breed of life scientists sought to further elucidate the essence of life by narrowing their focus to organisms so rudimentary and so simple as to be maximally immune from the mystifying and recalcitrant chaos of complex organisms. Their strategy led to the study of life in test tubes and petri dishes populated by bacterial cultures of E. coli and their viral parasites, the bacteriophage, forms of life that seemed simple enough to preserve the linearity of simple codes and telegraph-like messages.

Such arch reductionism had ample support in the 1950s and not only from the unprecedented status of physics: it had the cultural support of a massive, almost visceral, reaction against the horrific associations with the revival of holism that had flourished in Germany in the 1930s and 1940s, with what Anne Harrington calls the German "Hunger for Wholeness."[18] Molecular biologists sought to build a new biology in clear and often loudly voiced opposition to an older, organismic biology. Even more fully than Morgan and his colleagues had succeeded in doing, they would irrevocably rid descriptions of the organism of residual vitalism and functionalism; especially, they would expunge from biological language those conspicuously teleological notions of purpose, organization, and harmony.

But elsewhere, in response to the exigencies of war,[19] cyberscientists were avidly appropriating these older preoccupations for their own uses, leaning heavily on the very images, language, and even conceptual models of organicism that were now being so vigorously discredited in biology. Like Kant, they wished to develop new paradigms of circular (or reciprocal) causality, but unlike (and even contra) Kant, they did so with explicitly practical ends in mind. Their mission was to build a new kind of machine that could put the principles of life to work in the world of the nonliving. The end of the war did little to dampen this mission; by then, it had taken on a life of its own.

Body and mind: self-organization redux

Even more conspicuously than in Wiener's visionary musings of the 1940s, this appropriation of an older organicism for a new mission is evident in the concerted efforts of the late 1950s and 1960s to understand and to build "self-organizing systems." Self-organizing systems would marry the ultramodern with the premodern, entirely bypassing the successes of molecular biology. Between 1959 and 1963, a flurry of conferences and symposia were convened under this mandate, mostly under the aegis of the information systems branch of the Office of Naval Research (ONR) and drawing from a wide array of scientific disciplines both in the United States and in England.[20] A major impetus behind these conferences was the Navy's hope of building a new kind of machine, more adept and more versatile—indeed, more autonomous—than the digital computer developed during the war years.[21] If one wanted to build a machine that could handle truly complex problems, that could learn from experience, the trick, they argued, would be to study the natural systems (i.e., living organisms) that did so with such obvious success.[22] For the ONR, Frank Rosenblatt, at Cornell's Astronautical Laboratory, provided the direct inspiration. Rosenblatt, who had in turn been inspired by Donald Hebb's claim that an ensemble of neurons could learn if the connection between neurons were strengthened by excitation, attempted to build a machine which could learn in just that way. The perceptron—a simple neuronlike learning device—is now seen as the first connectionist machine: it should be able, Rosenblatt argued, to "perceive, recognize and identify its surroundings without human training or control." Inspired by the work of von Neumann in the late 1940s on self-reproducing automata, he went on to suggest that, "in principle, it would be possible to build Perceptrons that could reproduce themselves on an assembly line and which would be 'conscious' of their existence."[23]

Kant's term, *self-organization*, had been resurrected in the 1930s by von Bertalanffy to characterize the central feature of organismic development and by Gestalt psychologists to describe the way humans process experience. At one and the same time, a single word denoted the key feature of minds and of organisms. And it was to these authors that the new cyberneticians referred. But a crucial difference had now become evident: The hope of building a "self-organizing" machine required a new order of precision for notions of organization and purpose. In a review of the first three conferences, all held in 1959, Rapaport and Horvath wrote:

> The notion of the organism . . . is taken to be central in most holistic expositions of method . . . [But] how does placing the organism at the center of one's conceptual scheme help one get away from "limited mechanistic concepts"?. . . . What . . . can a philosopher mean when he suggests that the method of biology may have more to offer than physics to the science of the future, the science of organized complexity? Can he mean that the modern scientist should place more reliance on those aspects of biology which are carry-overs from a previous age? Such a position would be difficult to defend. Yet a defense of the "holist's" position can be made, provided we can spell out just how the older schemes . . . can fit into the new modes of analysis. . . .
>
> The question before us is whether we can . . . extend systematic

rigorous theoretical methods to "organized complexity," with which the holists are presumably concerned. There are at least two such classes of concepts discernible, and both are reminiscent of the ways of thinking of the biologist, . . . namely, teleology and taxonomy.

Of course, as the authors go on to explain, "When we say teleology, we mean teleology in its modern garb, made respectable by cybernetics," defining cybernetics as "the theory of complex inter-locking 'chains of causation,' from which goal-seeking and self-controlling forms of behaviour emerge."[24]

It is striking, however, that postwar molecular biologists had little interest in and even less patience for such an agenda. Their own discipline was represented neither in the earlier Macy conferences nor in the conferences on self-organization. They may have been happy to appropriate and adapt to their own uses some of the lexicon of information theory, but they sharply distanced themselves from the focus on complexity, organization, and purpose. Their interests and their agenda lay elsewhere. But the ears of many who were still interested in the problems of embryogenesis did perk up recognizing here a new kind of scientific ally, an ally who was willing for whatever reasons to restore the organism-as-a-whole to conceptual primacy, and even eager to acknowledge its apparent "mind of its own." And what's more, an ally promising to bring clarity to the meaning of *organization*.

As J. H. Woodger, the philosopher of biology responsible for introducing von Bertalanffy to English speaking audiences, had written in 1929:

> If the concept of organization is of such importance as it appears to be it is something of a scandal that biologists have not yet begun to take it seriously but should have to confess that we have no conception of it. . . . Biologists in their haste to become physicists have been neglecting their business and trying to treat the organism not as an organism but as an aggregate.[25]

In 1950, much the same could still be said—and it was. Edmund W. Sinnott, for example, wrote:

> The word *organism* is one of the happiest in biology, for it emphasizes what is now generally regarded as the most characteristic trait of a living thing, its *organization*. . . . But this central stronghold, we must ruefully admit, has thus far almost entirely resisted our best efforts to break down its walls.
>
> Organization is evident in diverse processes, at many levels, and in varying degrees of activity. It is especially conspicuous in the orderly growth which every organism undergoes and which produces the specific forms so characteristic of life.

Like embryogenesis or like mind. Indeed, Sinnott asserts "that biological organization . . . and psychical activity . . . *are fundamentally the same thing.*"[26]

Now, however, a new breed of scientist appeared who was eager to help out. A direct response to the challenge can be found in the work of a young German refugee by the name of Gerd Sommerhoff who credits Uexküll for his inspiration. In the same year, Sommerhoff published a work called *Analytical Biology* which begins by endorsing the primacy accorded by holists and organicists to organization and purpose as the key features of life, but then adds that "none of these schools manage to tell us in precise scientific terms what exactly is meant by 'organization.'" Sommerhoff undertakes to fill this gap, to create an "Analytical Biology," that is, "to form a set of deductively employable concepts which will enable theoretical biology to deal with the really fundamental characteristics of observed life, viz. its apparent purposiveness."[27]

In the mid-1950s, Sommerhoff joined the ARTORGA group in England (named for the study of artificial organisms), and his book becomes one of the most frequently cited sources for participants of the conferences on self-organization in the late 1950s. Years later, W. Ashby described Sommerhoff's formulation as "of major importance, fit to rank with the major discoveries of science," for he had shown that "The problem of the peculiar purposiveness of the living organism, and especially of Man's brain" could be represented with "rigour, accuracy, and objectivity."[28] To be sure, it was the properties of mind that were of primary importance to these would-be architects of self-organizing systems, but, like Sinnott, they viewed the design principles of biological organization and of mind as one and the same thing. And even C. H. Waddington seemed to concur: "The behaviours of an automatic pilot, of a target-tracking gunsight, or of an embryo, all exhibit the characteristics of an activity guided by a purpose."[29] The solution of a problem at once so difficult and so general clearly required ignoring minor disciplinary differences and pooling the resources of psychologists, embryologists, neurophysiologists, physical and mathematical scientists, and engineers.

Like the wartime architects of command, control, and communication systems, a chronic slippage is evident in these discussions between control and self-control. What distinguished them, and retrospectively earned them the appellation "second-order cybernetics," was paradoxically their greater emphasis on autonomy and, at the same time, their explicit acknowledgment that autonomy is meaningless. In the very first conference on self-organizing systems (convened in May 1959), Heinz von Foerster, a postwar émigré from Vienna who was soon to become the leading name associated with "self-organizing systems," opened with the provocative claim, "There are no such things as self-organizing systems."[30] Systems achieved their apparent self-organization by virtue of their interactions with other systems, with an environment. In the same vein, Robert Rosen noted the "logical paradox implicit in the notion of a self-reproducing automaton" (1959), and two years later, Ross Ashby (1962) reiterated the point: "No system reproduces itself," he wrote. What we call "self-reproduction" is a process resulting from the interaction between two systems, or between the part and the whole.[31] Another player in these early conferences, Gordon Pask, reminded his colleagues that the crucial point in the development of an organism ("a self-organizing system from any point of view") is not its components (e.g., nucleus and cytoplasm), but the interaction between and among them: "When we speak of an organism, rather than the chemicals it is made from, we do not mean something *described by* a control system. An organism *is a control system* with its own survival as its objective."[32]

For von Foerster, Ashby, and Pask, the goal was to understand the logic of the coupling (or "conversation") between subsystems that would generate the capacity of such a system to have its own objectives, that is, to be a natural system. Sometime in 1959 or 1960, von Foerster founded the Biological Computer Laboratory (BCL) at the University of Illinois at Urbana,[33] and for the next fifteen years, until von Foerster's retirement in 1975, the BCL (with support from the ONR, NIH, and the U.S. Air Force) served as hearth and home for those mathematicians, logicians, and biologists interested in the problems of self-organizing systems.

Thirty years later, Pask summarized the differences between the old and new cybernetics as a shift in emphasis from information to coupling; from the reproduction of "order-from-order" (Schrödinger) to the generation of "order-from-noise" (von Foerster);[34] from transmission of data to conversation; from stability to "organizational closure"; from external to participant observation—in short, from "command, control, and communication" to an approach that could be assimilated to Maturana and Varela's concept of "autopoiesis."[35] The term *second-order* is von Foerster's, introduced to denote the explicit preoccupation of the new cybernetics with the nature of self-reflexive systems.

What came of their efforts? As it happens, not very much, at least not in the short run. Throughout the 1960s, a number of developmental biologists attempted to make use of this new alliance, but by the end of that decade, their effort had collapsed, as, indeed, (at least in the United States) had cybernetics itself. The 1960s were the years in which molecular biology established its hegemony in biology and by the end of that decade its successes lent its agenda an authority that seemed impossible to contest. Even its appropriation of cybernetic language worked effectively to eclipse the inherently global questions of developmental biology—especially the question of what it is that makes an organism self-organizing. The cybernetic vision also collapsed in computer science, where it gave way to competing efforts from what Daniel Dennett has called "High-Church computationalism." The starting assumption of the latter was that, just like digital computers, minds are nothing more than physical symbol systems, and intelligent behavior could therefore be generated from a formal representation of the world.[36] A fierce funding war raged between the two schools in the early 1960s, championed respectively by the ONR and the U.S. Air Force; by the mid-1960s, with the award of more than 10 million dollars from ARPA to Marvin Minsky (through Project MAC, created in 1963) to develop artificial intelligence (AI) at MIT, it was effectively over.[37] Funding for the perceptron and its requisite hardware (especially, for parallel processors) dried up,[38] and the ONR-supported conferences on self-organizing systems ceased. In 1969, Minsky and Seymour Papert published a devastating attack on the entire project, explaining their intent in clear terms: "Both of the present authors (first independently and later together) became involved with a somewhat therapeutic compulsion: to dispel what we feared to be the first shadows of a 'holistic' or 'Gestalt' misconception that would threaten to haunt the fields of engineering and artificial intelligence as it had earlier haunted biology and psychology."[39]

Shortly afterward (11 July 1971), Frank Rosenblatt, its guiding inspiration, died in a boating accident that some say was a suicide. But within the AI community, early connectionism had already met its demise.[40] In the absence of funding and the kinds of concrete successes in embodying biological principles in nonbiological systems that might have generated new alliances and new sources of funding, the union promised between organisms and computers failed to materialize. Neural net models, which had been the heart of Rosenblatt's effort, had to wait another fifteen years to be revived; interactionist models of development have had to wait even longer. Evidently, the soil of American science in the 1960s was not quite right for this largely European transplant to take root. Even so, as Hubert and Stuart Dreyfus write, "blaming the rout of the connectionists on an antiholistic bias is too simple."[41] In a rather expansive sweep, they fault the influence of an atomistic tradition of Western philosophy, from Plato to Kant; von Foerster, by contrast, blames both the funding policies of American science (the short funding cycle, the emphasis on targeted research) and the excessively narrow training of American students.[42] (Historians have yet to have their say on the matter.)

Today, however, the winds have shifted once again. With the explosive development of computer technology over the last fifteen to twenty years, the roots of computer culture have spread wide and deep. With the computational hardware needed to implement them now available,[43] neural net modeling has become a flourishing business, and thanks to the avidity of the media industry, complexity is becoming everyone's favorite buzz word. Fed the term *self-organization*, my search engine turns up more than a dozen books and conference proceedings over the past decade. One of these is a conference held in Bielefeld in 1988 on "Selforganization: Portrait of a Scientific Revolution." Heinz von Foerster emerges here as a retrospective hero of this revolution, credited not only as "one of the founders of modern self-organization theory" and the founder of the BCL, but also as the inspiration for Maturana and Varela's theory of autopoiesis. Other names—Manfred Eigen, Hermann Haken, Ilya Prigogine, Erich Jantsch—also figure prominently.[44]

The scope of this latest "revolution" is dizzying. The 1990s, we are repeatedly told, was the era of postreductionism. The notion of life as an emergent phenomenon has become a truism. As physicist P. C. W. Davies writes: "Because living organisms are nonlinear (for a start!), any attempts to explain their qualities as wholes by the use of analysis is doomed to failure. . . . This is not to denigrate the very important work of molecular biologists who carefully unravelled the chemical processes that take place in living systems, only to recognize that understanding such processes can not *alone* explain the collective, organisational properties of the organism."[45] In fact, the vision of molecular biology has itself changed radically since Jacob's early description of an "invisible guide" encoded in the sequence of nucleotide bases. For those now engaged in the molecular study of development, "the challenge is to link the genes and their products into functional pathways, circuits, and networks."[46] George Miklos and Gerald Rubin suggest that sequence information "will not, by itself, be sufficient to determine biological function." What is needed is "to unravel the multilayered networks that control gene expression and differentiation."[47] And even the aims of molecular biology have begun to converge with those of the first generation of biological computer scientists. As Miklos and Rubin conclude, "In the Post-Sequence Era, we may eventually be able to move beyond what evolutionary processes have actually produced and ask what can be produced. . . . That is, we may be able not only to discern how organisms were built and how they evolved, but, more importantly, estimate the potential for the kinds of organisms that can still be built."[48]

Moreover, just as the early cyberneticians envisioned, in this new age of complexity life is not limited to biological organisms. Davies continues his commentary by noting, "The most cogent argument against vitalism is the existence of self-organizing processes in nonbiological science. . . . Mathematically we can now see how nonlinearity in far-from-equilibrium systems can induce matter [here quoting Charles Bennett (1986)] to 'transcend the clod-like nature it would manifest at equilibrium, and behave instead in dramatic and unforeseen ways, molding itself for example into thunderstorms, people and umbrellas.' "[49] Indeed, for some, life is not even limited to physicochemical materiality. Artificial Life—as its supporters put it, "life as it could be"—can exist as pure simulation.

Conclusion

Two hundred years ago, biologists invoked the notion of organization to distinguish the living from the nonliving, replacing an earlier tripartite division of the world (animal, vegetable, and mineral) with a bipartite division (animate and inanimate); today that same notion is employed to unify the animate and inanimate, arranging its objects not along a great chain-of-being but along a new succession ordered solely by complexity. Born of the exigencies of war and its need for better technologies of control, this marriage of premodern categories with ultramodern technology has given birth to a postmodern ontology. In the old chain of being, we knew clearly enough who or what defined the end point, but this new succession has no end point, no limit to the possibilities of being. It is an organismic ontology, to be sure, but one which needs neither the vital force of generation nor an invisible guide to direct its beings along their journey. What now is an organism? No longer a bounded, organic body (for some, not necessarily even a material body). Instead, it is a nonlinear, far-from-equilibrium system that can mindlessly (or virtually) transcend the clodlike nature of matter and emerge as a self-organizing, self-reproducing, and self-generating being. It might be green or gray, carbon- or silicon-based, real or virtual.

Also one other difference: first-order cybernetics, born of concrete military needs, gave rise to command-control systems; second-order cybernetics, bred from the new alliances of the cold war, fostered the birth of self-organizing systems. For both, systems of control (with all the ambiguities of the term in place) were the primary preoccupation. But the revival of self-organizing systems in the 1980s was celebrated with promises of transcendence, even of self-transcendence. In Erich Jantsch's vision of this New Age, we are told, "life, and especially human life, now appears as a process of self-realization, . . . the inner, co-ordinative aspect of which becomes expressed in the *crescendo* of an ever more fully orchestrated consciousness. . . . Each of us would then, in Aldous Huxley's (1954) terms, be Mind-at-large and share in the evolution of this all-embracing mind and thus also in the divine principle, in meaning."[50] Today, in the post–cold war era of the 1990s, a new breed of self-organizing systematists offer us "life at the edge of chaos," celebrating not command and control but life "out of control."[51] Meanwhile, less interested in meaning and more interested in products, an entire industry waits poised for the new technologies of design and construction given birth by their dreams.

If we are in fact in the midst of a revolution, the question is, what kind of revolution? In what domain is it taking place? How much of "life on the edge of chaos" is a biological, technological, cultural or even a political vision? And, finally, where does vision leave off and reality begin? Or does it not any longer—for self-organizing, self-steering systems like ourselves and our progeny—even make sense to ask?

Notes

1   Quoted in Georges Canguilhem, *A Vital Rationalist* (Cambridge, Mass.: MIT Press, 1994), 293.

2   François Jacob, *The Logic of Life* (New York: Vintage, 1976), 4.

3   Jacques Monod, *Chance and Necessity: An Essay on the Natural Philosophy of Modern Biology* (New York: Knopf, 1971).

4   Jacob, *Logic of Life*, 74.

5   Immanuel Kant, *Critique of Judgement* (1790), in *Great Books*, vol. 39 (Chicago: Encyclopedia Brittanica, 1993), 461.

6   Ibid., 558, para. 66.

7   Ibid., 557, para. 65; italics in original.

8   Ibid.

9   Timothy Lenoir, *The Strategy of Life* (Chicago: University of Chicago Press, 1982), 2.

10 See, e.g., Robert Brain and M. Norton Wise, "Muscles and Engines: Indicator Diagrams in Helmholtz's Physiology," in Lorenz Krüger (ed.), *Universalgenie Helmholtz: Ruckblick nach 100 Jahren* (Berlin: Akademie Verlag, 1994), 124–145.

11 Hermann Helmholtz, "The Interaction of Natural Forces" (1854), quoted in David Cahan (ed.), *Science and Culture: Popular and Philosophical Essays of Hermann von Helmholtz* (Chicago: University of Chicago Press, 1995), 37.

12 Claude Bernard, *Introduction à l'étude de la médicine expérimental* (Paris: Librairie J. B. Ballière, 1865), pt. 2, ch. 2, sec. 1, quoted in Canguilhem, *Vital Rationalist*, 297.

13 See, e.g., Evelyn Fox Keller, *Refiguring Life: Metaphors of 20th Century Biology* (New York: Columbia University Press, 1995).

14 Thomas Hunt Morgan, "The Rise of Genetics: II," *Science* 76 (1932): 285.

15 Thomas Hunt Morgan, *Embryology and Genetics* (New York: Columbia University Press, 1934), 6–7. Driesch seems to have concurred. In 1912, he abandoned experimental research and joined the philosophy faculty at Heidelberg.

16 Keller, *Refiguring Life*, chapter 3.

17 Lily E. Kay, "Who Wrote the Book of Life?," in M. Hagner and H.-J. Rheinberger (eds.), *Experimente in den Biologische-Medizinischen Wissenschaften* (Berlin: Akademie Verlag, 1996); also Lily E. Kay, *Who Wrote the Book of Life: A History of the Genetic Code* (Stanford, Calif.: Stanford University Press, 2000).

18 Anne Harrington, *Hunger for Wholeness* (Princeton, N.J.: Princeton University Press, 1995).

19 See, e.g., Peter Galison, "The Ontology of the Enemy: Norbert Wiener and the Cybernetic Vision," *Critical Inquiry* 21 (1994): 228–266.

20 Especially from a group formed in England in the mid-1950s under the name "ARTORGA," for the study of artificial organisms. For the ONR, see, e.g., Marshall Yovits and Scott Cameron (eds.), *Self-Organizing Systems* (New York: Pergammon Press, 1960).

21 C. A. Muses (ed.), *Aspects of the Theory of Artificial Intelligence* (New York: Plenum, 1962). As organizer of one of the early conferences on self-organizing systems, Muses suggested a slightly different aim, one he described as "man's dominating aim," namely "the replication of himself by himself by technological means" (114). Indeed, he introduced the published volume with the suggestion that "Man Build Thyself?" might be a better title (v).

22 The U.S. Army's interest in bionics had a similar basis. In 1960 Harvey Saley of the Air Force addressed a bionics symposium as follows: "The Air Force, along with other military services, has recently shown an increasing interest in biology as a source of principles applicable to engineering. The reason clearly is that our technology is faced with problems of increasing complexity. In living things, problems of organized complexity have been solved with a success that invites our wonder and admiration"; quoted in Geof Bowker, "How to Be Universal: Some Cybernetic Strategies, 1943–1970," *Social Studies of Science* 23 (1993): 118.

23 Frank Rosenblatt, *Principles of Neurodynamics: Perceptrons and the Theory of Brain Mechanisms* (Washington: Spartan Books, 1962); see also Donald O. Hebb, *The Organization of Behavior* (New York: Wiley, 1949).

24 Anatol Rapaport and William J. Horvath, "Thoughts on Organization Theory and a Review of Two Conferences," *General Systems* 4 (1959): 90.

25 J. H. Woodger, *Biological Principles* (New York: Harcourt, Brace, 1929), 291.

26 Edmund W. Sinnott, *Cell and Psyche: The Biology of Purpose* (Chapel Hill: University of North Carolina Press, 1950), 22, 48.

27 Gerd Sommerhoff, *Analytical Biology* (Oxford: Oxford University Press, 1950), 13–14. The book was originally intended as his doctoral dissertation for the biology department at Oxford but was rejected (according to Sommerhoff) because it was "too mathematical" for biology. The work interested others, however, and, instead of a degree, it earned him a research fellowship in neuroscience (personal communication, 12 April 1996).

28 W. Ashby, unpublished recommendation, holdings of Gerd Sommerhoff.

29 C. H. Waddington, *Biology, Purpose, and Ethics* (Worcester, Mass.: Clark University Press, 1971), 20.

30 See Heinz von Foerster, "On Self-Organizing Systems and Their Environments," in Yovits and Cameron, *Self-Organizing Systems*, 31–50. Trained in physics and engineering, von Foerster spent the war years employed in top secret research on electromagnetic wave transmission in Berlin and Liegnitz. But his interests were far broader: after the war, he introduced himself to Warren McCulloch with a short volume he had written on quantum mechanics and memory. McCulloch was impressed; he sponsored von Foerster's emigration in 1949, recruited him to edit the Macy conference proceedings, and helped him obtain a faculty position in electrical engineering at the University of Illinois.

31 In Heinz von Foerster and George W. Zopf (eds.), *Principles of Self-Organization* (New York: Pergammon Press, 1962), 9.

32 Gordon Pask, *An Approach to Cybernetics* (New York: Harper and Brothers, 1961), 72.

33 The exact date is unclear: von Foerster's papers suggest that the official date of its origin, 1 January 1958, was decided retroactively—probably based on the starting date for his first major contract from the ONR for "the Realization of Biological Computers." Once named, however, a wide assortment of books and papers written by visitors to his lab from 1957 on could be claimed as products of the BCL.

34 Gordon Pask, "Different Kinds of Cybernetics," in Gertrudis Van de Vijver (ed.), *New Perspectives on Cybernetics* (Dordrecht: Kluwer Academic Publishers, 1992), 24–25; also von Foerster, "On Self-Organizing Systems and Their Environments," 31–50.

35 See, e.g., Humberto Maturana and Francisco Varela, *Autopoiesis and Cognition: The Realization of the Living* (Dordrecht: Reidel, 1980). Maturana was a frequent visitor at the BCL during the 1960s and after 1968 so was Varela; indeed Varela thanks von Foerster for his "role in the gestation and early days of the notion of autopoiesis," in Francisco Varela, "The Early Days of Autopoiesis: Heinz and Chile," *Systems Research* 13(3) (1996): 407. Both Maturana and Varela have remained von Foerster's close personal friends ever since.

36 As Allen Newell and Herbert Simon asserted, "A physical symbol system has the necessary and sufficient means for general intelligent action"; quoted in Herbert L. Dreyfus and Stuart E. Dreyfus, "Making a Mind versus Modeling the Brain: Artificial Intelligence Back at a Branchpoint," *Daedalus* 117(1) (1988): 16.

37 Minsky recalls that their "original funding stream was created by J. C. R. Licklider, who went to ARPA in order to realize his dream of a nationwide network of computer research and application. He arranged the MIT collaboration with Prof. R. M. Fano at MIT, who became the first director of MAC. . . . I should add that I had been a student and worshipper of Licklider from my sophomore days at Harvard in 1947" (personal communication, 26 March 1997).

38 Yovits says "it was starved to death" (personal communication, 1997).

39 Marvin Minsky and Seymour Papert, *Perceptrons* (Cambridge: MIT Press, 1969), 19.

40 See, e.g., Dreyfus and Dreyfus, "Making a Mind," 24. Von Foerster, however, was able to keep the BCL going for another five years, largely by forging new alliances (with, e.g, the National Institutes of Health and the Department of Education).

41 Dreyfus and Dreyfus, "Making a Mind," 24.

42 Interview by Stefano Franchi, Güven Güzeldere, and Eric Minch, *Stanford Humanities Review* 4.2 (1995).

43 Thanks in good part to private corporations like Thinking Machines, Inc.

44 Wolfgang Krohn, Gunter Kuppers, and Helga Nowotny (eds.), *Selforganization: Portrait of a Scientific Revolution* (Dordrecht: Kluwer Academic Publishers, 1990), 3. For a valuable overview, see N. Katherine Hayles, *How We Became Posthuman: Virtual Bodies in Cybernetics, Literature, and Informatics* (Chicago: University of Chicago Press, 1999).

45 P. C. W. Davies, "The Physics of Complex Organisation," in Peter Saunders and Brian C. Goodwin (eds.), *Theoretical Biology: Epigenetic and Evolutionary Order from Complex Systems* (Edinburgh: Edinburgh University Press, 1990), 101–111.

46 William F. Loomis and Paul Sternberg, "Genetic Networks," *Science* 269 (4 August 1995): 649.

47 George L. Gabor Miklos and Gerald M. Rubin, "The Role of the Genome Project in Determining Gene Function," *Cell* 86 (1996): 521.

48 Ibid., 527. For an update on the meaning of genetic sequencing, see Evelyn Fox Keller, *The Century of the Gene* (Cambridge: Harvard University Press, 2000).

49 Davies, "The Physics of Complex Organisation," 111.

50 Erich Jantsch, *The Self-Organizing Universe* (Oxford: Pergammon Press, 1980), 397–398.

51 *Out of Control* is the title of a recent popular account notable for its celebratory (and perhaps obfuscatory) moral. The author concludes, "To be a god, at least to be a creative one, one must relinquish control and embrace uncertainty. . . . To birth the new, the unexpected, the truly novel—that is, to be genuinely surprised—one must surrender the seat of power to the mob below"; see Kevin Kelly, *Out of Control: The New Biology of Machines, Social Systems, and the Economic World* (New York: Addison-Wesley, 1994), 257.

7 Immunology and the enigma of selfhood

Alfred I. Tauber

Immunology has appropriated the task of defining the organismal self. This is not a typical definition of the discipline, and I freely admit a heterodox orientation. The general view is that this science, since its very beginnings during the last twenty-five years of the nineteenth century, was committed to discerning those mechanisms by which the "self" discriminates host elements from the foreign. On this view, the latter are destroyed by immune cells and their products, whereas the normal constituents of the animal are ignored. In other words, the identity of the host organism was given or assumed. But such neat divisions or boundaries were adopted or at best were drawn with a certainty that remained problematic. Aside from competing theoretical formulations, early discrepancies accompanied the full embrace of a mode for discriminating between self and nonself to explain immune function. Concomitant to demonstrating the beneficial effects of immunity, these same defensive functions were shown to be the cause of much of tissue damage. But inflammation only broadened the conceptual arena of immune mechanisms and was soon incorporated as the necessary untoward effects of the defense system. However, the immune system not only was found to inflict damage as the price of cleansing; it also was capable of apparently capricious assault on its host. So-called autoimmune reactions were described at the turn of the century and later determined to cause autoimmune disease, but because the entire orientation of the science was to see immunity as a mediator of host defense, these findings were viewed as a pathological aberrancy. Arising from an unregulated killer system gone awry, autoimmunity, on this view, could hardly be regarded as part of an expected continuum of normal immune function. In each case, "ideal" immunity was the agent of the Self, and although immunity might behave inconsistently regarding that mandate, the basic structure of immunology demanded articulation of a model for identifying and protecting organismal identity. In short, by the mid-twentieth century, a formal theory of discrimination between self and non-self was articulated that would attempt to clearly demarcate the host organism and the foreign by an immune system that under normal conditions discerned the self and protected it. This basic formulation has served as the foundation of the contemporary science.

We can trace this theme that immunology's history resulted from discoveries elucidating the bacterial etiology of infectious diseases. This orthodox narrative draws together twin disciplines: microbiology, the study of the offenders, and immunology, the examination of host defense. Thus typical accounts of immunology's development begin with relating how, on the one hand, German immunochemists, led by Robert Koch's epochal bacteriological studies, discerned specific antibody protection of infected animals, and on the other hand, Elie Metchnikoff, at the Pasteur Institute, championed cellular immune mediators for a similar purpose. These apparently disparate versions of immunity eventually merged and a unified theory of host defense was outlined. On this view, both humoral and cellular agents were used for one purpose, to kill pathogens, and the argument seemed confined to questions of mechanism. Other autoimmune behaviors were marginalized as aberrancies, and the evolutionary origins of the immune system were postulated only in regard to the ostensible defensive functions discerned in vertebrate animals. So by 1908, when Metchnikoff and Paul Ehrlich, a stalwart immunochemist, shared the Nobel Prize, the discipline had quickly reached an apparent theoretical maturity amid great clinical fanfare for its promise to combat infectious diseases.

Thus, in this pathological context, immunology began as the study of how a host animal reacts to pathogenic injury and defends itself against the deleterious effects from such microbial insult. This is the historical account of immunology as a clinical science, a tool of medicine, and as such it focused almost exclusively on the role of immunity as a defender of the infected. And note, the paradigmatic host is the patient, an infected "self" which I believe is a critical element for the power of this view. Explaining the development of the field by erecting such an edifice presupposes that a definable self exists that might be defended. After all, are we not selves? I have contested that view at its foundations. For me the history of immunology is precisely the very attempt to define such an entity. And the reading I have offered contends that it is the problem of a self that has besieged the theoretical fortress immunology has attempted to erect. Briefly put, the issue has been neatly divided between those who adopt a modernist notion of the self, namely that there indeed is a self, and the contesting postmodernist argument that the self is an artifice, a conceit, a model at best. On this latter view, immunology cannot accept organismal identity as a given but must pose the self as its core problematic. This latter opinion is a minority voice, but a nagging one that cannot be ignored, for it articulates the unsettled foundations of the science.

The clinical orientation, which assumes a given entity—the self, is obviously a dominant organizing perspective, but it has been challenged by a formulation that works by an altogether different theoretical construction, dispensing with the self altogether. This modernist assumption has dominated the histories of immunology, but such narratives omit the deeper perplexities of the science, which remain largely implicit and are thus ignored. The historical examination of immunology is in its infancy, and perhaps it is too early to discern schools of interpretation, but since the view I have embraced is nevertheless heterodox, I believe some accounting is due to explain my vantage and how it was developed.

My essay explores the philosophical underpinnings of the discipline's theoretical construction, and I have adopted a firsthand recapitulation here to clearly expose the interpretative aspects of my work. In so doing, I wish to sharply contrast my perspective with other histories of immunology. Specifically, I am not referring to the accounts given by the scientists themselves as they seek to recount the growth of their discipline,[1] but to the interpretations and historiographic strategies adopted by my peers such as Arthur Silverstein, Ann Marie Moulin, Thomas Soderqvist, Warwick Anderson and his colleagues, Pauline Mazumdar, Alberto Cambrosio and Peter Keating, and Ilana Lowy.[2] As different as their respective interests might be, they hold in common what I shall call "the presence of the self," and this posture necessarily affects their histories. So beyond summarizing the conceptual issues that have intrigued me, the historiography employed to elucidate these questions, and the rationale for predicting a significant turn in the theoretical orientation arising from the early responses to the quandary now faced, I hope to illustrate how my different orientation has grappled with immunology's theoretical agenda. In short, on my view, immunology does not at heart seek to discern the basis of discriminating between self and nonself as current dogma would have us believe, but rather it is concerned with establishing organismal identity. This issue has been present since the birth of the field but effectively subordinated to the history of immunological specificity. No doubt this latter problem was a dominant concern, but it rested upon an unstable theoretical construction. By excavating at deeper depths, the history of immunology assumes a different pattern than that typically depicted.

Nineteenth-century origins of immunology

If we begin our inquiry with a careful examination of the earliest immunologists—Elie Metchnikoff, the German microbiologists who debated him, and the immunochemists who built on his formulation—we may clearly discern the origins of the immune selfhood problem.[3] Metchnikoff came to the nascent field of immunology from an unexpected theoretical and methodological perspective. He was an embryologist who sought to discover genealogical relationships in the context of Darwinian problematics. Intrigued with the problem of how divergent cell lineages were integrated into a coherent, functioning organism, Metchnikoff was thus preoccupied with the problems of development as process, which he regarded as Darwinian: cell lineages were inherently in conflict to establish their own hegemony, and he thus hypothesized that a police system was required to impose order, or what he called "harmony" on the disharmonious elements of the animal.[4] He found such an agent in the phagocyte, which retained its ancient phylogenetic eating function, to devour effete, dead, or injured cells that violated the phagocyte's sense of identity.

Thus the phagocyte was initially viewed as a purveyor of identity, and when Metchnikoff entered the field of infectious diseases in the early 1880s (with almost twenty years of comparative embryology to his credit), he was poised to apply his understanding of phagocytes to the duty of protecting the organism from pathogens (i.e., maintaining integrity).[5] It was a grand scheme, which he presented in a series of public lectures in Paris in the spring of 1891. Later published as *Lectures in the Comparative Pathology of Inflammation*, Metchnikoff argued that the phagocyte had preserved its most ancient phylogenetic function: in simple animals it served as the nutritive organ (eating resident microbes) and in animals with a gut it continued to eat but now for defense. Thus in Metchnikoff's theory, immunity was a particular case of physiological inflammation, a normal process of animal economy. But there was a more subtle message: (1) immunity was an active process with the phagocyte's response seemingly mounted with a sense of independent arbitration, and (2) organismal identity was a problem bequeathed from a Darwinian perspective that placed all life in an evolutionary context. This last point cannot be overemphasized: Darwinian conflict extended to the competing cell lineages of the individual animal. The quality of agency in his argument and the radical sense of self-definition reflected major Nietzschean themes, a parallel I have attempted to make explicit elsewhere.[6]

Metchnikoff had been brushed aside by his German detractors as a hopeless Romantic, with outdated teleological precepts that caricatured phagocytes as possessing volition and intention and thus vitalist independence. His polemics with the Germans were complicated by both political and personal issues,[7] but the conceptual differences should dominate this history. Specifically, we must explicate his stance against the strong reductionist program of contemporary immunochemists and the basis of his holistic orientation as the infrastructure of a comprehensive biological theory of organismal identity. Later historians, however, have generally followed the initial German assessment and discounted Metchnikoff's role in the development of the science, in large measure because they correctly perceived that immunology was dominated by serology during the first half of this century. But this reading does not recognize how Metchnikoff's theory has underpinned the discipline, laying dormant until it was rediscovered after World War II and explicitly activated.[8]

If one considers the documentation of Metchnikoff's Nobel archives, public testimony, and a careful contrast of his views with other scientists involved in similar research,[9] we must recognize that his scientific posture employed emergent and dynamic thinking appropriate to his organismal orientation as a biologist keenly aware of the problem of identity in a post-Darwinian age.[10] Metchnikoff, with perhaps unique insight, deeply comprehended the Darwinian revolution and his approach critically complemented the reductionists. He maintained that throughout the organism's life, it experiences changing environments, new insults, and encounters with novel challenges, and it is the organism's immune adaptability and versatility that determines its overall success. His concerns were different from those who approached immunology from the more narrow perspective of infectious diseases; he was concerned with the grand challenge of Darwinism writ large. He sought the primary lesson of evolutionary biology, and he held a radically different conception of the organism from that offered in the pre-Darwinian era. Prior to *On the Origin of Species*, the organism was a given. It was viewed as essentially unchanging and stable, a view that contrasts the dynamic image implicit in Metchnikoff's formulation. For him the organism is in a dialectical relationship with its world. In an ever-changing set of relationships, at many different levels of engagement, the animal responds to its environment, and it was this *active* component that was so revolutionary. Moreover, because of the phagocyte's apparently independent volition (or agency), which could hardly be described by physicochemical laws, he drew the ire of those committed to establishing the chemical basis of the immune reaction.

In recognizing Metchnikoff's own holistic view of the organism, his theory-driven, integrated, and comprehensive approach to biology, and the neo-Romanticism of his extrapolated biological thought, he may be placed within the broader intellectual currents of the period. From this perspective, we can discern at the heart of his theory the question of identity and more specifically the notion of disharmony, the idea of struggle turned inward that characterized Metchnikoff's phagocytosis theory. More to the point, Metchnikoff's case presaged a science yet to come, the foundation of current immune theory. As an embryologist, he understood that the organism was continuously developing, and a dialectical approach is required to understand his theory.[11] Without firm boundaries and structure, this vision of organismal identity was decentered and emergent.[12] This formulation framed twentieth century theoretical immunology.

The problem of the self

In seeking a contemporary voice, or at least a resonance to Metchnikoff's formulation, I stumbled upon a Santa Fe Institute publication, *Theoretical Immunology*.[13] The institute was founded in 1984 to examine complex systems from a multidisciplinary perspective, and it quickly became a dominant organ for applying computer modeling to nonlinear dynamical analyses. Interestingly, this text (two volumes) was the first systematic analysis of immunology in the institute's series of publications, which continues to give immunology a high billing. Of particular note, the contribution by Varela, Coutinho, and their colleagues immediately struck a responsive chord to the issues raised by the study of Metchnikoff.[14] They espoused a self-determinism closely related to Metchnikoff's dialectical vision of the organism: "The self is not just a static border in the shape space, delineating friend from foe. Moreover, the self is not a genetic constant. It bears the genetic make-up of the individual and of its past history, while shaping itself along an unforeseen path." This dynamic vision of immune identity had a complex and revealing history.

Immunology during the first half of the twentieth century was preoccupied with the chemical questions of immune specificity, and the biological questions concerning immune identity were set aside.[15] But after World War II, transplantation and autoimmunity became increasingly relevant both to basic immunologists and to clinicians. It was at this juncture that Macfarlane Burnet introduced the "self" into the immunological lexicon, and upon that metaphor erected a theory of immunological tolerance that would dominate the field to this date.[16] Immunology today defines itself as the science of discriminating between self and nonself, and Burnet's clonal selection theory (CST), by which selfhood is understood, "with only slight modification . . . has passed from the status of theory to that of paradigm," according to one textbook. For those uncomfortable with such sweeping notions as "paradigms," another textbook contends that there is still a consensus that CST "is no longer a theory but a fact."[17] The immune self has indeed become dogma, and the self-versus-other axis has organized the thinking of the entire discipline.

Burnet was a virologist by original training, so he came to immunology from a biological perspective quite different from the immunochemists then dominating the field. He was ambitious to integrate developmental biology, genetics, and immunology into a cohesive theoretical whole, and he did so by drawing upon both Metchnikoff and later ecological theory to devise a view of the immune system as the purveyor of organismal identity.[18] But if this history is regarded through a prism similar to that used to dissect and reconstruct the Metchnikovian saga, an ethos that appreciated the dynamical and hierarchal properties of biological systems, one must conclude that Burnet was not the best champion to carry the Russian's mantle. The self is a complex construction, and immunologists have had different visions of selfhood as borrowed from various philosophical and psychological formulations. In *The Immune Self, Theory or Metaphor?*,[19] the case was made that the self concept was developed along a continuum, stretching from "punctual" (i.e., defined, demarcated) to "elusive" views of identity. The dominant view among immunologists is that a self exists and that it has borders defined by a genetic signature. The immune system is designed to react against the "foreign" and not against the host. When the immune system would attack the body, immunologists generally regarded it as pathological autoimmunity, a condition Paul Ehrlich called "dysteleological in the highest degree"[20] and which generations of immunologists believed to be true. Not surprising, Metchnikoff thought that autoimmunity was expected because the immune system was always sensing the animal's inner environment, seeking abnormal cells to destroy, whether originating from the host or invading pathogens. Burnet, assuming the Ehrlich precept, sought a firm definition of the immune self. This goal has been the guiding principle of immunology, an orientation that I have critiqued at length and in summary.[21]

Burnet's theory, in brief, is that the animal during prenatal development exercises a purging function of self-reactive lymphocytes (the cells responsible for synthesizing reactive antibodies and mediating so-called cellular reactions) so that the immune system would ignore all antigens (substances that initiate immune responses) encountered during this period. The hypothesis, first presented in 1949, was later developed into the clonal selection theory, which maintains that lymphocytes with reactivity against host constituents are destroyed during development, and only those lymphocytes that are nonreactive are left to engage the antigens of the foreign universe.[22] These potentially deleterious substances would select lymphocytes with high affinity for them, and through clonal amplification a population of lymphocytes would differentiate and expand to combat the offending agents. The theory was assumed to be proven in the 1970s, but this vision of immune identity should be regarded with considerable skepticism. Bountiful evidence in recent years has suggested that autoimmunity is a normal finding, and immune reactivity, rather than functioning only in an "other-directed" mode, is in fact bidirectional:

> Clearly, one can define "self" from a biochemical or genetic or even a priori basis. But from our vantage point, the only valid sense of immunological self is the one defined by the dynamics of the network itself. What does not enter into its cognitive domain is ignored (i.e. it is nonsense). This is in clear contrast to the traditional notion that IS [immune system] sets a boundary between self in contradistinction to a supposed non-self. From our perspective, there is only self and its slight variations.[23]

The critical turning point was to appreciate that the immune system in fact recognizes selfness as natural autoimmunity and that such host-directed reactivity is physiologically normal. The significance of this orientation has taken some time to sink into the collective consciousness of the discipline, and its ramifications are still not widely appreciated. As Coutinho and Michel Kazatchkine later wrote,

> During this century, the evolution of concepts on autoimmunity could be summarized by "never, sometimes, always." Thus from the early "horror autotoxicus" [Ehrlich] to the 1960s, immune autoreactivity was simply not considered. . . . With the first identification of autoreactive antibodies in patients and the subsequent conceptual association with autoaggressive immune behaviors, the "sometimes" phase was entered, necessarily equated with disease. By this time, immunology had laid its foundation on the clonal selection theory, which forbids autoreactive clones in normal individuals. Immunologists thereafter devoted 30 years discovering ways by which autoreactive lymphocyte clones can be deleted and why they fail to be deleted in autoimmune patients. . . . In the 1970s at least three sets of observations and ideas began to alter this course of events and to herald the "always" period.[24]

This position contrasts with the "one-way" definition of selfhood, where there is a genetic self whose constitutive agents see the foreign, and immune reactivity arises from this polarization with attack directed only against nonself. Varela drew upon a definition of immune selfhood as analogous to the mind, which has no firm genetic boundaries but arises from experience and self-creative encounter. Not surprisingly, emphasizing the cognitive nature of immune function, Coutinho and his colleagues argued that the global properties of the immune system cannot be understood from analysis of component parts alone; "emergent properties," nonlinear network or complex systems, "global cooperativity," and other terms borrowed from the neurosciences emphasized the affinity of methods they employed with those already adopted for describing other complex cognitive systems.

Coutinho and others have been committed to the notion of selfhood, but such a designation ruling immune function is becoming increasingly problematic. There are at least half a dozen different conceptions of what constitutes the immune self,[25] models that might be situated on a continuum between an extreme genetic reductionism and a complex construct employing different organizing principles. With so much dispute surrounding the definition of self, a growing counter position suggests that the self might be better regarded as only a metaphor for the immune system's silence, that is, its nonreactivity. The theory built upon the self now appears to have many ad hoc caveats and paradoxes. Perhaps the evolution of the original metaphor into theory is now yielding again to another metaphorical construction. So what could the immune self be? An answer was offered by Niels Jerne, the father of the paradigm shift—if in fact there was one.

Niels Jerne and the deconstruction of the self

Niels Jerne's novel concept of immune regulation went well beyond the current notion of the immune network composed of lymphocyte subsets secreting immunostimulatory and -inhibitory substances (essentially a simple mechanical model with interlaced, first-order feedback loops). His idiotypic network theory was born in his attempt to model the immune system as analogous to the nervous system. The agenda of this theory was, from its very inception, a complex amalgam of fitting together the pieces of the regulation puzzle, with an overriding desire to understand the immune system as a cognitive enterprise. Exhibiting deep similarity with the brain and manifesting behaviors analogous to the mind, the idiotypic network elucidates the principles of self-organization that regulate the immune response. But beyond that accomplishment lurks the larger, metatheoretical desire to define the immune system as a cognitive entity or process.

Jerne had embarked on his theoretical odyssey as early as 1960, when he embraced the cybernetic enthusiasm of the period, writing of the antibody-producing system as being "analogous to an electronic translation machine."[26] By the mid-1960s, he had dealt explicitly with the metaphors of immunological "memory" and "learning." Before he formally proposed the network hypothesis, he noted how immunologists used metaphors such as "recognition" that were obviously derived from psychology, and he drew even more explicit comparisons and contrasts with the nervous system.[27] Jerne saw that each system has a history of encounters with the world that remain present both in the form of irreversible changes and in the form of memories which always affect the next response. Thus, both the immune and nervous systems change with, and learn from, experience. Over the next decade, Jerne continued to draw explicit parallels. Fundamentally, on his view, the immune system is like the central nervous system, with its cognitive functions and capacities of recognition, learning, and memory.

Jerne's idiotypic network hypothesis, as he extensively presented it in 1974 and proposed in outline earlier, initially had no explicit homology or analogy with language. But a decade later he developed a linguistic model for immune function as an important explicative ploy for his theory.[28] Jerne was intrigued by the possibilities of a language composed of ready-made constituents, construed as something akin to "words" that are sorted into sentences. He regarded the antibodies as forming a highly complex interwoven system, where the various specificities "referred" to each other. Under the general rubric of "cognition," Jerne conceived of the immune system as self-regulating, where antibody not only recognizes foreign antigen, but is also capable of recognizing self constituents as antigens (the so-called idiotopes). The "recognized" and the "recognizer" are essentially the same, since any given antibody might serve either or both functions. In other words, immune regulation is based on the reactivity of antibody (and later lymphocytes) with its own repertoire, forming a set of self-reactive, self-reflective, self-defining immune activities. There is no "self" and "other" for the immune system; according to Jerne's theory the system is complete unto itself, consisting of interlocking recognizing units: each component reacts with certain other constituents to form a complex network or lattice structure. When the system is perturbated by the introduction of a substance that is "recognized" (i.e., it reacts with a member[s] of the system), this disturbance initiates immune responsiveness. Thus foreignness per se does not exist in this formulation.

Jerne's theory presents a radically altered view of immune selfhood. If the biological world could be so easily divided between "host" and "foreign" constituents, as Burnet's original tolerance theory held, then anything an antibody (or lymphocyte) encountered would be suitable for destruction. In that simplified world of discrimination between a self and a nonself, the immune system learned such distinctions, generated an army of reactive antibody and lymphocytes, and acted accordingly when "antigen" was encountered. But Jerne coupled the simple antibody-antigen interactions to the far more complex and nondiscriminatory functions of the immune system built upon *self*-recognition. Thus "autoimmunity" became the organizational rule to explain immune function. The idiotypic network was fundamentally "dynamic" and "self-centered," generating anti-idiotypic ("self-reactive") antibodies to its own antibodies, which he thought constituted the overwhelming majority of antigens present in the body. Strikingly, Jerne proposed no explicit mechanism for discriminating between self and nonself, and this apparent lacuna would serve as the nexus of critiques.[29] But for Jerne, the need to define the self as distinct from the other seemed to recede from his primary theoretical concerns, and this posture was to have important repercussions.

When the immune system is regarded as essentially self-reactive and interconnected, the meaning of immunogenicity, that is, *reactivity*, must be sought in some larger framework. Antigenicity, then, is only a question of degree, where the self evokes one kind of response, and the foreign another, not because it is intrinsically foreign, but rather because the immune system sees that foreign antigen in the context of invasion or degeneracy. Foreignness per se does not exist because if a substance were truly foreign, it would not be recognized; the immune system would have no image by which to engage it. So the "foreign" becomes perturbation of the system; as observers, we record the ensuing reaction, and only as third parties do we designate "self" and "nonself." From the immune system's perspective, it only knows itself, and thus reaction to the foreign becomes secondary, or perhaps a by-product of this central self-defining function. Jerne's hypothesis helped to reformulate the entire question of how the immune system is organized. If there is a "self" in Jerne's theory, it is the entire immune system as it "senses" itself. Jerne's theory thus appears radically different from the dominant theories of interlocking inhibitory and stimulatory activities that described immune function built from Burnet's original self-nonself dichotomy. Jerne's theory commanded a reaction to the entire conception of the dichotomy self/nonself.

Jerne's legacy

The challenge posed by Jerne's network theory was twofold: most generally, it demanded an assessment of an inward-directed self-seeking process; its critical weakness was lacking a stable reference for defining selfness. Later models arising in reaction to his network theory attempted to circumvent this latter problem by demoting the problem of selfhood altogether. It is in this fundamental reorientation—if one takes it seriously—that we may perceive a decisive shift in immunology's theoretical foundations.

Not surprisingly, in the wake of Jerne's contributions much discussion ensued about whether the immune system might be regarded as semiotic,[30] and more generally, about how the immune system might be understood as a cognitive faculty. On this view, the immune system represents a complement to the nervous system; indeed, contemporary theorists represent immune function using models similar to those for understanding neural cognition. To engage its targets, the immune system must first perceive them and then, in a sense, decide whether to react. This is a cognitive model, where the immune and nervous systems are regarded analogously. Each has perceptive properties; each can discern both internal and external universes. Information processing is central to both the nervous and immune systems, and thus their perceptive properties are linked to effector systems.

Beyond the functional analogy, there is growing evidence that the nervous and immune systems are highly integrated with one another.[31] They share many of the same messenger molecules, have close developmental histories both in phylogeny and ontogeny, and intersect biochemically to achieve a common purpose. But beyond these interdependencies, there is increasing appreciation of a strong parallel in how these complex systems might be organized; increasingly, systems analyses applicable to one discipline are carefully examined for their applicability to the other based on the assumption that as a cognitive apparatus the immune system's structure may well mimic the architecture of the nervous system. Models based on neural networks, complete with analogous computer program simulations, suggest new research directions,[32] but there is little evidence to suggest that any of these efforts have either led immunologists to productive experimental strategies or predicted any research outcomes. If such applications will be successful, it is still unclear in which domain we might expect their utility.

Irrespective of particular success, the cognitive formulation has become an explicit mode for organizing theoretical discussion. For instance, Irun Cohen built on this general notion by declaring that a new "cognitive paradigm" has eclipsed Burnet's clonal selection theory,[33] and he organized the first conference dedicated to exploring this theoretical shift. Although these recent discussions have not explicitly built on Jerne's idiotypic theory, it is apparent that his ideas have filtered into the immunological community in diverse ways. The problem of meaning—how does an antigen become antigenic and evoke a response?—serves as the nexus of theoretical discussion. To formulate the matter, I have used an analogy with language based on Shanon's provocative cognitive study.[34] The "representational" sense of an antigen, that is, that it carries its meaning with an intrinsic property, has been replaced with the notion that meaning is derived from the antigen's *context*. In linguistic terms, each word has a spectrum of definitions; the context in which the word is used confers the meaning. When I say, "Let's go to the bank," do I mean to go to a building and get some money or to the river and go fishing? Multiple contextual elements confer specificity to words, and those supporting structures delimit possible interpretations.

For me, contextual meaning seemed to hold together modern immunology, as in Jerne's ideas of the network being "perturbed" and in the dominant model concerning lymphocyte activation. This latter view holds that specific recognition of antigen by a lymphocyte receptor is not sufficient for activation, and that additional signals are necessary to determine whether a cellular response follows. In short, an antigen is neither self nor nonself except as it attains its meaning within a broader construct.[35] Orthodox immune theory encompasses this idea in the so-called two-signal model, which does not require any of Jerne's hypotheses to fulfill its agenda.[36] But there are more radical readings of the "contextualist" setting by which antigens are sensed. Debate concerning what constitutes the milieu of antigenicity and ensuing reaction have spawned certain provocative and potentially important models of immune regulation. The more radical responses inspired by Jerne's theory herald a shift in how immune regulation might be understood. At this point, I detected a postmodern ethos.

The entire contra-Burnetian perspective rests on recognizing the relativity of perspectives, for the context of the immune encounter is paramount in conferring meaning on any antigen.[37] Once the self-nonself structure is weakened, more radical perspectives may be entertained. For instance, Zvi Grossman and William Paul have proposed that under the cover of immunological tolerance, autoreactive T lymphocytes may actively participate in classical protective responses as well as in nonclassical physiological activities, such as stabilizing the differentiation profiles of other cells.[38] Grossman has argued that "the immune system is not 'devised for aggression against foreign antigens' more than it is devised to manifest tolerance, or [a more] complex relationship, to self or foreign antigens; recognition of antigen is necessary for both aggression and tolerance but is not sufficient for either."[39] This credo is developed most explicitly by Polly Matzinger (based on the suggestions of Ephraim Fuchs), whose model arises from rethinking the lymphocyte activation hypothesis rather than as an evolution of Jerne's theorizing.[40] On her view, the immune system decides what is insultive to the organism, that is, what causes distress, destruction, or nonprogrammatic death, and signals of such aberrancy initiate immune reactions. Selfhood per se recedes as the basis of immune definition; immunity becomes organismally driven (i.e., functional), and immunocytes become dependent on extraimmune factors and context.

In Matzinger's theory, discrimination between self and nonself has thus been replaced with a contextualist scheme based on responses to danger and destruction. From this perspective, both the self and its immune system have been deconstructed. Rather than a specialized system that patrols the rest of the body, the immune system becomes extended and intricately connected with every other cell of the organism so that tolerance and reactivity are governed by the cooperation of lymphocytes, antigen-presenting cells, and other tissues. Matzinger's model is fundamentally a process-driven, functionally conceived model, and interestingly, it builds on the antigen-presenting cell (likely a close relative of Metchnikoff's original phagocyte) as the arbiter of immune reaction, a cell that cannot distinguish self from nonself in traditional terms of lymphocyte recognition. Thus, to varying degrees some contemporary immunologists are moving beyond the self as a governing principle, and replacing discrimination between self and nonself with a functional, process-dominated conception based on the context in which antigen is encountered.

The differences in the way the cognitive or contextual orientation takes shape can be linked to the different elements with which the various theorists develop the analogy of the immune system to the nervous system, each assuming a distinctive slant from Jerne's own formulation. Irun Cohen has singled out the imprinted knowledge of the external world, a knowledge stored developmentally in sets of molecules and their complementarities and, to a limited extent, in simple regulatory networks. Coutinho and Varela have focused on the distributed nature of information processing and self-definition; Grossman has emphasized that the rate of change of lymphocyte stimulation requires intracellular integration, and at higher levels of organization, that coordination of complex physiological activities of many cells across tissues cannot be strictly preprogrammed or restricted to structures exclusively confined to the immune system. Matzinger dispenses with the self concept altogether, focusing instead on "damage control." In their scrutiny as to how well the "one-way" (i.e., self against the other) paradigm might account for selfhood, these critics decided that orthodox theory required radical revision. The self no longer commands center stage.[41]

By the mid-1990s, the New York Times had discovered that the Burnetian paradigm was being threatened. Reporting on three different experimental scenarios appearing in a single issue of Science, the general public was now alerted to the apparent failure of what were heretofore well-accepted discriminatory boundaries between self and nonself.[42] The implications of what was regarded by some as a major challenge to the self-nonself paradigm quickly spread well beyond the esoteric musings of a few investigators. These opinions extended from the enthusiasm of a palace revolt ("We're challenging 50 years of immunological thought" [Paul Lehrmann]), to its equivocation ("In a way, the new studies undermine what has been taken as a pillar of the self-nonself model. That doesn't mean the model is necessarily wrong. But the reports undermine its foundations" [Albert Bendalac]), to its denial ("This is being blown so far out of proportion. . . . I don't think the studies fundamentally challenge the self-nonself theory" [Alfred Singer]), to a cautious middle ground ("I think the work is an extension of the theory rather than a direct contradiction" [Charles Janeway, Jr.]). At a minimum, whether the self-nonself paradigm falls or stands, these new findings have highlighted paradoxes that demand explanation, and irrespective of the final verdict, the challenge to the self model has become big news.[43]

The question of postmodernism

Elsewhere I have discussed the contextualist theories of other immunologists, presented arguments that favor a significant shift in immunology's fundamental theoretical structure, offered an extensive philosophical analysis of the self metaphor in immunology, and suggested a linguistic model by which we might understand these various theoretical approaches based on representational versus contextualist models of language.[44] These efforts have been made to support a postmodernist vision of the organism, and in a recent publication I have explicitly drawn parallels between immunological theory and modernist versus postmodernist notions of selfhood.[45] There I argued that the immune self taps into an imprecise lexicon that carries a complex cultural construction, some of whose elements possess postmodern literary and social critical overtones. One might contest the extent of correspondence, but the analogies are, at the very least, highly suggestive. This is not to contend that immune theory supports particular postmodern tenets, only that I believe deep resonances exist between certain hypotheses regarding immune models and the language—with its cultural meanings attached—used to describe those theories. My orientation might be summarized by the following selection from my "Postmodernism and Immune Selfhood":

> The defining characteristic which corresponds with the spirit of postmodernism is the understanding of the nature of the organism as decentered and indeterminant. Thus in contrast to the modernist vision of the self, the post-modern view stresses the dynamic, if not dialogical character of the organism. . . . There are those immunologists who resist defining immune activity as based upon the mutually affecting presence of the other. They are wedded to a modernist notion of the self as a given entity, neatly defined and entailed by its own "selfness." This self is guarded by the immune system, which is then conceptualized as a sub-system fully (i.e., firmly) determined by a genetic prescription. I believe this is fundamentally a simple mechanical conception, where well-integrated parts function together like Descartes' clock: If only we understood its workings better, we would per-

ceive the mechanical order and operative causal relationships of this complex system. This . . . view is limited in its explanatory power. . . . Such "entities" cannot be characterized by metaphorical approximations, but require organizational principles to describe new kinds of mechanistic models, perhaps best described by non-linear logic, complexity theory, and self-organizational precepts of various ilk.[46] In immune theory this general view is represented by the cognitive paradigm. . . .

The argument rests on recognizing that Cartesian reductionism is a major pillar of the modernist strategy, and irrespective of its historical appearance in biology, the reductionist physiologists pursued deterministic and mechanical ideals alien to post-modernism's emphasis on holism, chance, emergence and, most important, process. Barely submerged beneath these concepts are vague notions of organismic contingency and self-actualization. These are easily found in the postmodern vocabulary, but not generally regarded as suitable "scientific" terms, but they lurk, hardly hidden in our general understanding of the organism defined by its immune system.[47]

Parallels with postmodern tenets are useful for limited demonstrative purposes of this sort and should be regarded as no more than suggestive devices in discerning the theoretical structure of the science. I invoke such an orientation to see the immune self as a powerful metaphor in need of major revision. Interpretation requires metaphors and borrowed vocabularies to gain a perspective outside the strict confines of scientific discourse. We must, in some fashion, get outside the subject. But even within the boundaries of a strict scientific grammar, immunology may well apply for new ideas to understand the dynamics and organization of its object of study, its use of probabilistic causality, nonlinear-dynamical and self-organizational constructions. The immune self would thereby be radically altered.

Non-Cartesian mechanics have yet to appear in immunology's normal practice. On the other hand, such approaches beckon as they seductively do in the neurosciences for theorists who seek better representations of these complex systems. The shared interests of theoreticians who respectively seek designs of the immune and neural systems reside in their common concern with modeling cognition, a parallel Jerne posited as early as the 1960s when he dropped vague notions concerning their possible similarities. Later theorists have been more explicit in their attempts to apply modeling from one discipline to the other. I have already mentioned some of those efforts and the general consensus that little substantive progress has been made in this effort. Though the attempts have become more sophisticated, the utility of this approach either in directing or predicting research is highly problematic. So while such models have a certain seductive appeal, I do not see a meaningful dialogue between theorists who embrace the computer and experimentalists who are committed to going directly from a to b with simple mechanical models of molecular or cellular behavior. Quite simply, theoretical papers are virtually unknown in experimental immunology journals and rarely found outside their own specialty sources. This is not to say that a union of these two communities is not likely, or even imminent, but as the millennium turned, no dramatic advances were made to propel the engagement to a meaningful marriage.

So if one is to assign a postmodern label to immunology based on a non-Cartesian model grounded on principles of chaos or complexity, then I believe we are still awaiting a significant contribution from that speculative sphere. Given these fundamental caveats, I do, however, see glimpses of a postmodern orientation in regards to immunology's metatheoretical concerns, when growing numbers of experimental immunologists are questioning the utility of a firm definition of immune identity, the so-called immune self, and have eroded this modernist construction with one based on a contextualist understanding of immune function.[48] Although this may prove to be a tantalizing first glimpse of a postmodern sensibility, it cannot be regarded as a dominant trend in the discipline and even less so a proven experimental strategy. I have given a great deal of credence to this movement against the staunch self-nonself proponents, but clearly this is a minority report.

Scarce indeed is another scientist who might write a similar history, for he or she would, as is often seen, offer an interpretation based on the dominant thought style. The decenteredness or even abandonment of the immune self, notwithstanding the New York Times's excitement, is generally regarded as frivolous by the immunological citizenry.[49] And well they might, for my espousal of a postmodern conception is based on a paucity of firm evidence, much speculation, a few tentative models, and finally, a prejudice that complex systems must behave in a way approximated by these nascent notions of immune function formulated by Jerne and his intellectual progeny. I must leave to later history whether this interpretation confirms that we are in fact witnessing a major turn in immunology's theory.

Notes

1   E. Metchnikoff, Immunity in Infective Disease, trans. F. G. Binnie (Cambridge: Cambridge University Press, 1905); K. Landsteiner, The Specificity of Serological Reactions, 2d ed. (Cambridge: Harvard University Press, 1945); P. M. H. Mazumdar (ed.), Immunology 1930–1980: Essays on the History of Immunology (Toronto: Hall and Thompson, 1989); R. B.

Gallagher, J. Gilder, G. J. V. Nossal, and G. Salvatore, Immunology: The Making of a Modern Science (London: Academic Press, 1995).

2   A. M. Silverstein, A History of Immunology (San Diego: Academic Press, 1989); A. M. Moulin, Le dernier langage de la médecine: Histoire de l'immunologie de Pasteur au Sida (Paris: Presses Universitaires de France, 1991); T. Soderqvist, "How to Write the Recent History of Immunology—Is the Time Really Ripe for a Narrative Synthesis?," Immunology Today 14 (1993): 565–568. W. Anderson, M. Jackson, and B. Rosenkrantz, "Toward an Unnatural History of Immunology," Journal of the History of Science 27 (1994): 575–594; P. M. H. Mazumdar, Species and Specificity: An Interpretation of the History of Immunology (New York: Cambridge University Press, 1995); A. Cambrosio and P. Keating, Exquisite Specificity: The Monoclonal Antibody Revolution (Oxford: Oxford University Press, 1995); I. Löwy, Between Bench and Bedside: Science Healing and Interleukin-2 in a Cancer Ward (Cambridge: Harvard University Press, 1996).

3   A. I. Tauber and L. Chernyak, Metchnikoff and the Origins of Immunology (New York: Oxford University Press, 1991).

4   Ibid.; see also A. I. Tauber, "The Immunological Self: A Centenary Perspective," Perspectives in Biology and Medicine 35 (1991): 74–86; Tauber, "Introduction: Speculations Concerning the Origins of Self," in Tauber (ed.), Organism and the Origins of Self (Dordrecht: Kluwer Academic, 1991), 1–39.

5   A. I. Tauber, The Immune Self: Theory or Metaphor? (New York: Cambridge University Press, 1994), 20, 62–63.

6   A. I. Tauber, "The Organismal Self: Its Philosophical Context," in L. Rouner (ed.), Selves, People, and Persons, Boston University Studies in Philosophy and Religion, vol. 13 (Notre Dame: Notre Dame University Press, 1992), 149–167; Tauber, "A Typology of Nietzsche's Biology," Biology and Philosophy 9 (1994): 25–44; Tauber, Immune Self.

7   A. I. Tauber, "A Case of Defense: Metchnikoff at the Pasteur Institute," in P. A. Cazenave and G. P. Talwar (eds.), L'Immunologie: L'Heritage de Pasteur (New Delhi: Wiley Eastern Limited, 1991), 21–36.

8   Tauber, Immune Self, 40ff.

9   Ibid., and A. I. Tauber, "The Birth of Immunology: III. The Fate of the Phagocytosis Theory," Cellular Immunology 139 (1992): 505–530.

10 Tauber, "Immunological Self"; also Tauber, "Introduction."

11 R. Levins and R. Lewontin, The Dialectical Biologist (Cambridge: Harvard University Press, 1985).

12 L. Chernyak and A. I. Tauber, "The Idea of Immunity: Metchnikoff's Metaphysics and Science," Journal of the History of Biology 23 (1990): 187–249; also Tauber and Chernyak, Metchnikoff and the Origins of Immunology.

13 A. S. Perelson (ed.), Theoretical Immunology: The Proceedings of the Theoretical Immunology Workshop Held June, 1987, in Sante Fe, New Mexico, 2 vols. (Redwood City, Calif.: Addison-Wesley, 1988).

14 F. J. Varela, A. Coutinho, B. Dupire, and N. N. Vaz, "Cognitive Networks: Immune, Neural, and Otherwise," in Perelson, Theoretical Immunology, 2:359–375.

15 Silverstein, History of Immunology.

16 Tauber, Immune Self. A. I. Tauber, "The Molecularization of Immunology," in S. Sarkar (ed.), The Philosophy and History of Molecular Biology: New Perspectives (Dordrecht: Kluwer Academic, 1996), 125–169.

17 E. S. Golub and D. R. Green, Immunology, a Synthesis, 2d ed. (Sunderland, Mass.: Sinauer, 1991); J. Klein, Immunology (Boston: Blackwell Scientific Publications, 1990).

18 Tauber, Immune Self; see also A. I. Tauber and S. H. Podolsky, "Frank Macfarlane Burnet

and the Declaration of the Immune Self," *Journal of the History of Biology* 27 (1994): 531–573.

19 Tauber, *Immune Self*.

20 Ibid., 114.

21 Tauber, *Immune Self*; Tauber, "Molecularization of Immunology"; A. I. Tauber, "The Elusive Self: A Case of Category Errors," *Perspectives in Biology and Medicine* 42 (1999): 459–474; Tauber, "Moving Beyond the Immune Self?," *Seminars in Immunology* 12 (2000): 241–248.

22 F. M. Burnet and F. Fenner, *The Production of Antibodies*, 2d ed. (Melbourne: Macmillan, 1949); see also Burnet, *The Clonal Selection Theory of Acquired Immunity* (Nashville: Vanderbilt University Press, 1959).

23 Varela et al., "Cognitive Networks."

24 A. Coutinho and M. Kazatchkine, "Autoimmunity Today," in Coutinho and M. Kazatchkine (eds.), *Autoimmunity: Physiology and Disease* (New York: Willey Liss, 1994), 3–6.

25 P. Matzinger, "Tolerance, Danger, and the Extended Family," *Annual Review of Immunology* 12 (1994): 991–1045.

26 N. K. Jerne, "Immunological Speculations," *Annual Review of Microbiology* 14 (1960): 341–358.

27 N. K. Jerne, "Antibody Formation and Immunological Memory," in J. Gaito (ed.), *Macromolecules and Behavior* (New York: Appleton Century Crofts, 1966), 151–157; see also N. K. Jerne, "Antibodies and Learning: Selection versus Instruction," in G. C. Quarton, T. Melnechuk, and F. O. Schmitt (eds.), *Neurosciences* (New York: The Rockefeller University Press, 1967), 200–205.

28 N. K. Jerne, "What Precedes Clonal Selection?," in *Ontogeny of Acquired Immunity: A Ciba Foundation Symposium*, new ser., no. 5 (Amsterdam: Elsevier, 1972), 1–15; N. K. Jerne, "The Immune System," *Scientific American* (July 1973): 52–60; Jerne, "Towards a Network Theory of the Immune System," *Annals of Institute Pasteur/Immunology* 125C (1974): 373–389; Jerne, "Idiotypic Networks and Other Preconceived Ideas," *Immunological Review* 79 (1984): 5–24; Jerne, "The Generative Grammar of the Immune System," EMBO [European Molecular Biology Organization] *Journal* 4 (1985): 847–852.

29 M. Cohn, "Conversations with Niels Kaj Jerne on Immune Regulation: Associate versus Network Recognition," *Cellular Immunology* 61 (1981): 425–436; also Cohn, "What Are the 'Must' Elements of Immune Responsiveness," in R. Guillemin, Cohn, and T. Melnichuk (eds.), *Neural Modulation of Immunity* (New York: Raven Press, 1985), 3–25; Cohn, "The Concept of Functional Idiotype Network for Immune Regulation Mocks All and Comforts None," *Annals of Institute Pasteur/Immunology* 137C (1986): 64–76; Cohn, "The Ground Rules Determining Any Solution to the Problem of the Self/Nonself Discrimination," in P. Matzinger, M. Flajnik, H.-G. Rammensee, G. Stockinger, T. Rolink and L. Nicklin (eds.), *The Tolerance Workshop* (Basle: Editiones, 1987), 3–35.

30 E. E. Sercarz, F. Celada, N. A. Mitchison, and T. Tada (eds.), *The Semiotics of Cellular Communication in the Immune System*, NATO ASI [Advanced Study Institute] Series, vol. H23 (Berlin: Springer, 1988).

31 R. Ader, D. L. Felten, and N. Cohen (eds.), *Psychoneuroimmunology*, 3d ed. (San Diego: Academic Press, 2001).

32 Perelson, *Theoretical Immunology*; I. R. Cohen and H. Atlan, "Network Regulation of Autoimmunity: An Automaton Model," *Journal of Autoimmunity* 2 (1989): 613–625; also F. Celada and P. E. Seiden, "A Computer Model of Cellular Interactions in the Immune System," *Immunology Today* 13 (1992): 56–62; J. Stewart, F. J. Varela, and A. Coutinho, "The Relationship between Connectivity and Tolerance as Revealed by Computer Simu-

lation of the Immune Network: Some Lessons for an Understanding of Autoimmunity,"
Journal of Autoimmunity 2(Suppl.) (1989): 15–23; F. J. Varela, A. Coutinho, and J. Stewart,
"What Is the Immune Network For?," in W. D. Stein and F. J. Verla (eds.), *Thinking About
Biology: An Invitation to Current Theoretical Biology*, Sante Fe Institute Studies in the Sci-
ences of Complexity, Lecture Notes, vol. 3 (Redwood City, Calif.: Addison-Wesley, 1993),
3:215–230.

33 I. R. Cohen, "The Cognitive Paradigm Challenges Clonal Selection," *Immunology Today*
13 (1992): 441–444.

34 B. Shanon, *The Representational and the Presentational: An Essay on Cognition and the Study of
Mind* (London: Harvester-Wheatsheat, 1993).

35 S. H. Podolsky and A. I. Tauber, *The Generation of Diversity: Clonal Selection Theory and the Rise
of Molecular Immunology* (Cambridge: Harvard University Press, 1997); also Tauber, "Historical and Philosophical Perspectives on Immune Cognition," *Journal of the History of
Biology* 30 (1997): 419–440.

36 Cohn, "Conversations with Niels Kaj Jerne"; Cohn, "What Are The 'Must' Elements";
Cohn, "Concept of Functional Idiotype Network"; Cohn, "Ground Rules."

37 Tauber, "Moving Beyond the Immune Self?"

38 Z. Grossman and W. E. Paul, "Adaptive Cellular Interactions in the Immune System: The
Tunable Activation Threshold and the Significance of Subthreshold Responses," *Proceedings of the National Academy of Science* 89 (1992): 10365–10369.

39 Z. Grossman, "Cellular Tolerance as a Dynamic State of the Adaptable Lymphocyte,"
Immunological Reviews 133 (1993), 47.

40 Matzinger, "Tolerance, Danger, and the Extended Family"; E. Fuchs, "Two Signal
Model of Lymphocyte Activation," *Immunology Today* 13 (1992): 462; Fuchs, "Reply from
Ephraim Fuchs," *Immunology Today* 145 (1993): 236–237.

41 Cohen, "Cognitive Paradigm"; also Varela et al., "Cognitive Networks"; Z. Grossman,
"Recognition of Self, Balance of Growth and Competition: Horizontal Networks Regulate Immune Responsiveness," *European Journal of Immunology* 12 (1982): 747–756; Grossman, "Cellular Tolerance"; Grossman and W. E. Paul, "Self-tolerance: Context Dependent Tuning of T Cell Antigen Recognition," *Seminars in Immunology* 12 (1992): 197–203;
C. C. Anderson and P. Matzinger, "Danger: The View from the Bottom of the Cliff,"
Seminars in Imunology 12 (2000): 231–238; T. E. Starzl and A. J. Demetris, "Transplantation Milestones Viewed with One- and Two-Way Paradigms of Tolerance," *Journal of the
American Medical Association* 173 (1995): 876–879.

42 G. Johnson, "Findings Pose Challenge to Immunology's Central Tenet," *New York Times*,
26 March 1996, C-1 and C-3. See T. Forsthuber, H. C. Yip, and P. V. Lehmann, "Induction
of $T_H 1$ and $T_H 2$ Immunity in Neonatal Mice," *Science* 271 (1996): 1728–1730; also J. P.
Ridge, E. J. Fuchs, and P. Matzinger, "Neonatal Tolerance Revisited: Turning on Newborn T Cells with Dendritic Cells," *Science* 271 (1996): 1723–1726; and M. Sarzotti, D. S.
Robbins, and P. M. Hoffman, "Induction of Protective CTL Responses in Newborn Mice
by a Murine Retrovirus," *Science* 271 (1996); 1726–1728.

43 E. Pennisi, "Teetering on the Brink of Danger," *Science* 271 (1996), 1665–1667.

44 Tauber, *Immune Self*; also A. I. Tauber, "Postmodernism and Immune Selfhood," *Science in
Context* 8 (1995): 579–607; Podolsky and Tauber, *Generation of Diversity*; Tauber, *Immune
Self*; Tauber, "Historical and Philosophical Perspectives."

45 Tauber, "Postmodernism and Immune Selfhood."

46 H. Atlan, *L'Organization Biologique et la Theorie de l'Information* (Paris: Hermann, 1992); also
S. A. Kaufmann, *The Origins of Order: Self-Organization and Selection in Evolution* (New York:
Oxford University Press, 1993); W. Krohn, G. Kuppers, and H. Nowotny, *Selforganization:*

Portrait of a Scientific Revolution (Dordrecht: Kluwer Scientific, 1990); A. R. Peacocke, *An Introduction to the Physical Chemistry of Biological Organization* (Oxford: The Clarendon Press, 1983); D. Pines (ed.), *Emerging Syntheses in Science* (Redwood City, Calif.: Addison-Wesley, 1988); F. E. Yates, *Self-Organizing Systems: The Emergence of Order* (New York: Plenum Press, 1987).

47 Tauber, "Postmodernism and Immune Selfhood," 588–589.

48 Ibid.

49 R. Langman, "Self-Nonself Discrimination Revisited," *Seminars in Immunology* 12 (2000): 159–344; also R. E. Vance, "A Copernican Revolution? Doubts about the Danger Theory," *Journal of Immunology* 165 (2000): 1725–1728.

8 Immunology and AIDS: growing explanations and developing instruments

Ilana Löwy

AIDS can be viewed as immunology's "demonstrative disease." This pathology can be simultaneously presented as the apotheosis and the apocalypse of the immune system because the apocalyptic results of the immune system's failure dramatically display its central importance.[1] Physicians who in 1981 observed patients suffering from severe infections, especially "opportunist infections," that is, infections induced by germs which are as a rule harmless in healthy individuals (such as *Pneumocystis carinii* pneumonia, buccal candidiasis, toxoplasmosis, or cryptosporidosis), rapidly made the link between pathological phenomena observed in previously healthy individuals (often young males), and those observed in immunosuppressed individuals (persons with severely dysfunctional immune mechanisms) such as children with inborn defects of the immune system, cancer patients suffering from hematological malignancies or treated with cortisone (a drug which blocks the activity of the immune system), and transplant patients who have undergone immunosuppressive therapy in order to prevent graft rejection. The first name of the new disease, GRID (gay-related immunodeficiency) stigmatized a sexual minority. The gay connection rapidly weakened, however, when the same symptoms were observed in other groups, leaving acquired immunodeficiency as the main characteristic of the mysterious syndrome. Accordingly, the new disease was named AIDS (Acquired ImmunoDeficiency Syndrome).

An infectious immunodeficiency

From the very beginning (1981) clinicians had shown that patients suffering from the new syndrome had an unusually low level of T4 (or CD4+) lymphocytes in their blood. (CD4+ lymphocytes are a subset of white blood cells which play a central role in numerous immune mechanisms; they are a part of a larger category of T-lymphocytes.)[2] The first hypotheses on the causes underlying the disappearance of these lymphocytes and the subsequent dysfunction of the immune mechanisms variously posited infectious, genetic, and toxic causes. Some explanations combined several elements—for example, "overload of the immune system" with "hereditary susceptibility" and "consumption of recreational drugs." The balance rapidly shifted, however, in favor of infectious hypotheses. In 1982 the new syndrome was described in hemophiliacs treated with clotting factors produced from pooled sera (including children with hemophilia), in recipients of blood transfusions, and in intravenous drug users. The epidemiological pattern of the new syndrome took a familiar pattern: that of hepatitis B, a disease which, in Western countries, is transmitted mainly through sexual contact—and for this reason frequently found among homosexuals who had a large number of sexual partners—and through blood and blood products, hence its frequent occurrence among hemophiliac patients, intravenous drug users, and occasional infections of transfused individuals and of health care professionals.

The similarity between epidemiological patterns of AIDS and hepatitis B led, in late 1982, to a shared conviction that the new syndrome was induced by a virus (probably belonging to the group of retroviruses) found in the blood and in other body fluids.[3] Candidate viruses were presented in 1983, and in 1985 a consensus was achieved that a previously unknown retrovirus was the etiology of AIDS.[4] The virus was finally named HIV or Human Immunodeficiency Virus—the first time an immunological function gave a name to an infectious disease and to its etiologic agent. The widespread agreement that AIDS is a viral disease was coupled with the agreement that the pathogenic effect of the putative "AIDS agent" was mediated through the destruction of a specific subpopulation of cells involved in immune reactions, the CD4+(T4) lymphocytes, an event which was invariably perceived as perturbing a complicated regulatory network.

AIDS and immunology: the elusive complexity

Immunology, its historians have pointed out, oscillated from its very early stages on, between explanations which stressed simplicity, complementary structure of chemical compounds, single-hit reactions, and war metaphors, and explanations which stressed complexity, multiplicity of interactions, and physiological images. To put it simply, while selected immunologists (and in some periods and places, the majority of immunologists) developed reductionist approaches, these approaches never completely dominated immunological investigations. Moreover, from the 1960s on, the focus of immunological investigations shifted toward cellular immunology, that is, toward the study of multilevel interactions among cells and molecules.[5]

Physicians traditionally used expressions such as *constitution* or *terrain* to explain the great variety of individual reactions to an encounter with a pathogenic microorganism. The insistence on the diversity of these reactions is aptly summed up in the expression (attributed to Claude Bernard) "le microbe n'est rien; le terrain est tout." Investigators such as Alexis Carrel and Claude Richet proposed circa 1910 to investigate the chemical and physiological basis of biological individuality.[6] In the early twentieth century, such investigations were not feasible, however, and most immunologists turned instead to studying specific antibodies in the serum and their uses for diagnosing and treating infectious diseases. The practical importance of the latter studies diminished with the development of antibiotics. In the post–World War II era, immunologists began to investigate immunopathological phenomena such as allergy and autoimmunity (diseases induced by dysfunction of the immune system) and graft rejection (mediated by immune mechanisms). These investigations led to a new understanding of immunity. According to this new view, immune mechanisms were responsible for distinguishing between self and nonself and reacting to nonself. Briefly, according to this view, immune mechanisms, previously considered capable of merely reacting to invading pathogens, acquired a much broader physiological significance as devices able to recognize the self (the unique chemical constitution of the individual) and to react to every nonself structure, be it a bacterium, a virus, a foreign protein, or even a synthetic molecule.[7] The task of distinguishing self from non-self and reacting to the latter is conducted by a complex network of cells and molecules, the *immune system*. The theoretical elaboration of the immune system concept and, in parallel, of the clonal theory of antibody formation in turn facilitated an impressive "proliferation of immunologists" (the expression is borrowed from the immunologist Niels Jerne).[8]

Network-based explanations and immunodeficiency

An approach which stressed regulatory circuits, complicated networks of cells and molecules, interactivity, and the great diversity of reactions was especially well adapted to describe complex pathological phenomena and to promote the translation of these phenomena into a new biological language. In the 1970s and 1980s, individual differences between patients, once presented as the variability of the terrain, could be adapted to biochemical, genetic, and molecular biological idioms, and presented, for example, as differences in major histocompatibility complex (MHC) antigens (a family of proteins on the surface of human cells which play an important role in self–nonself recognition), in the activity of intercellular mediations such as cytokins (molecules secreted by one subset of immuno-competent cells which stimulate another subset[s] of such cells), or in expression of specific receptors on cell surfaces. Immunological explanations stressed cascades of reactions, feedback loops, and complicated networks of cells and molecules. Cellular immunologists developed new techniques that allowed them to visualize and to trace antibodies by labeling them with radioisotope-marked or fluorescent stains and methods that made possible the study of interactions between immunocompetent cells in the test tube, such as mixed lymphocyte reactions. These methods were combined with new biochemical approaches (electrophoresis, chromatography), and later with molecular biological techniques. Together these new methods enabled immunologists to dissect the mechanisms of immune reactions, and they consolidated the conceptual framework of cellular immunology. AIDS was at first defined and conceptualized within this framework.

Clinicians who first described opportunistic infections and/or Kaposi's sarcoma in homosexuals correlated their findings solely with a sharp decrease in the T4 (later CD4+) subpopulation of lymphocytes. Immunologists, who rapidly became interested in the new syndrome, studied other parameters of the immune system and concluded that the abnormally low number of CD4+ lymphocytes was but one of numerous immune-mechanism anomalies found in AIDS. The Working Group on AIDS of the Clinical Immunology Committee of the International Union of Immunological Societies/World Health Organization, published in 1984 an exhaustive list of immunological abnormalities found in AIDS patients.[9] These abnormalities included a decrease in the global number of lymphocytes, a selective reduction in the T4 subset of lymphocytes, decreased or absent skin reactions to antigens, elevated levels of immunoglobulins (a fraction of serum proteins which include the antibodies), altered functions of other white blood cells such as monocytes, and increased spontaneous secretion of antibodies by individual cells. Numerous AIDS patients showed also a decrease in cellular responses to antigens and other stimulants, in the ability of immunocompetent cells to kill tumor cells, in cell-mediated immune responses, and in the ability to mount an antibody response to a new antigen, as well as an increase in the levels of molecules having a non-specific regulatory role in the immune system (such as acid-labile alpha-interferon, of beta-microglobulin and of alpha-1 thymosin). Occasionally these patients also produced antilymphocyte antibodies (antibodies able to destroy one's own white blood cells).

The data provided by the Working Group pointed to an unexpected multiplicity and diversity of defective immune mechanisms associated with AIDS. Immunologists explicitly correlated this proficiency of immune abnormalities with the complexity of immunological regulatory networks, and more specially, with the putative cascade effects perturbing the activity of one subset of cells active in immune reactions on other elements of the immune system. The viral hypothesis was prominent in the Working Group on AIDS's conclusions, but the study's authors stressed that a viral infection might have multiple effects on the immune system. In 1984, the interpretative situation was anything but clear:

> These viruses appear to have a positive tropism for activated T-lymphocytes. . . . In theory, the virus could destroy its target cell, or its precursor, or both, interfere with the growth or the cognitive function of its target cell, or interfere with the biosynthesis of molecules involved in its function. The etiologic virus could interact directly with B-lymphocytes and cause polyclonal activation, or it may be the consequence of the primary effect of the virus on T4 cells. In addition to affecting T-lymphocytes, the virus may also interfere with the functions of antibody-presenting cells. Alternatively, impaired antigen-presenting cell function due to down regulation of Ia antigens expression may be secondary to the T4 cell defect. Many of the immunological abnormalities observed in AIDS may be secondary to the host response, although some could be direct effects of the retrovirus infection.[10]

AIDS as an autoimmune disorder

Between 1983 and 1986, researchers consolidated the viral hypothesis of AIDS causation, isolated and cultivated the virus, decrypted its sequence, followed the variability of virus strains, and studied in detail viral antigens and antibodies against these antigens. The improved understanding of viral structure and antigenic variability did not lead, however, to a better understanding of the ways HIV infection leads to the disease AIDS. At first, scientists believed that, following the identification of the etiologic agent of AIDS, "we will rapidly learn much more about the relations of the virological features of this agent to the natural history and pathogenesis of AIDS."[11] This prediction was not fulfilled. The rapid progress in unraveling the structure of HIV was not matched by an equally rapid progress in understanding the physiological effects of HIV infection and the great diversity of its physiopathological and clinical manifestations.[12] In 1988 Jay Levy pointed to the persistence of "AIDS mysteries": lack of precise information about the mechanisms by which the virus infects susceptible cells (especially cells lacking CD4 receptors), insufficient explanation of the genetic variability of the virus, and, above all, lack of understanding of the mechanisms of HIV-induced pathology and of factors which influence progress to disease. One of the major difficulties, Levy explained, was that scientists found very low numbers of viral particles in the blood of HIV-infected individuals. It was not clear how such a low-level infection could induce important pathological effects (the "pre-AIDS' syndrome"), and why it changed into full blown AIDS.[13] Similarly, David Baltimore and Mark Feinberg explained in 1989 that "we are rapidly learning about the role of each of HIV's approximately 10.000 nucleotides, but remain largely ignorant of rudimentary aspects of the process underlying the development of AIDS in humans."[14]

In 1986 John Ziegler and Daniel Stites proposed that AIDS be considered an autoimmune disease (a disease in which immune mechanisms turn, by mistake, against the self and destroy components of the body).[15] The HIV infection, they argued, should be seen merely as a triggering stimulus, while the true pathology resulted from the body's exaggerated reaction to this stimulus.[16] They grounded their explanation in the network theory of immunoregulation according to which immune responses are kept in check by a complex regulatory network that has a "damping" effect on the system and limits its infinite activation. An error in a single site of this network can, however, lead to catastrophic effects because it can start a chain of uncontrollable events. Ziegler and Stites pointed to the fact that the CD4 molecule of a T4 (or CD4+) lymphocyte—the receptor which makes possible the entrance of the HIV virus into the cell—is a molecule which plays an important role in self–nonself recognition (it belongs to class II of MHC antigens). When the HIV attaches itself to this important regulatory molecule, they proposed, it "confounds" the body's immune mechanisms because it makes the very site responsible for recognizing the self (the CD4 molecule) into a locus recognizing the nonself. The virus can thus induce an attack of cytotoxic (cell-killing) cells on CD4+ lymphocytes, block the communication between these lymphocytes and other cells of the immune system, severely disturb the normal patterns of recognizing foreign antigens, induce the body to progressively deregulate numerous immune mechanisms, and provoke a dysfunctional immune response of catastrophic proportions.[17] The advantage of this hypothesis, its authors stressed, was its ability to explain the contradiction between the very low load of HIV in the blood and the extreme severity of the pathological phenomena observed in AIDS patients and to account for the array of immunological phenomena that accompany disease progression.[18]

The autoimmune hypothesis of AIDS's origin remained attractive but unproven until 1991, when two papers for the first time proposed experimental evidence to support it. In one study, mice injected with T cells from another mouse strain (but not with HIV) developed anti-HIV antibodies. These findings were explained in the framework of a hypothesis (similar to the one developed earlier by Ziegler and Stites) that the HIV virus induced antibodies which attack the immune system itself because the viral gp120 protein has structural similarities with the class II MHC proteins, which are central to self–nonself recognition.[19] A second study indicated that monkeys immunized with normal T cells (a control group, which was compared with monkeys immunized with T cells infected with SIV, a simian virus akin to HIV) were unexpectedly also protected from SIV infection, indicating that normal T cells and SIV share molecular structures. Together these results were presented by Nature's editor, John Maddox, as "AIDS research turned upside down" and as an indication that the solution to the AIDS problem might well come from immunologists. Possibly, Maddox explained, instead of hunting for an elusive anti-AIDS vaccine or an efficient and nontoxic antiretroviral compound, researchers should look for ways to make people tolerant to HIV proteins.[20]

Another immunological explanation of the pathogenesis of AIDS, also proposed in 1991, was the hypothesis of a "superantigen." Several research groups independently proposed that HIV envelope proteins are superantigens—that is, antigens that initially produce a massive stimulation of the immune system, so massive that it leads ultimately to the cell's dysfunction and death. In 1988 Asher and Sheppard proposed that the generalized lymphadenopathy (enlargement of lymph nodes) observed in HIV-infected individuals and the disparity between the dramatic effects of HIV infection and the low number of viral particles detected in the blood of infected individuals might be explained if one assumed that HIV induces a nonspecific hyperstimulation of the immune system. Such hyperstimulation might be mediated by HIV-infected macrophages (white blood cells involved in the uptake of antigens) which express viral components on their surface. The reaction of these macrophages with CD4-expressing T lymphocytes might result in the progressive loss of capacity of hyperstimulated CD4+ cells to react normally to antigenic stimuli and, in fine, to the disappearance of these cells.[21] The superantigen hypothesis adopted the concept of hyperstimulation and explained this hyperstimulation by the presence of superantigens.[22] Neither the autoimmune nor the superantigen hypotheses was formally disproved, but they were not confirmed either, and between 1992 and 1996 this approach was quietly shelved, perhaps awaiting a new resurrection.

The practical uses of immunology in AIDS

While the attempts to explain AIDS through immune theories remained inconclusive, scientists nevertheless attempted to apply the immunological explanatory framework to the therapy and diagnosis of AIDS. Let's look at therapy first. The definition of AIDS as a disorder of the immune system rapidly led to attempts to develop immunotherapies for AIDS. Between 1982 and 1984, these therapies included bone marrow transplantation, lymphocyte plasmapheresis (transfusion of concentrated white blood cells), and administration of molecules which modulate immune functions: recombinant alpha-interferon, natural product gamma-interferon, natural product and recombinant interleukin-2, and thymic factors. Alpha-interferon induced partial remissions of Kaposi's sarcoma and later of AIDS-associated lymphomas (this molecule is an accepted therapy for several hematological tumors). Other immunotherapies had no detectable therapeutic effects.[23] In the late 1980s, following the announced success of adaptive immunotherapy of cancer with interleukin-2 activated lymphocytes (the patient's lymphocytes were activated outside the body with interleukin-2, then reinjected into the bloodstream) researchers attempted to cure AIDS through the injection of purified and activated CD8 + lymphocytes (a subpopulation of T lymphocytes, involved in the suppression of immune reactions).[24] Ronald Herberman explained the rationale behind this therapy: "thus far, most of the treatment for patients with HIV infection or with AIDS has been focused on an antiviral approach. It is clear, however, that AIDS and HIV infection is primarily a disease affecting the immune system. . . . If one could select an effector cell population that would have reactivity against the HIV-infected cells and possibly also against opportunistic infections, this would be a good rationale for treatment."[25] Laboratory investigations indeed indicated that activated CD8+ lymphocytes have anti-HIV activity in the test tube.[26] The clinical results, however, were disappointing. The injected CD8+ cells, the researchers had shown, maintained their anti-HIV activity, but the clinical improvement obtained by this method was slight if any, and it did not justify the use of technically complicated and costly therapy. Attempts to develop immunotherapies of AIDS have continued, but they have often been presented as an effort to develop a complement to antiretroviral treatments, not as independent therapeutic approaches.

Immunological approaches played a more important role in the development of diagnostic tests for AIDS. One should distinguish here between diagnostic tests aiming at the detection of HIV infection (seropositive status) and those destined to monitor the clinical status of HIV-infected individuals. The latter include immunological tests which aim at evaluating the clinical status of a given patient, predicting the rate of progress from asymptomatic infection to full-blown AIDS, and monitoring the effects of therapies. The diagnosis of HIV infection based on the presence of specific anti-HIV antibodies in the serum was rapidly stabilized. This stabilization was facilitated by the use of techniques borrowed from virologists and molecular biologists to confirm a positive diagnosis. In contrast, the use of immunological parameters to follow the clinical evolution of HIV-infected individuals remained problematic.

In the years 1985 and 1986, several tests evaluating the immune mechanisms' level of activation, such as assessing a specific cellular response to HIV antigen, the production of gamma-interferon by antigens, the number of circulating lymphocytes recognized by CD8 and Leu7 monoclonal antibodies, lymphocyte proliferation following the stimulation with pokeweed mitogen (a nonspecific stimulus), and antilymphocyte antibodies in the serum, seemed to have prognostic significance in HIV infection. A *Lancet* editorial stated in 1986 that no one of these tests was sufficiently simple to perform and to interpret that it could become an efficient predictive test and then expressed the belief that the present rapid progress of understanding of immunological aspects of HIV infection would be followed "with astonishing speed" by the development of more adequate testing.[27] Four tests emerged in the mid-1980s as candidates for predicting progression to AIDS: assessing the CD4+ lymphocyte count, antibodies against p24 core antigen of HIV (a viral protein), and levels of two proteins which influence immune reactions, beta-microglobulin (in the serum), and neopterin (in the serum and in urine).[28] In addition, researchers examined titers of antibodies against other HIV proteins and continued to follow the general decrease in cellular immune functions.[29] Measuring levels of CD4+ lymphocytes in the blood (the first immunological defect described in AIDS), although an expensive and relatively complicated technique, emerged as the major test used to evaluate the clinical status of HIV positive individuals. This development paralleled the distribution of an instrument—the FACS (flourescein-activated cell sorter)—which automated the measurement of lymphocyte subpopulation levels in the blood. The official redefinition of AIDS patients as HIV infected individuals who have less than 200 CD4+ lymphocytes per ml further impelled the widespread use of the FACS. It stimulated efforts to increase the instruments' reliability and to standardize results obtained in different settings.[30] The quantification of CD4+ lymphocytes indeed became more reproducible and more reliable, but the exact meaning of the results obtained using better instruments and techniques remained unclear and was not improved by wider diffusion of these techniques. Just the opposite is true: the accumulation of data on the levels of CD4+ lymphocytes in patients and attempts to correlate CD4+ lymphocyte data with other immunological parameters partly undermined the meaning attributed to specific results.[31]

One of the problems with using immunological parameters as "surrogate markers" for progression to AIDS (biological measures which can be substituted for the observation of clinical developments) was the physiological variability of these markers. Thus the CD4+ lymphocyte count was found to be influenced by the effects of exercise, tobacco, and the consumption of alcohol and caffeine.[32] Moreover, their use became more problematic following the introduction of AZT into AIDS therapy. The use of AZT, the first chemical compound found to be capable of affecting HIV infection in humans, increased the need to monitor patients' therapeutic progress. The quantification of CD4+ lymphocytes became more frequent, but at the same time agreement on the predictive value of the CD4+ lymphocyte count decreased. To be more precise, nearly everybody agreed that an important fall in the level of CD4+ lymphocytes indicated a degradation of an HIV-infected individual's clinical status, but smaller variations were of lesser—if any—predictive value. The association between the levels of CD4+ lymphocytes and progression to AIDS was especially weak during the first four months of AZT therapy—thus precisely during the period in which it was crucial to find out if the treatment was beneficial or if the patient was being unnecessarily submitted to toxic (and costly) treatment.[33] Between 1985 and 1995 AIDS continued to be defined as "acquired immunodeficiency," but the complexity of the immunological reactions in HIV-infected individuals, coupled with the multiplicity of causes influencing these reactions, hampered the development of reproducible and easy to interpret tests which measured this immunodeficiency. In the mid-1990s, monitoring HIV-infected persons gradually shifted (at least in major hospitals and research centers) to measuring viral load—that is, on a test which evaluates not immunodeficiency, but infection. This shift was made possible by adapting PCR (polymerase chain reaction) techniques to measuring retroviral load in infected individuals. Beyond their use in monitoring patients, these techniques dramatically changed the perception of HIV infection and modified therapeutic approaches to it.

AIDS and virology: technical developments and industrial strategies

From 1983 on, scientists affirmed that the etiological agent of AIDS is a retrovirus. The long latency period between initial infection with HIV (primoinfection) and development of full-blown AIDS, the fact that a retrovirus may be concealed as an inactive provirus in the cell's DNA, and the observation that HIV-infected individuals have a low concentration of viral particles in their blood favored representing HIV as a hidden virus which, after a short period of rapid multiplication during the primoinfection, remains for many years concealed within receptive cells (CD4+ lymphocytes, macrophages) and then, for unknown reasons, is "awakened" and induces a full-blown immunodeficiency. This view of the natural history of AIDS led to the "network amplification" hypothesis, which attributed the pathological effects of HIV infection to a cascade of events initiated by the retrovirus but later independent of viral presence.[34]

The data on the scarcity of viral particles in the blood of HIV-infected individuals came from investigations made with traditional virological techniques such as limited dilution tests and with the first generation PCR tests. In 1989 David Ho and his collaborators used a newer and more sensitive PCR test to quantify the number of retroviral particles in the blood. They found that in asymptomatic, HIV-positive patients, one in about 20,000 peripheral blood cells was infected with HIV, while in patients with full-blown AIDS, about one in 500 peripheral white blood cells contained the virus—data which were a two to three log increase over previous estimates.[35] The new results, Baltimore and Feinberg explained, "should dispel any lingering doubts about whether HIV is the true culprit in AIDS."[36]

From latent virus to occult proliferation

Investigations made in the early 1990s using these more sensitive PCR methods indicated that one of the virus's hiding places during the (clinically) latent phase of HIV infection was the infected individuals' lymph nodes. Specific categories of cells within the lymph nodes, such as the follicular dendritic cells, served as a reservoir for the virus. This finding was not surprising. The enlargement of lymph nodes (lymphadenopathy) was described in "pre-AIDS" patients even before the description of the virus and, in fact, HIV (then called LAV) was first isolated from the enlarged lymph nodes of a pre-AIDS patient. What was new was the finding that patients in the latent phase of AIDS carried many more viral particles than was previously believed.[37] The development of in situ PCR techniques led in 1993 to the finding that during the latent phase the virus was actively and rapidly proliferating in the lymph nodes. The "silent period" of the disease was, the researchers argued at that point, silent only when the viral activity was measured in the (readily accessible) peripheral blood, but it was proliferating rapidly in the lymphatic tissue. HIV infection was "semi-latent" at best.[38] This finding was analyzed in the framework of the putative complexity of immunological events following HIV infection. The ongoing proliferation of HIV in lymph nodes was related to the dysfunction of CD4+ lymphocytes, the histological phenomenon of syncythia (multinuclear cell) formation in infected lymph nodes, autoimmune mechanisms, the loss of immune reactivity in immunocompetent cells, the presence of superantigens, and programmed cell death (apoptosis).[39] Researchers interested in the role of lymph nodes in HIV infection stressed that "the pathogenic mechanisms of HIV disease are multifactorial and multiphasic," and expressed the hope that detailed studies of cellular events in the lymphoid tissue during the early stages of the HIV infection might give clues to the design of an efficient therapy.[40]

From semilatency to "total war"

In 1993, the persistent replication of HIV in the lymph nodes was presented as the most important feature of HIV infection. Viral replication "generates conditions that promote its continued growth. In other words, HIV eventually becomes its own opportunistic infection. . . . The emerging view is . . . that a progressive HIV burden involving first activation and eventual destruction of the immune system is what lies behind AIDS."[41] These findings did not put an end, however, to inquiries concerning the mechanisms of HIV-induced pathology. The presence of HIV in one in 500 cells (in advanced stages of HIV infection), even less its presence in one in 50,000 cells (in an asymptomatic infection), could not account by itself for the massive destruction of the immune system in AIDS patients, while the proposal that the virus induces a chain of events within the immune system, while plausible, lacked an experimental basis. A 1993 survey on the status of AIDS research organized by *Science* and published under the heading "AIDS Research: The Mood Is Uncertain," named as the most important question still to be solved, "What causes the immune system to collapse in AIDS?"[42]

Two years later, the mood had shifted from uncertainty to a new perception of HIV-infection as a frontal and global attack by the virus. In 1995 scientists described the extreme rapidity of viral turnover during the infection's supposedly "silent" stage. The new understanding of HIV infection's dynamics was made possible because improvements in the PCR technique (branched DNA signal-amplification technique, in situ ultrasensitive quantitative PCR) increased its ability to produce (trustworthy) measures of viral replication. Applying these techniques to studies of the dynamics of HIV infection reversed HIV's original image as a "lentvirus" which induces a "latent infection." HIV, the new data indicated, was anything but slow.[43] As Ho and his collaborators stressed, "AIDS is primarily a consequence of the continuous, high-level replication of HIV-1, leading to virus- and immune-mediated killing of CD4+ lymphocytes."[44] The rapid turnover of HIV could also explain why indirect methods such as endpoint dilution detected low amounts of this virus in the blood. At any given moment the quantity of the virus in the blood is indeed low because viral particles released to the bloodstream are immediately eliminated by immune mechanisms—especially in asymptomatic individuals without major impairment of these mechanisms. This finding could not, however, be interpreted as pointing to a low level of viral activity.

Immunologists rapidly integrated the new vision into data on the dynamics of HIV-induced transformations in lymph nodes. The interest in complicated cellular interactions was replaced by investigations—using PCR in situ—of the rate at which specific cellular populations in the blood and in the lymph nodes were infected: T-cells, macrophages, and dendritic cells. In parallel, researchers compared data on viral load (seen as indicating the true status of infection) with measurements of the number of CD4+ cells and with lymph node biopsies and concluded that the level of CD4+ cells in the blood is a less precise indicator of the infection's evolution than morphological changes in lymph nodes, which were found to be relatively well-correlated with PCR data.[45] The shift from an interest in cellular regulations of anti-HIV response to the emphasis on the dynamics of cell destruction by the retrovirus illustrates the changes brought by new findings on rapid multiplication of HIV.[46] In 1995, Simon Wain-Hobson likened this infection to a total war: "that billions of virions and infected cells can be destroyed every day vividly illustrates the very hostile environment created by the immune system—the meanest of streets are nothing in comparison."[47]

That year, the transition occurred from perceiving AIDS as (above all) a complicated multilevel "immunodeficiency" to perceiving it as "global war" between rapidly multiplying viruses and rapidly multiplying CD4+ lymphocytes. In 1993 and 1994 investigators from leading groups in AIDS research (such as the teams directed by David Ho and Anthony Fauci) attempted to correlate events in the lymphoid tissue with new data on viral proliferation. Researchers in these groups studied the role of cellular immune responses in the early stages of HIV infection, the relations between cellular dynamics in the lymph node and patterns of retroviral multiplication.[48] Two years later, the same groups were involved in a very different enterprise, investigating molecules and receptors on lymphoid cells which may convey resistance to infection with HIV. The focus had shifted from studying complicated cellular interactions to investigating molecules (coreceptors) which control the HIV's entry into a cell and thus from interest in complicated networks of cells to interest in specific molecules. This shift did not abolish all the differences in the ways HIV infection was perceived. David Ho and Giuseppe Pantaleo (Anthony Fauci's collaborator) presented contrasting views of HIV infection's pathogenicity at the Eleventh International Conference on AIDS (1996). Ho insisted on the predominant role of viral factors (viral kinetics, rate of viral mutation) in the development of AIDS, while Pantaleo stressed the role of host factors which modulate the organism's response to HIV.[49] These host factors were not presented, however, in terms of complicated and unpredictable networks of cells but in terms of single-hit interactions between well-defined molecules.

Redefinition of the host's response to AIDS in terms of receptors which make possible the entrance of HIV into cells and molecules which interact with these receptors emerged in a long series of coordinated studies published in 1995 and 1996. Briefly, scientists observed that a category of immunocompetent cells, the CD8+ subset of T lymphocytes, slow down HIV infection. They had isolated HIV suppressive factors secreted by these lymphocytes—chemokines (regulatory molecules)—and then found that these molecules block the entry into cells of macrophage-tropic HIV strains (a virus which selectively infects macrophages) but not T-cell tropic (a virus which selectively infects T lymphocytes).[50] At about the same time, researchers who had studied the mechanism of entry of HIV into cells proposed that a coreceptor, named "fusin," is necessary (in addition to the previously known HIV receptor, the CD4 molecule) for the infection of T cells with T-cell tropic strains.[51] Experts then concluded by analogy that a similar coreceptor should be found in macrophages too, and proposed that the newly described inhibitory molecules may react with this coreceptor. Both suppositions were confirmed. No less than five large groups (a telling testimony to the density of interactions in this area and to perception of the subject as an especially "hot" topic) published articles almost simultaneously that identified a macrophage-specific coreceptor of HIV virus, named CKR-5, and showed that this coreceptor reacted with inhibitory chemokines.[52] In parallel, investigators studied individuals who, despite multiple exposures to HIV virus through sexual contacts, remained infection free, and displayed a correlation between the presence of mutant CKR-5 and resistance to HIV infection.[53]

Viral dynamics and therapeutic approaches

Shifting the representation of HIV infection from a latent infection with a hidden provirus to an active and continuous viral attack on the immune system had direct implications for therapeutic strategies. If the HIV infection was a pitiless and dramatic war between rapidly multiplying viruses and millions of white blood cells, winning this war would require a concentrated attack on the rapidly proliferating and mutating virus: "monotherapies could not succeed. Only a combination of drugs have the potential to outgun the virus."[54] When scientists believed that clinical latency was synonymous with low viral burden and microbiological latency, they hesitated about starting an aggressive treatment with antiretroviral agents.

The pharmaceutical industry had an obvious interest in AIDS. At the early stages of the epidemic, AIDS patients had sought "parallel" or natural drugs, while the industry invested in developing therapies for "opportunistic infections" in AIDS patients. The marketing of AZT (zidovudine) by the Burroughs-Wellcome Company (in 1988) was a turning point in AIDS therapy. For the first time, a chemical compound was found to be effective against the HIV virus, and it helped to improve the clinical status of many patients. The AIDS movement, previously opposed to the medical establishment and major drugs companies, started an (often uneasy) collaboration with these companies and with drug testing agencies, while pharmaceutical firms launched an intensive search for other antiretroviral compounds.[55] AZT and other molecules (ddI developed by Bristol-Myers, ddC developed by Hoffman Roche, 3TC developed by Glaxo) which inhibited retroviral multiplication were, however, toxic and often could not be tolerated for a long time. Some experts proposed that administering a combination of drugs could increase their efficacy, but other specialists argued that it could also enhance toxicities, leaving the physician without a possibility of further therapeutic intervention. Many doctors preferred not to use these therapies too early and to reserve antitretroviral drugs for an acute stage of AIDS when the patients had "nothing to lose." Moreover, clinicians feared that treating patients early would lead the virus to mutate into drug-resistant strains and compromise the chances of later therapeutic intervention.[56] When more sensitive molecular techniques indicated that a relatively high load of HIV is present in the lymphoid tissue during all stages of infection, they strengthened the rationale for early and vigorous treatment.[57] In 1993, however, Anthony Fauci warned that the advantages of such early treatment had to be weighed against the possible disadvantages of disrupting a delicate physiological equilibrium.

> Any comprehensive therapeutic strategy must consider the complexities of the pathogenic mechanisms of HIV disease. . . . Certain stages of the disease may benefit more than others from an intensive regimen of antiretroviral drugs. . . . Certain types of intervention may be appropriate at one stage of the disease and contraindicated at another. . . . It is only through the process of carefully conducted clinical trials based on the expanding knowledge of HIV pathogenesis that answers to these important questions will be forthcoming.[58]

Two years later, researchers were no longer arguing that the subtle dynamics of HIV infection had to be investigated before efficient anti-HIV therapies could be developed. In 1995 David Ho, for example, stated that the only efficient way to deal with an HIV infection is to hit HIV early and hard.[59] The new generation of PCR tests, which could more precisely measure viral load, helped doctors to rapidly estimate the efficacy of drug combinations in the organism. These new techniques demonstrated an impressive drop in the virus load—an average of 99 percent and in some cases a 99.99 percent drop—after therapy was initiated.[60] Such a rapid drop was important because, the experts argued, the success of a given antiretroviral therapy depended to a large extent on its capacity to act faster than the virus: "current protocols for monitoring the acute antiviral activity of novel compounds should be modified to focus on the first days following drug initiation."[61]

The latter proposal was attractive for pharmaceutical companies which produce anti-HIV drugs. It was also attractive to AIDS patients who wished to shorten the time for testing promising therapies. In the late 1980s researchers first hoped that the clinical trials of anti-HIV therapies would improve patients' survival rates. Later, they could shorten clinical trials of antiretroviral compounds by introducing "biological markers" (or "surrogate markers") to indicate the efficacy of a given therapy. Several months are necessary, however, to document changes in "biological markers" such as the number of CD4+ lymphocytes. By contrast, changes in virus load measured with PCR can be observed after only a few days, dramatically reducing the length (and the cost) of clinical assays and thus the expenses linked with developing new therapeutic compounds. For some experts, immunosuppression, which gave its name both to the disease and to its etiological agent, became a secondary issue, and they sought to use AIDS therapy to eliminate HIV. As Joep Lange from the World Health Organization explained, "The virus is the real thing. Clinical end points are the surrogates."[62] Some AIDS activists agreed. Thus Xavier Rey-Coquais, from the French group Actions–Traitement, explained that "the aim of a clinical trial should not be to verify if a given combination of drugs increases the level of CD4+ lymphocytes or diminishes the frequency of opportunistic infections, but to check if it decreases the viral load and maintains it at the lowest possible level."[63]

In January 1996 came the remarkable announcement that a combination of three anti-HIV drugs (two nucleotide analogues and an antiprotease), tested for several weeks in HIV-infected individuals, eliminated the virus from their blood. This led to a wave of enthusiasm for the new cure and to its availability to important groups of AIDS patients—those who could afford it or who were covered by an insurance system (national or private) willing to pay for it.[64] The early clinical findings were rapidly confirmed. New drugs, especially combinations of antiproteases with other antiretroviral molecules such as the newly tested T4D, eliminated HIV from blood circulation.[65] The slogan of the Eleventh International Conference on AIDS, held in Vancouver in July 1996, was "one world, one hope." (French observers, socialized in a more skeptical culture, added a question mark after this slogan.) It was centered on the promise of "victory on AIDS" brought about by the new therapeutic approaches. This promise was indeed fulfilled when the number of AIDS-related deaths in Western countries dropped sharply in 1997.[66] New approaches to AIDS therapy developed from the progress of methods for quantifying viral particles in the blood and from finding that the new tests are much better surrogate markers for the clinical status of the HIV-infected individuals than studies of CD4+ lymphocytes.[67]

Tri- and quadritherapies were indeed a true breakthrough in AIDS history in Western countries. The organizers of the Vancouver conference were nevertheless overoptimistic when they equated the disappearance of a pathogenic element from the body (or rather its disappearance from the clinicians' sight) with a cure. The slogan carried by New York city buses in late 1997, "AIDS = 0.000% cure," remained valid for the next five years. Between 1995 and 2002 the standard treatment of Western AIDS patients, highly active antiretroviral therapy (HAART) allowed long-term survival with HIV infection but did not completely eliminate the virus from the body. AIDS remains a chronic disease, complicated by development of drug-resistant HIV strains and by the high toxicity and bothersome secondary effects of HAART.[68] The inability of antiretroviral drugs to eradicate the reservoir of HIV within CD4+ T lymphocytes led to renewed interest in immunotherapy of HIV infection.[69] Numerous clinical trials were set to study the effects of interleukin-2 (IL-2) and other immunomodulators on AIDS.[70] These clinical trials have been framed in radically different ways, however, from similar attempts in the 1980s and early 1990s. From the mid-1990s on, AIDS was redefined as a cascade of pathological events following retroviral multiplication. Accordingly, manipulation of the immune system is seen as an adjuvant therapy that may allow doctors to reduce the doses of toxic antiretroviral compounds or to temporarily interrupt HAART.[71] In the 1980s AIDS was perceived as a complex, multilevel deregulation of the immune system triggered by a viral infection; in the early twenty-first century AIDS is seen as a viral infection which can be (perhaps) attenuated by stimulating the immune system.[72]

Representing lymphocytes and inhibiting viruses

Immunological explanations of AIDS, which defined this disease and shaped its image in the 1980s, were grounded in the concept of the "immune system" developed in the late 1950s. This concept shaped immunological thinking from the 1960s on. The complexity of immune interactions and regulation through networks had, however, been postulated much earlier. Between 1898 and 1904 leading experts on immune reactions (such as Ehrlich, Morgenroth, Metchnikoff, and Bordet) studied "anti-antibodies." Metchnikoff's collaborator Alexandre Besredka affirmed that anti-antibodies neutralize harmful autoantibodies (antibodies against self-components). In parallel, Ehrlich and Morgenroth proposed that auto-antibodies are neutralized when they are combined with the self-antigen (a "lateral chain" liberated into the bloodstream) against which they are directed. This self-antigen acts therefore at the same time as an anti-auto-antibody (a logical extension of Ehrlich's lateral chains theory).[73] While these early proposals did not achieve the degree of complexity of immune theories developed in the 1970s and 80s and were not linked with the concept of self, they do exhibit some of the characteristics of later immu-nological theories, such as the presence of complicated regulatory chains postulated in Jerne's idiotypic network theory (which proposes, among other things, that immune responses are regulated by anti-antibodies, anti-anti-antibodies, etc.).[74] Perhaps the very nature of pathological investiga-tions, especially those which involve interactions between two distinct organisms, the parasite and the host, may favor "growing explanations." A parallel may thus exist between the unpredictable growth of organisms within organisms—or, to use Ludwik Fleck's expression "a complicated revolution within the complex life unit"[75]—and the nature of explanations stimulated by this growth.

In the mid-1990s, explanations of AIDS stressing complexity were, how-ever, increasingly replaced by explanations viewing the multiple physio-logical effects of HIV-infection as secondary effects of a single event: the "struggle" between retroviruses and immunocompetent cells. Adapting the PCR technique to studies of the dynamics of viral multiplication, to inves-tigation of virus in tissues, and to quantitative evaluation of viral load in the blood, together with the practical pressure to develop antiretroviral drugs, changed the perception of AIDS from a "hidden infection" that triggered a cascade of multilevel effects on immune mechanisms, to an army of rapidly multiplying viruses that directly confronts an army of CD4+ lymphocytes. New developments in studies of "resistance to HIV infection" (translated into molecular terms), along with the new perception of this infection's natural history (expressed as a change in viral load measured by PCR), may point to an intensifying "technological drive" in AIDS studies.

Since the mid-1990s, a new image of AIDS resulting from this "techno-logical drive" has become dominant in the scientific community of retro-virologists and has spread throughout the organized community of AIDS activists and the media. One should not, however, confuse highly visible research trends (which tend to change rapidly with changing fashions) with the daily practices of scientists and physicians who study HIV infection and who monitor HIV-infected individuals. Those practices often represent long-term commitments to experimental systems, a disciplinary matrix, and long-standing research interests, and they are therefore less prone to rapid changes. The new perception of AIDS as a "total war" between retroviruses and CD4+ lymphocytes has not explained all of the pathological phenomena observed in AIDS patients, and physicians and scientists remain interested in the chain effects caused by the destruction of CD4+ lymphocytes on other immunocompetent cells. Similarly, despite the introduction of new techniques to evaluate viral load and the claim that viral load more accurately predicts the progression to full-blown AIDS and better indicates the therapeutic value of antiretroviral drugs, doctors still use CD4+ lymphocytes as a major indicator of HIV-infected individuals' clinical status.

The finding that retroviral load is a better indicator of a given patient's clinical status than the patient's level of CD4+ lymphocytes, is not sufficient by itself to eliminate the use of tests measuring CD4+ cells to determine clinical status in HIV infection. The choice between routine testing based on using FACS and lymphocyte counts, and routine testing based on quantifying viral load, will probably depend, in fine, on the capacity of manufacturers to develop relatively cheap and reliable tests (e.g., kits for the determining viral load), to establish satisfactory quality control of these tests, and to integrate them harmoniously into the complex division of medical labor. In complicated enterprises such as modern biomedicine, Andrew Abbott explains, the "organizational efficacy" of an action (i.e., its ability to articulate domains of activity), may be more important than its "technical efficacy" (i.e., its ability to provide specific results).[76] In 1997, the organizational efficacy of the quantitative PCR (its ability to articulate the interests of virologists, clinicians, industrialists, and AIDS patients) seemed to be higher than that of CD4+ lymphocyte counts. Other elements, such as cost, facility of use, and reproducibility, may influence the fate of tests based on the quantification of viral particles in the blood and may still shift the balance in favor of CD4+ lymphocyte-based tests in routine evaluations, perhaps with the occasional use of viral-load measures to confirm a trend or to clarify an uncertain situation.

To sum up, the relative success since the mid-1990s of the image of AIDS as "total war" reflects the central role of new techniques (such as the development of quantitative PCR and of methods to visualize viruses in situ), the importance of practical considerations (such as the need to find efficient "surrogate markers" for the progress of HIV), and the influence of the pharmaceutical industry (interested in rapid clinical testing of new molecules in order to decrease the cost of developing new drugs). In addition, however, its success mirrors a well established bacteriological and virological tradition. In the second half of the nineteenth century, the shift from the search for sufficient causes of diseases (which define the conditions leading to the appearance of a given pathology, or "if A then B") to the search for necessary causes of diseases (the conditions without which an infectious disease cannot appear, or "if not A, not B") led to the recognition that microorganisms were the etiologic agents of infectious diseases.[77] Thus physicians uncovered numerous conditions (poverty, poor lodging conditions, lack of sunlight, insufficient nutrition, "constitution") that were closely correlated with tuberculosis, but bacteriologists showed that in the absence of the Koch bacillus, the worst living conditions did not induce, by themselves, the disease's characteristic pathological lesions. AIDS, described as "an epidemic of signification," perhaps epitomizes postmodernity and reveals our complicated relationships with nature and with new technology, but it is above all an infectious disease.[78] If pathology favors studies of complexity and "growing explanations" (a tendency linked with a different, and equally well established bacteriological and virological tradition, the one which investigates the multiple levels of complicated host–parasite relationships), then therapy is above all the search for efficient solutions. And in our medical tradition, the most efficient way to cure an infectious disease is still to eliminate or control its "necessary cause," the causative microorganism.

A final remark. New methods such as quantitative PCR favored reductionist approaches and redefined AIDS as a struggle between retroviruses and white blood cells. This was not, however, the only consequence of introducing these techniques. Methods which in the early 1990s led to a simplified view of AIDS as the battle between retroviruses and lymphocytes, were later employed to investigate the details of host–parasite interaction.[79] New molecules and new techniques have thus had a double effect. They favored the standardization of diagnostic and prognostic tests (such as measuring viral load) and at the same time led to an increasingly fine-grained understanding of the dynamics of HIV infection.[80] The result is the emergence of two images of AIDS: the investigative image, which maps a mosaic of differences, and a pragmatic image, which defines zones of uniformity that guide therapeutic interventions.

Notes

1   Anne Marie Moulin, Le dernier langage de la médecine: L'histoire de l'immunologie de Pasteur au Sida (Paris: Presses Universitaires France, 1991), 423.
2   M. S. Gotlieb, R. Schroff, H. M. Shanker et al., "Pneumocystis carinii Pneumonia and Mucosal Candidiasis in Previously Healthy Homosexual Men," New England Journal of Medicine 305 (1981): 1425–1431; Henry Masur, Marry-Ann Michelis, Jeffrey B. Greene et

al., "An Outbreak of Community-Acquired *Pneumocystis carinii* Pneumonia: Initial Manifestation of Cellular Immune Dysfunction," *New England Journal of Medicine* 305 (1981): 1431–1438; David T. Durack, "Opportunistic Infections and Kaposis's Sarcoma in Homosexual Men," *New England Journal of Medicine* 305 (1981): 1465–1467; F. P. Siegal, C. Lopez, G. S. Hammer et al., "Severe Acquired Immunodeficiency in Male Homosexuals, Manifested by Chronic Perianal Ulcerative Herpex Simplex Lesions," *New England Journal of Medicine* 305 (1981): 1439–1444.

3   Jean L. Marx, "New Disease Baffles Medical Community," *Science* 217 (1982): 618–620.

4   "The Chronology of AIDS Research," *Nature* 326 (1987): 43–45.

5   Arthur Silverstein, *A History of Immunology* (San Diego: Academic Press, 1988); Moulin, *Le dernier langage de la médecine*; Ilana Löwy, "The Immunological Construction of the Self," in Alfred Tauber (ed.), *Organism and the Origins of Self* (Dordrecht: Kluwer Academic, 1991), 43–75.

6   Alexis Carrel, "Remote Results of the Transplantation of the Kidneys and the Spleen," *Journal of Experimental Medicine* 12 (1910): 12, 146–150; also Charles Richet, *L'Anaphylaxie* (Paris: Felix Alcan, 1911), 250–251.

7   Löwy, "Immunological Construction of the Self."

8   Niels Jerne, "Summary: Waiting for the End," *Cold Spring Harbour Symposia on Quantitative Biology* (1967): 601.

9   Maxime Seligman, Leonard Chess, John L. Fahey et al., "AIDS—An Immunological Reevaluation," *New England Journal of Medicine* 311 (1984): 1286–1292.

10 Seligman, Chess, Fahey et al., "AIDS—An Immunological Reevaluation," 1289.

11 Thomas C. Marigan, 'What Are We Going to Do about AIDS and HTLV-III/LAV Infection?," *New England Journal of Medicine* 311 (1984): 1311–1313.

12 P. Volberding, "AIDS—Variations on a Theme of Cellular Immune Deficiency," *Bulletin de l'Institut Pasteur* 85 (1987): 87–94.

13 Jay A. Levy, "Mysteries of AIDS: Challenges for Therapy and Prevention," *Nature* 333 (1988): 519–522.

14 David Baltimore and Mark B. Feinberg, "HIV Revealed: Towards the Natural History of the Infection," *New England Journal of Medicine* 321 (1989): 1673–1675.

15 Isolated manifestations of autoimmunity were found in AIDS patients; see Raphael B. Stricker, Donald I. Abrams, Laurence Corash and Marc A Shuman, "Target Platelet Antigen in Homosexual Men with Immune Thrombocytopenia," *New England Journal of Medicine* 313 (1985): 1375–1380.

16 Related explanations were advanced to explain why a person infected with a "flu" virus (influenza virus, rhinovirus) feels febrile, aching, tired, and miserable: the symptoms of the disease are not induced by the virus itself, but by molecules (lymphokines) secreted by the body as an often exaggerated reaction to the multiplication of this virus.

17 John L. Ziegler and Daniel P. Stites, "Hypothesis: AIDS Is an Autoimmune Disease Directed at the Immune System and Triggered by a Lymphotropic Retrovirus," *Clinical Immunology and Immunopathology* 41 (1986): 305–313.

18 These phenomena include hypergammaglobulinemia, allergic hypersensitivity, autoimmune manifestations, production of acid-labile interferon and suppressor substances, and dysfunction of monocytes and natural killer cells (two important classes of immunocompetent cells); see Ziegler and Stites, "Hypothesis: AIDS Is an Autoimmune Disease," 311.

19 Geofrey W. Hoffman, Tracy A. Kion, and Michal D. Grant, "An Idiotypic Network Model of AIDS Immunopathogenesis," *Proceeding of the National Academy of Sciences USA* 88 (1991): 3060–3064.

20 John Maddox, "AIDS Research Turned Upside Down," *Nature* 353 (1991): 297.

21 Michael S. Asher and Haynes W. Sheppard, "AIDS as Immune System Activation: A Model for Pathogenesis," *Clinical and Experimental Immunology* 73 (1988): 165–167; see also Sheppard, Asher, Brian McRaye et al., "The Initial Response to HIV and Immune System Activation Determine the Outcome of HIV Disease," *Journal of AIDS* 4 (1991): 704–712.

22 Jean Marx, "Clue Found to T Cell Loss in AIDS: A 'Superantigen' Encoded by the AIDS Virus May Cause Progressive Immune Cell Depletion that Leads to the Collapse of Patients' Immune System," *Science* 254 (1991): 798–800.

23 Seligman, Chess, Fahey et al., "AIDS—An Immunological Reevaluation," 1290.

24 Ronald B. Herberman, "Adoptive Therapy with Purified CD8(+) Cells in HIV Infection," *Seminars in Hemotherapy*, 29 (1992), 35–40; also Nancy G. Klimas, "Clinical Impact of Adoptive Therapy with Purified CD8(+)Cells in HIV Infection," *Seminars in Hemotherapy*, 29 (1992), 40–44.

25 Herberman, "Adoptive Therapy," 35.

26 C. Walker, D. Moody, D. P. Stites et al., "CD8(+) Lymphocytes Can Control HIV Infection in Vitro by Suppression of Virus Replication," *Science* 234 (1986): 1563–1566.

27 Editorial, "Who Will Get AIDS," *The Lancet*, 2 (1986), 953.

28 A.R. Moss, "Prediction: Who Will Progress to AIDS," *British Medical Journal* 297 (1988): 1067–1069.

29 B. Frank Polk, Robin Fox, Ron Brokmeyer et al., "Predictors of the Acquired Immunodeficiency Syndrome Developing in a Cohort of Seropositive Homosexual Men," *New England Journal of Medicine* 316 (1987): 61–66; also F. De Wolf, Joep M. A. Lange, Jose T. M. Houveling et al., "Appearance of Predictors of Disease Progression in Relation to the Development of AIDS," *AIDS* 3 (1989): 563–569; John F. Fahey, Jeremy M. G. Taylor, Roger Detels et al., "The Prognostic Value of Cellular and Serologic Markers in Infection with Human Immunodeficiency Virus Type 1," *New England Journal of Medicine* 322 (1990): 166–172.

30 Alberto Cambrosio and Peter Keating, "Interlaboratory Life: Regulating Flow Cytometry," in Jean Paul Gaudillière and Ilana Löwy (eds.), *The Invisible Industrialist* (London: Macmillan, 1998), 250–295.

31 A. Lafeuillade, C. Tamalet, P. Pellegrino et al., "Correlation between Surrogate Markers, Viral Load, and Disease Progression in AIDS," *Journal of AIDS* 7 (1994): 1028–1033.

32 Janet M. Raboud, Lawrence Haley, Julio G. Montaner et al., "Quantification of the Variation Due to Laboratory and Physiological Sources in CD4+ Lymphocyte Counts of Clinically Stable HIV-Infected Individuals," *Journal of AIDS* 10 (1995): S67–S73.

33 Paul Cotton, "HIV Surrogate Markers Weighted," *Journal of the American Medical Association* 265 (1991): 1357–1359; also Sungub Coi, Stephen W. Lagakos, Robert T. Sholey, and Paul A. Volberding, "CD4+ Lymphocytes Are an Incomplete Surrogate Marker for Clinical Progression in Persons with Asymptomatic HIV Infection Taking Zidovudine," *Annals of Internal Medicine* 118 (1993): 674–680.

34 See, e.g., Mirko D. Grmek, *Historie du sida* (Paris: Payot, 1989), 128–136; Margaret A. Hamburg and Anthony S. Fauci, "AIDS: The Challenge to Biomedical Research," in Stephen R. Graubard (ed.), *Living with AIDS* (Cambridge: MIT Press, 1990), 43–64.

35 David Ho, Tarsem Mougdil, and Masud Alam, "Quantification of Human Immunodeficiency Virus Type I in the Blood of Infected Persons," *New England Journal of Medicine* 321 (1989): 1621–1625.

36 David Baltimore and M. J. Feinberg, "HIV Revealed: Towards a Natural History of the Infection," *New England Journal of Medicine* 321 (1989): 1675.

37   Giuseppe Pantaleo, Cecilia Graciosi, Luca Buttini et al., "Lymphoid Organs Function as Major Reservoirs for Human Immunodeficiency Virus," *Proceedings of the National Academy of Sciences, USA* 88 (1991): 9838–9842; also Cecil H. Fox, Klara Tenner-Racz, Paul Racz et al., "Lymphoid Germinal Centers Are Reservoirs of Human Immunodeficiency Virus," *Journal of Infectious Diseases* 164 (1991): 1051–1057; Hans Spigel, Herman Hebst, Gerald Niedobitek, Hans-Dieter Foss, and Harald Stein, "Folicular Dendritic Cells Are a Major Reservoir for Human Immunodeficiency Virus Type 1 in Lymphoid Tissues, Facilitating Infection of CD4+ T-Helper Cells," *American Journal of Pathology* 140 (1992): 15–22.

38   Giuseppe Pantaleo, Cecilia Graciosi, James E. Demerest et al., "HIV Infection Is Active and Progressive in Lymphoid Tissue During the Clinically Latent Stage of Disease," *Nature* 362 (1993): 355–358; also Janet Emberson, Mary Zupanic, Jorge L. Ribas et al., "Massive Covert Infection of Helper T Lymphocytes and Macrophages by HIV During the Incubation Period of AIDS," *Nature* 362 (1993): 359–362.

39   Giuseppe Pantaleo, Cecilia Graciosi, and Anthony Fauci, "The Immunopathogenesis of Human Immunodeficiency Infection," *New England Journal of Medicine* 328 (1993): 327–335; also Pantaleo, Graciosi, and Fauci, "The Role of Lymphoid Organs in the Immunopathogenesis of HIV Infection," *AIDS* 7 (1993): S19–S23.

40   Giuseppe Pantaleo, Cecilia Graciosi, James F. Demerest et al., "Role of Lymphoid Organs in the Pathogenesis of HIV Infection," *Immunological Review* 140 (1994): 125.

41   Robin A. Weiss, "How Does HIV Cause AIDS?," *Science* 260 (1993): 1277–1278.

42   Jon Cohen, "AIDS: The Mood Is Uncertain," *Science* 260 (1993): 1254–1256.

43   David D. Ho, Avidan U. Neumann, Alan S. Perelson et al., "Rapid Turnover of Plasma Virions and CD4 Lymphocytes in HIV-I Infection," *Nature* 373 (1995): 123–126; also Xiping Wei, Sajal K. Ghosh, Maria E. Taylor et al., "Viral Dynamics in Human Deficiency Virus Type I Infection," *Nature* 373 (1995): 117–122.

44   Ho, Neumann, Perelson et al., "Rapid Turnover of Plasma Virions," 126.

45   Giuseppe Pantaleo, Oren G. Cohen, Douglas J. Schwartzentrouber et al., "Pathogenic Insights from Studies of Lymphoid Tissue from HIV-Infected Individuals," *Journal of AIDS* 10 (1995): S6–S14.

46   Martin A. Novak, "AIDS Pathogenesis: From Models to Viral Dynamics in Patients," *Journal of AIDS* 10 (1995): S1–S5.

47   Simon Wain-Hobson, "Virological Mayhem," *Nature* 373 (1995): 102.

48   Richard A. Koup, Jeffrey T. Safrit, Yunzhen Cao et al., "Temporal Association of Cellular Immune Responses with the Initial Control of Viremia in Primary Human Deficiency Virus Type I Syndrome," *Journal of Virology* 68 (1994): 4650–4655; also Giuseppe Pantaleo, Oren J. Cohen, Douglas Schwartzentrouber et al., "Pathogenic Insights from the Study of Lymphoid Tissue from HIV-Infected Individuals," *Journal of AIDS* 10 (1995): S6–S14.

49   Laurent de Villepin, Gilles Pialoux, and Yves Souterand, "Deux mondes, deux espoirs," special issue on Vancouver AIDS conference, July 1996, *ANRS-Journal du Sida-Transcriptase* (fall 1996): 3–4.

50   Florenza Cocchi, Anthony L. DeVico, Alfredo Garziano-Demo et al., "Identification of RANTES, MIP 1-Alpha and MIP 1-Beta as the Major HIV Supressive Factors Produced by CD8+ T Cells," *Science* 270 (1995): 1811–1815; also Benjamin J. Doranz, Joseph Rucker, Yanjie Yi et al., "A Dual-Tropic Primary HIV-I Isolate that Uses Fusin and the Beta-Chemokine Receptors CKR-5, CKR-3, and CKR-2b as Fusion Cofactors," *Cell* 85 (1996): 1149–1158. All these articles were published between 20 and 29 June 1996.

51   Yu Feng, Christopher C. Border, Paul E. Kennedy et al., "HIV-Entry Co-Factor: Functional C-DNA Cloning of a Seven Transmembranase, G-Protein Coupled Receptor," *Science* 272 (1996): 872–877.

52 Ghalib Alkhatib, Christophe Combadiere, Christopher C. Border et al., "CC CKR-5: A RANTES, MIP 1-Alpha and MIP 1-Beta Receptor as a Fusion Cofactor for Macrophage-Tropic HIV-1," *Science* 272 (1996): 1955–1958; also Hong Kui Deng, Rong Liu, Wilfried Ellmeier et al., "Identification of a Major Co-Receptor for Primary Isolates of HIV-1," *Nature* 381 (1996): 661–666; Tatjana Dragic, Virginia Litwin, Graham P. Allaway et al., "HIV-1 Entry into CD4+ Cells is Mediated by the Chemokine Receptor CC-CKR-5," *Nature* 381 (1996): 667–673; Hyeryun Choe, Michale Farzan, Ying Sun et al., "The Beta-Chemokinine Receptors CCR-3 and CCR-5 Facilitate Infection by Primary HIV-1 Isolates," *Cell* 85 (1996): 1135–1148; Liu, William A Paxton, Sunny Choe et al., "Homozygous Defect in HIV-1 Co-Receptor Accounts for Resistance in Some Multiply-Exposed Individuals to HIV Infection," *Cell* 86 (1996): 367–377. The signs CC-CKR-5, CC-CCR-5, CCR-5, and CKR-5 describe the same molecule.

53 Michel Samson, Fréderic Libert, Benjamin J. Doranz et al., "Resistance to HIV-1 Infection in Caucasian Individuals Bearing Mutant Alleles of the CCR-5 Chemokine Receptor Gene," *Nature* 382 (1996): 722–725; also Patricia da Souza and Victoria A. Harden, "Chemokines and HIV-1 Second Receptors: Confluence of Two Fields Generates Optimism in AIDS Research," *Nature Medicine* 2 (1996): 1293–1201; Anthony S. Fauci, "Resistance to HIV-1 Infection: Its in the Genes," *Nature Medicine* 2 (1996): 966–967.

54 Wain-Hobson, "Virological Mayhem."

55 Peter S. Arno and Karyn L. Feiden, *Against the Odds: The Story of AIDS Drugs Development, Politics, and Profits* (New York: Harper Collins, 1992); also James Harvey Young, "AIDS and the FDA," in Caroline Hannaway, Victoria A. Harden, and John Parascandola (eds.), *AIDS and the Public Debate* (Amsterdam: IOS Press, 1995), 47–66; Steven G. Epstein, *Impure Science: AIDS Activism and the Politics of Knowledge* (Berkeley: University of California Press, 1996).

56 Paul Cotton, "Controversies Continue as Experts Ponder Zidovudine's Role in Early HIV Infection," *Journal of the American Medical Association* 263 (1990): 1605–1609. The "latency" observed in the 1980s might have been an artifact of observation—or if one prefers, of the tendency of physicians and biomedical scientists to avoid using invasive techniques such as lymph node biopsies. At first researchers looked for viral replication in an easily accessible site, the blood, and only later they turned to the lymph nodes.

57 The Concorde trial—a multicenter British and French study which evaluated the effects of treating asymptomatic HIV-infected individuals with AZT—did not show clinical benefit of that therapy; see Concorde Coordinating Committee, "Concorde: MRC/ANRS Randomised Double-Blind Controlled Trial of Immediate and Deferred Zidovudine in Symptome-Free HIV Infection," *The Lancet* 1 (1994): 871–881. Other (and more restricted) studies indicated that an aggressive therapy with AZT may be efficient if initiated in the very first stages of HIV infection; see S. Kinloch de Löes, B. J. Hirschel, B. Hoen, et al., "A Controlled Trial of Ziduvidine (AZT) in Primary HIV Infection," *New England Journal of Medicine* 333 (1995): 408–413; L. Perrin and Kinloch de Löes, "Therapeutic Interventions in Primary HIV Infection," *Journal of AIDS* 10 (1995): s69–S76.

58 Anthony S. Fauci, "Multifactorial Nature of Human Immunodeficiency Virus Disease: Implications for Therapy," *Science* 262 (1993): 1011–1118.

59 David D. Ho, "Time to Hit HIV Early and Hard," *New England Journal of Medicine* 333 (1995): 450–451.

60 Wei, Ghosh, Taylor et al., "Viral Dynamics in Human Deficiency Virus," 122.

61 Ho, Neumann, Perelson et al., "Rapid Turnover of Plasma Virions," 126.

62 Quoted in Paul Cotton, "Many Clues, Few Conclusions on AIDS," *Journal of the American Medical Association* 272 (1994): 753–756.

63   Interview with Xavier Rey-Coquis, *Le Journal de la* MGEN (May 1996): 41. Actions-Traitement is a French association specialized in the follow up of clinical trials of AIDS therapies. On the history of this and related associations, see Janine Barbot, *Les Malades en mouvements: La médecine et la science . . . l'épreuve du sida* (Paris: Balland, 2002).

64   Frank Fontenay and Jean François Chambon, "La conference de Washington: L'indavir et le ritonavir relancent l'interet pour le tritherapies," *Journal du Sida* 82 (1996): 4–11; also Jon Cohen, "Results on New Anti-AIDS Drugs Bring Cautious Optimism," *Science* 271 (1996): 755–756.

65   Alan S. Perlson, Avidan U. Neumann, Martin Markowitz, John M. Leonard, and David D. Ho, "HIV-1 Dynamics in Vivo: Virion Clearance Rate, Infected Cells Life-Span, and Viral Generation Time," *Science* 271 (1996): 1583–1586.

66   De Villepin, Pialoux, and Souterand, "Deux mondes, deux espoirs?"

67   John W. Mellors, Charles R. Rinaldo, Phalguni Gupta et al., "Prognosis of HIV-1 Infection Predicted by Quantity of Virus in the Plasma," *Science* 272 (1996): 1167–1170; also Henri Aguti, "La charge virale en première ligne," special issue on Vancouver AIDS conference, July 1996, ANRS-*Journal du Sida-Transcriptase* (fall 1996), 57–58.

68   S. Vella and L. Palmisano, "HIV Therapy in the Post-Eradication Era," *Journal of Biological Regulation and Homeostatic Agents* 14 (2000): 34–37.

69   G. P. Rizardi, A. Lazzarin, and G. Pataleo, "Potential Role of Immune Modulation in the Effective Long-Term Control of HIV-1 Infection," *Journal of Biological Regulation and Homeostatic Agents* 16 (2000): 83–90.

70   R. T. Mitsuyasu, "The Potential Role of Interleukin-2 in HIV," *AIDS* 16 (2002): S22–S27; also M. Dybul, B. Hidalgo, T. W. Chun et al., "Pilot Study of the Effects of Intermittent Interleukin-2 on HIV-Specific Immune Responses in Patients Treated During Recently Acquired HIV Infection," *Journal of Infectious Diseases* 185 (2002): 61–68; R. Paredes, J. C. Lopez Benaldo de Quiros, E. Fernandez-Cruz et al., "The Potential Role of Interleukin-2 in Patients with HIV Infection," *AIDS Review* 4 (2002): 36–40; A. K. Pau and J. A. Tavel, "Therapeutic Uses of Interleukin-2 in HIV-Infected Patients," *Current Opinions in Pharmacology* 2 (2002): 433–439. On the history of interleukin-2, see Ilana Löwy, *Between Bench and Bedside: Science Healing and Interleukin-2 in a Cancer Ward* (Cambridge: Harvard University Press, 1996), 128–137.

71   I. Sereti and H. C. Lane, "Immunopathogenesis of Human Immunodeficiency Virus: Implications for Immune-Based Therapies," *Clinical Infectious Diseases* 32 (2001): 1738–1755.

72   M. M. Lederman, "Is There a Role for Immunotherapy in Controlling HIV Infection?," *AIDS Reader* 10 (2000): 209–216; S. L. Pett and S. Emery, "Immunomodulators as Adjunctive Therapy for HIV-1 Infection," *Journal of Clinical Virology* 22 (2001): 289–295.

73   Silverstein, *History of Immunology*, 262–270. The theories of autoantibodies were dismissed circa 1905, when scientists had found that experiments which apparently showed the widespread existence of anti-antibodies were technically flawed.

74   Niels Kay Jerne, "Towards a Network Theory of the Immune System," *Annales de l'Institut Pasteur* 125C (1974): 373–389.

75   Ludwik Fleck, *Genesis and Development of a Scientific Fact*, trans. Fred Bradley and Thaddeus J. Trenn (Chicago: University of Chicago Press, 1979), 61.

76   Andrew Abbott, *The System of Professions: An Essay on the Division of Expert Labor* (Chicago: University of Chicago Press, 1988), 184–195.

77   K. Codell-Carter, "Ignatz Semmelweis, Carl Meyerhofer, and the Rise of Germ Theory," *Medical History* 29 (1985): 33–53.

78 Paula A. Treichler, "AIDS, Homophobia, and Biomedical Discourse: An Epidemic of Signification," *Cultural Studies* 1 (1987): 263–305.

79 See, e.g., Liu, Paxton, Choe et al., "Homozygous Defect in HIV-1 Co-Receptor," 367–377; T. Tasaren, M. O. Hottinger, and U. Hubsher, "Functional Genomics in HIV-1 Virus Replication: Protein-Protein Interactions as a Basis for Recruiting the Host Cell Machinery for Viral Propagation," *Biological Chemistry* 382 (2001): 993–999.

80 See, e.g., M. Emerman and H. Malim, "HIV-1 Regulatory/Accessory Genes: Keys to Unraveling Viral and Host Cell Biology," *Science* 280 (1998): 1880–1884.

Artificial life

9   Artificial life support: some nodes in the Alife ribotype

Richard Doyle

Panic . . . the sudden, intolerable knowing that everything is alive.
—William S. Burroughs, *Ghost of Chance*

Everyone knows that in 1953 Watson and Crick diagrammed the structural and functional characteristics of the double helical molecule deoxyribonucleic acid. Much of the rhetoric of this remarkable achievement suggested that Life's secret had finally been uncovered, and that Life was therefore localized in the agency of genes. Watson, writing in his autobiography named for a molecule, put it this way: "In order to know what life is, we must know how genes act."[1] The action and manipulation of nucleic acids became the hallmark of a molecular biology that no longer analyzes organisms but—in symbiosis with stock markets and venture capitalists—transforms them, generating life-forms without precedent.

From agents to events: distributing life

Localizing life onto genetic actors—"what life is"—has also enabled an astonishing distribution of vitality, one that allows us to speak of "artificial life," simulacra which are not simply models of life but are in fact instances of it. In short, life is no longer confined to the operation of DNA but is instead linked to the informatic events associated with nucleic acids: operations of coding, replication, and mutation. I will argue that the emergence of artificial life signals more than the liberation of living systems from carbon—it also maps a transformation of the scientific concept of life itself, a shift from an understanding of organisms as *localized agents* to an articulation of living systems as *distributed events*. Rather than located in any particular space and time, living systems are in this view understood as unfolding processes whose most compressed descriptions are to be found in the events themselves—growing explanations.[2] This essay will focus primarily on the challenges such a transformation poses for our means of representing life in both scientific and extrascientific texts. I analyze both rhetorical milieus not because such realms of discourse are equivalent, but because they can sometimes fruitfully borrow from each other. Indeed, as I hope to make clear, the rhetorical practices involved in such representations of life are potent allies in the transformation of life, as rhetorical "softwares" become elements in the very network through which artificial life becomes lively.

Alife makes me nervous

Alife, I must admit, makes me nervous. This wracking of my nerves is not the deep anxiety of any fear and loathing of machines, those alleged alienators of souls, labor, depth, bodies—in short "Technology." No, Alife is like a joke that I just can't seem to get. You know the feeling . . . Oh, I'll admit that some of the creatures are, well, "cute," that they scamper across my screen, seemingly out of control, on the same drugged electricity as the Eveready Energizer Bunny. Sometimes I'll get all enthused about a simulation, sitting at my desk while I simulate work, and fetch a colleague from across the hall. "Look!" I'll say, "Look! It's alive!," and point at the swirling, flickering, flocking pixels. "It's *alive!*"

There just seems to be no convincing him. I've even tried the old rhetorical ploy of implicating my neighbor in a simulation. With LifeMaker, a cellular automata program available on the Web, I spelled out his name in "cells" and put the simulation on ultrafast. The cells swirled, flickered, and dissolved the writing and produced something more akin to a proliferating growth than a name. My colleague, a Joyce scholar, just looks at the screen, then at me, and deadpans, "It's just language, Rhetoric Boy."

In these situations I realize that I am being called upon to justify my expertise, so I lean back in my chair, do my best impression of a *professor*, and provide some historical and cultural context.

"Just *information*, you mean." I say this as if I have trumped him, as if the distinction is itself so stuffed with information that the scales should fall from his eyes. "But since Erwin Schrödinger's articulation of the genetic substance as a code-script in 1943, life itself has gradually been conflated with information. The trajectory is long and complex—from George Gamow's 1954 discussion of the diamond-code scheme for the translation of DNA into proteins, Jacques Monod and François Jacob's research on induction and the genetic program, to the recent human genome initiatives and their mapping and decoding of the Book of Life—but suffice to say that from the perspective of many contemporary biologists, life is just an interesting configuration of information. Biologist Richard Dawkins, whom you may have read about in the *New Yorker*, has claimed that we are nothing but 'lumbering robots,' vehicles for the propagation of DNA. We're like the host; it's the parasite. Or maybe it's the other way around."

"I still don't get it. Which part is alive?" At this point I despair because, I have to admit, it starts to feel like I am explaining a joke, a task that is neither fun nor funny. I can't really pull off the knowing, laughing "Oh, you don't get it. Well, if you don't *get it*, I can't really explain it. It would just take too long." At least, I can't *say* any of that. So all I can really do is imply it, suggest through my silence that it's a generational, theoretical thing. Perhaps he doesn't have the secret poststructuralist Alife decoder ring, I suggest through my silence. "I'll give you a copy of this segment I am writing about Alife when I am done," I say. "Maybe that'll help."

So Alife makes me nervous because it seems to be inarticulable in some way. And periodically, I don't get the joke. But it also makes me nervous because I can't simply write it off. Because there is something uncanny about Alife. It's a creepy doubling of something that no longer appears—"Life."

Life, as a scientific object, has been *stealthed*, rendered indiscernible by our installed systems of representation. No longer the attribute of a sovereign in battle with its evolutionary problem set, the organism its sign of ongoing but always temporary victory, life now resounds not so much within sturdy boundaries as between them. The very success of the informatic paradigm—in fields as diverse as molecular biology and ecology—has paradoxically dislocated the very object of biological research. "Biologists no longer study life today," writes Nobel Prize–winning molecular biologist François Jacob, "they study living systems."[3] This postvital biology is, by and large, interested less in the characteristics and functions of living organisms than in sequences of molecules and their effects. These sequences are themselves articulable though databases and networks; they therefore garner their effects through relentless repetitions and refrains, connections and blockages rather than through the autonomous interiority of an organism. This transformation of the life sciences in the twentieth century, while hardly homogeneous and not univocal, marks a change in kind for biology, whose very object has shifted, has become "distributed."

Consider, for example, the Boolean networks of Stuart Kauffman's research. Kauffman argues in his work that such nets—networks of buttons threaded to each other in a random pattern—display autocatalytic configurations after a phase transition when the ratio of threads to buttons reaches 0.5, after which "all of a sudden most of the clusters have become cross-connected into one giant structure."[4] Kauffman and others find that this deep principle of autocatalytic nets suggests another transition—at the origin of life: "The rather sudden change in the size of the largest connected cluster of buttons, as the ratio of threads to buttons passes 0.5, is a toy version of the phase transition that I believe led to the origin of life."[5]

The paradox of such a formulation emerges around the dependence of this remarkable model on the "sudden" quality of the transition. Precisely the power of such a network—a persuasiveness that would link surprise to the very emergence of life—renders a difficult problem for a rhetoric that could narrate such a sudden change—It's *alive!* How to articulate such a parallel event, an event in which difference emerges not in a serial, one-after-another story, but all at once, "all of a sudden"? To visualize this change in kind, Kauffman's own visual rhetoric has recourse to a flicker or a "blinking" when he constructs a Boolean net composed of bulbs that light up based on their logical states: "I will assign to each light bulb one of the possible boolean functions[,] . . . [the] AND function to bulb 1 and the OR function to bulbs 2 and 3. At each tick of the clock each bulb examines the activities of its two inputs and adopts the state 1 or 0 specified by its boolean function. The result is a kaleidoscopic blinking as pattern after pattern unfolds."[6]

Crucial to representing such a networked understanding of life is a semiotic state that is neither here nor there—the flicker of such a "kaleido-scopic blinking" signals less the location of any orderly pattern than its status as a fluctuation. This configuration emerges precisely between loca-tions rather than in any given node of the network. Neither on nor off but the difference between on and off, the blinking signals the capacity of each specific type of network, such as Kauffman's K=2 networks, where each bulb is connected to two other nodes, for orderly patterns. In short, even in models of that allegedly singular event, the origin of life, articulations of life involve a multiplicity—multiple nodes—whose liveliness emerges be-tween locations.[7] This notion of life relies less on classical conceptions of autonomy than on a rigorous capacity for connection, orderly ensembles representable only as transformations—flickering patterns unfolding in space and time.

This makes Alife's claims concerning the vitality of virtual organisms all the more perplexing. For if Life seems to have disappeared as a sover-eign entity and joined the ranks of all those other relational attributes—economic value, for example—then it seems odd that it should reappear so visibly on my screen.[8] It's enough to make one believe in time travel. It's as if computers, with the right softwares, could travel back to a past when life was an autonomous attribute of organisms, capture it, and display it on the screen.

Heterodox theoretical biologist Marcello Barbieri has argued in this context that organisms have always already been networks. Moving beyond Johannsen's rendering of the organism into the duality of genotype and

phenotype, Barbieri argues persuasively that a living system must be understood as a tripartite ensemble of *genotype*, hereditary information, primarily
although not exclusively born by DNA; *ribotype*, the swarm of translational
apparatuses that transform DNA into the tertiary structures of folded proteins; and *phenotype*, the dynamic embodiment of these informations and
their transformations.[9] Crucial to Barbieri's argument is the recognition
that DNA information is necessary but not sufficient for the emergence of
life; yet another translational actant is needed to transform the immortal
syntax of nucleic acids into the somatic semantics of living systems.

By analogy, I want to suggest that Alife, too, emerges only through the
complex of translational mechanisms that render it articulable as "lively."
The ribotype that transforms the coded iterations and differences of Alife
softwares into the lifelike behavior of artificial life is composed of, among
other things, "rhetorical softwares." These rhetorical formulations—as
simple as a newly coined metaphor or as complex as an entire discourse—
don't construct scientific objects so much as they discipline them, render
them available for scientific observation, analysis, and argument, much as
the flicker of bulbs images Kauffman's autocatalytic order. The rhetorical
challenge posed by life that emerges out of networks goes beyond the
ontological uncertainty that haunts artificial life—are they really alive?—
and becomes a problem of articulation: How can something that dwells
not in a place but in virtuality, a network, be rendered? Hence rhetorical
problems haunt not simply the status of Alife creatures—what they are—
but their location—where they are.[10]

It is perhaps due to this uncanny distribution of life that the parallel
rhetorical formulations of localization and ubiquity have particular force
on artificial life, even as these effects are in tension. Rhetorics of localization suggest that some particular organism "in" or "on" the computer is
"alive," thereby occluding the complex ecology of brains, flesh, code, and
electric grids that Alife thrives on and enabling the usual habits of narrative—an actor moving serially through a world—to flourish, as a more
recognizable and perhaps seductive understanding of organism as "agent"
survives. At the same time, rhetorics of ubiquity provoke the possibility
that, as in Burroughs's observation, *anything could be alive*, leaving the observer continually alert to the signs of vitality in the midst of machines. It is
this latter effect that produces much of the excitement of artificial life, a
sublime incapacity to render the sudden, ubiquitous complexity of a phase
transition into the sequential operations of narrative.[11]

All scientific practices are differently comported by their rhetorical softwares: my focus has been on the researches and insights enabled by articulations of organisms as extensions of code. But Alife is in a slightly different position with respect to its rhetorical components, as the actual

difference of artificial life, as life, is continually at stake. This crisis of vitality that pervades Alife is not simply due to Alife's status as a simulation; as I suggested earlier, Alife emerges from a context in which quite literally, *life disappears* as the life effect becomes representable through the flicker of networks rather than articulable and definable locales. As researcher Pierre Levy writes, "Virtualization comes as a shock to the traditional narrative."[12] My challenge here will be to determine the specific rhetorical mechanisms that enable the narration and instantiation of some versions of artificial life, a wetware ribotype that makes Alife creatures so lively at this moment.

As virtual organisms, Alife creatures are not fake. Like all simulacra, they are copies without original, producing an effect not of reference—what would they *refer* to?—but of provocation, the uncanny feeling of familiarity in the unfamiliar realm of the computer screen. They double and *fold* the organic into the virtual, a hybridization of machine and organism that, inevitably, makes one laugh. We laugh nervously because while we are not in a state of Burroughsian panic, all of the technological in Frankensteinian rebirth, full of life—no conspiracy involving a toaster, chainsaw, and a couple of CD players is in the cards—we nonetheless get the sense that indeed anything *could be alive.* This is the first element I'll assay in the ribotype of artificial life, a rhetorical ensemble that smears the borders between the computer and its environment, what we could call a silicon abduction.

Abducted by silicon

If artificial life creatures, as actualizations of information, enjoy the burdens and benefits of vitality, they do so through the operation of what Charles Sanders Peirce characterized as "abduction." Peirce—a nineteenth-century polymath who contributed to mathematics, semiotics, and philosophy—formulated his theory of abduction to supplement the more traditional logical categories of induction and deduction. Scientific thinking, Peirce held, didn't always proceed via the clean operation of these categories—Kepler's discovery of the laws of planetary motion was among Peirce's favorite examples of a scientific practice that differed from these logical frameworks. Abduction, as a category of reasoning, is characterized by its reliance on an *absence*: "An *abduction* is a method of forming a general prediction without any positive assurance that it will succeed either in the special case or usually, its justification being that it is the only possible hope of regulating our future conduct rationally, and that induction from past experience gives us strong encouragement to hope that it will be successful in the future."[13] A missing term—one that might arrive in the future—completes abduction's argument. The "possibility" that inheres in any specific abductive enterprise is tethered to the pathos of hope, an encounter with the future without grounds but with calculation, anticipation, and a bit of desperation—"the only possible hope of regulating our future conduct rationally." The past, too, offers itself up as a support to abductive reasoning but only in the "inductive" habit that Peirce identifies with sheer repetition and persistence, attributes that do little to aid in the evaluation of any future event.

Still, Peirce favored abduction because it seemed to be the only office of reasoning that allowed for the arrival of a novel, unprecedented thought. Induction, tied to habit, tends to subsume each event into the Same, and the logical necessity of the deductive syllogism relies on full knowledge of all the premises, knowledge which, by definition, is not available in exploratory scientific enterprise.[14] Thus Peirce sought to describe the persuasive force of abduction—what he sometimes called "hypothesis"—in terms other than those reserved for logic:

> Hypothesis substitutes, for a complicated tangle of predicates attached to one subject, a single conception. Now there is a peculiar sensation belonging to the act of thinking that each of these predicates inheres in the subject. In hypothetic inference this complicated feeling so produced is replaced by a single feeling of greater intensity, that belonging to the act of thinking the hypothetical conclusion. . . . We may say, therefore, that hypothesis produces the *sensuous* [element] of thought, and induction the *habitual* element.[15]

Peirce's rhetoric and semiotics sought, among other things, to materialize our understandings of language and conviction, so the "sensuous" character of abductive thinking should not be read through the lens of the Platonic and Aristotelian suspicions of pathos. Instead, Peirce offers it to describe the only mode of reasoning that seems fit for an encounter with the future, a deployment of persuasive force that gambles on the unprecedented. Such a gamble works through an intensive practice of substitution—a single concept is swapped for a tangle.

In Christopher Langton's 1987 manifesto for artificial life, precisely such an encounter with the future takes place. The missing term in the abductive transaction is "life." Writing of biology's need to expand its purview and look at material substrates other than carbon, Langton writes that the traditional, narrow context of biology (!) makes it impossible for researchers to "understand" life.[16] "Only when we are able to view *life-as-we-know-it* in the larger context of *life-as-it-could-be* will we really understand the nature of the beast." With this claim, Langton offers a formalist definition of life, one arguing that the phenomenon of living systems is tied to organizational and not material attributes. This hypothesis is not itself new:

Maturana and Varela's notion of "autopoietic" machines also argues that life is an organizational phenomenon, one that is perhaps independent of its material instantiation. Indeed, the rhetorical practices of information that have transformed the life sciences support such a dislocated conception of living systems, as information becomes mobile, capable of instantiation in contexts other than its origin. What is peculiar to Langton's abductive move is the claim that one can *only* understand life adequately on the basis of its instantiation elsewhere, "life writ large across all material substrates . . . is the true subject matter of biology." A larger sample size, so the argument goes, might help us understand both what is peculiar to carbon based life and what more general descriptions and laws govern living systems, a kind of metalife.

A tension or even a tangle emerges from this bold move into the "synthetic approach to biology" where organisms emerge *in silico*. What Langton proposes is as novel a shift for biology as the discovery and study of microscopic organisms with van Leeuwenhoek's deployment of the microscope. Instead of greater magnification that enables the representation of a new realm of the organic, the iterative capabilities of the computer make visible the lively and self-organizing capacities of an inorganic stratum. And yet, in its logical formulation, Langton's claim depends on the very knowledge that it seeks. The very understanding intended to orient the life sciences—"the true subject matter of biology"—is yet to come, bundled with an analysis of "life-as-it-could-be." In the meantime, our very criteria for identifying and studying living systems remain vague, operational definitions haunted by their character as simulacra. The very impetus for artificial life research—the lack of sufficient knowledge of the formal attributes of life—stymies what Langton will call the "big" claim of artificial life, a claim that defines *in silico* creatures as "alive." "The *big* claim is that a properly organized set or artificial primitives carrying out the same functional roles as biomolecules in natural living systems will support a process that will be alive in the same way that natural organisms are alive. Artificial Life will therefore be *genuine* life—it will simply be made out of different stuff than the life that has evolved here on Earth."[17] This claim, anchored as it is in an understanding of the "way that natural organisms are alive," begs the very question that Alife allegedly illuminates: In what way, precisely, are natural organisms alive?

But, as Peirce pointed out, the dependence of an abductive argument on a term that is yet to come, off the Earth—such as "life-as-it-could-be"—is not simply lacking in logical coherence. Such an opening toward the future allows for the contingent, even improbable arrival of precisely such a missing term. Indeed, Peirce writes of abduction, "Never mind how improbable these suppositions are; everything which happens is infinitely improbable."[18] For Langton, the supposition of a synthetic biology substitutes for, even doubles, yet another supposition: "Since it is quite unlikely that organisms based on different physical chemistries will present themselves to us for study in the foreseeable future, our only alternative is to try to synthesize alternative life-forms ourselves—Artificial Life: life made by man rather than by nature."[19] How might such organisms "present themselves" to humans? The enjoyment of Langton's joke depends on our capacity to abide the exchange of one improbability for another: it seems, for some reason, more likely that we can fabricate life than that aliens will arrive any time soon. Along with that other alien contact strategy SETI, however, Alife comes up against an operational difficulty: How to determine whether its object is "really" alive?[20]

To this query, Langton implicitly offers an intriguing answer. Discussing the transformation of an artificial life genotype or GTYPE into its phenotype or PTYPE, Langton traces out an irreducible contingency in artificial life and perhaps life:

> It is not possible in the general case to adduce which specific alterations must be made to a GTYPE to effect a desired change in the PTYPE. The problem is that any specific PTYPE trait is, in general, an effect of many nonlinear interactions between the behavioral primitives of the system. Consequently, given an arbitrary proposed change to the PTYPE, it may be impossible to determine by any formal procedure exactly what changes would have to be made to the GTYPE to effect that—and only that—change in the PTYPE. It is not a practically computable problem. There is no way to calculate the answer—short of exhaustive search—*even though there may be an answer.*[21]

Assuming that an attribute of PTYPE would be "living," then no inspection of any GTYPE can yield an understanding of any PTYPE's liveliness. What, then, would provide the Alife researcher with an understanding of the liveliness of a PTYPE?

"Trial and error," Langton claims, is "the only way to proceed in the face of such an unpredictability." On this abductive account—it can only make assumptions in the face of the future—Alife phenotypes, PTYPEs, can only emerge through the actual execution of an algorithm, whose translation and expression, are part of a process that is itself characterized by multiple levels of interaction: "It should be noted that the PTYPE is a multilevel phenomenon. First, there is the PTYPE associated with each particular instruction—the effect that the instruction has on the entity's behavior when it is expressed. Second, there is the PTYPE associated with each individual entity—its individual behavior within the aggregate. Third, there is the PTYPE associated with the aggregate as a whole."[22] Thus the actual status of any Alife creature cannot be inferred from its initial configu-ration—GTYPE—and its expression is characterized by levels, thresholds which are themselves the outcome of multiple nonlinear interactions. If life, as Langton claims, is not "stuff," but is instead an "effect," then the effects of liveliness can only be articulated after the execution, as it were, of the Alife code.

What I want to suggest is that at each level, translations occur. "Within" the screen, Alife organisms survive based on their interactions with both their virtual environments and other Alife creatures. So, for example, the strange dogs called Moofs that sometimes populate Simlife emerge as "translations" of their digital genomes, and one can tinker with the ge-nomes in the hope of tweaking Moof success and behavior, but the ecology even of a simple program like Simlife is sufficiently complex that one cannot predict the effect on the Moof phenotype, at least in terms of its behavior within the virtual ecology that it inhabits.

Representing life for a living

But once we shift our focus from the alleged interior of the computer, we encounter yet another translation practice. Moofs are inoculated into the virtual ecology of Simlife based on the preferences and habits of the hu-mans interacting with them. The overall success of an Alife world—the number of times that a mass of organisms sprout on silicon substrates all over the Earth—depends on their ability to *seduce humans.* That is, their liveliness—their ability to achieve reproductive success and other lifelike behaviors in the computer's virtual ecology—depends on their success in *representing* life to their human wetware. This would be simply tautological were it not for the fact that our definitions of life are themselves, at best, recursive—ongoing feedback loops whose origin and destination are quite simply non sequiturs.

Even the definition of life emerges less from a sequence than a tangle, a complex of interactions that conjoins rather than demarcates life from its alleged others, such as machines and avatars. Gaia, for example, both names the emergence of the Earth as a superorganism and announces the immense capacities of "matter" to be imbricated and infiltrated with life: It's *alive!* Lynn Margulis's analyses of symbiogenesis and Gaia both suggest a strange capacity to be invaded characteristic of both the Earth and the nucleated cell. In this frame, the eukaryotic cell emerges not only through a parasitic invasion by bacterial DNA on a prokaryotic cell, or, conversely, via a predator's consumption of prey DNA, but from the supple and specific capacities to be invaded coevolving with the prokaryotic cell. This capacity to survive an inhabitation is just as important as the bacterial "tolerance for their predators" that allows select bacteria to live off the cell's hospitality and avoid consumption.[23]

According to Margulis, it was probably the evolution of a protective membrane that allowed early prokaryotes to survive the "poisonous guests." Thus, vital to Margulis's account of the bacterial deterritorialization through which a new form of cell emerged was a form of blockage—a blockage of its DNA.[24] What would be the analogous operation in Alife, if there is one?

Perhaps it is in the blockage of the *self* that Alife proceeds. As familiars, Alife offers itself only through the production of indiscernibility—the sudden hailing of a life-form, a creation of complicity among humans and machines. This complicity—as with the long term embrace of mitochondria and the eukaryotic cell—emerges out of a liaison whose very operation entails a temporary blockage or even suspension of the self, a self that would wield control over the machine.

This self must be actively blocked or distracted through visible and sometimes narratable tactics, some of which are more successful than others. Thus the success of Alife organisms in their virtual ecology is tied to their success in an actual interactive visual ecology, an ecology also populated by humans, in which, for example, one simply *cannot look away:* The promise of Alife is that something is always about to happen . . .

Hence the cute, perky vitality of most Alife organisms. Alife organisms need not "attract" in this same way, just as all flowering plants do not tempt the wasp as the orchid does.[25] But Alife creatures must indeed represent the life effect visibly and articulably to the mass of humans with whom they are interacting. Thus, at the level of PTYPE, where the speech act "It's alive!" emerges, Alife creatures require a ribotypical apparatus that will render them compelling to their human hosts. If an Alife organism falls in the forest, and no one is around to hear it, it only makes a virtual sound. Not because of any allegedly postmodern solipsism attributable to Alife creatures, but because actualized Alife organisms *represent life for a living.* The Alife organisms that achieve the most success in Darwinian terms are those that are most readily and remarkably narrated or otherwise replicated. This is as much an attribute of Alife behavior as their feeding habits, and at the level of the actual it is an obligatory passage point for success. Note that this attribute by definition cannot simply re-present life—what would such a representation look like?—but must instead provoke, seduce it into actualization. As Pierre Lévy, following Deleuze's discussion, writes, "the real *resembles* the possible whereas the actual *responds* to the virtual."[26]

I want to be clear that I am not claiming that Alife organisms are simply the result of human decisions or that it is *only* the rhetorical softwares bundled with Alife organisms that make them lively. Such a humanist understanding of Alife would overlook the fact that humans do not simply choose rhetorical practices; rather, they are persuaded by, respond to, them. The rhetorical softwares of information that transformed biology, for example, were less careful deployments of knowledge than contingent experiments in persuasion. And the success that Alife organisms have as *virtual* organisms is not fake; it is simply another level of PTYPE, another level of contingency, than the actual.

Philosopher's Gilles Deleuze and Félix Guattari articulate this distinction between the actual and the virtual as one based on "chaos," the sheer contingency of an unactualized event, a program that may be successfully run. Real but not actualized, the virtual is a consistency—such as a configuration of code or a spore—that remains to be executed. The virtual is not, therefore, unreal. Nor does it *lack* actuality—such a description would depend upon an abduction of the future, a retroactive understanding of the virtual in terms of its instantiation as actual. Nor is the virtual without the resistances and finitudes we often attribute to the real. It bears its own constraints—the capacity to be rendered into a virtual substrate of code, the materiality of its substrate, and perhaps most strangely, the capacity to encounter the difference of the future, to negotiate the catastrophic change in kind that is the movement from the virtual to the actual, program to instantiation, as in Kauffman's "We must instead simply stand back and watch the pageant." Feminist thinker Elizabeth Grosz distinguishes this characteristic of the relation between virtual and actual precisely in terms of the future's occurrence to the virtual, "what befalls it." "The movement of realisation seems like the concretisation of a preexistent plan or program; by contrast, the movement of actualisation is the opening up of the virtual to what befalls it."[27]

What befalls Alife creatures is itself multiple, even a crowd. The evolution and emergence of artificial life-forms occurs in relation to both the materiality of Alife's virtual ecology—my hard disk is getting full!—and its relationships to other Alife creatures. Crucial for Alife, though, is that one level of their actualization depends on their ability to be "befallen" by human wetware, an actual response to the virtual. It should be objected that the privilege of this last level of actualization—the level at which "It's alive!" emerges—is entirely enmeshed with the ecology of humans. Thinking the novelty and specificity of artificial life, however, demands that we encounter the crucial ways in which human corporeality is entwined with Alife's status as life, even if it does not *dominate* Alife.[28] We swarm with silicon-based as well as carbon-based allies. These silicon life-forms can be said to live off the states of mind wedded to human uses. In some sense that I will discuss in more detail at some instant in the future, actualizing Alife familiars as life is perhaps the least novel level in the Alife PTYPE, one that reterritorializes the strange contingencies of the virtual becoming actual into that old saw, life. Indeed, perhaps simulating life is but a ruse, a hoax or stealth tactic that enables the propagation of entities that dwell much more in alterity than life, novel entities that mime life as a tactic and not an essence.

Artificial life is not, for example, an operation of simple Darwinian artificial selection, where humans breed the liveliest Alife creatures, consciously or unconsciously. Instead, Alife creatures' very existence as actualized life, rather than interesting enterprises in computation, is enmeshed with the human phenotypes with which they interact. The original status of Alife as life, that which makes it replicable in the first place, is thoroughly bound up with the affect—Peirce's "sensuous character of thought"—of the humans that encounter them.[29] At an actualized level of PTYPE, that level at which Alife becomes replicable as life and spreads across the hard disks and networks of the infosphere, the liveliness of Alife creatures is contingent on the relations between their effects—such as reproduction— and their ability to be articulated as lively, an ability that does not simply reside in human narrators but actually provokes them.

Still, it is worth considering a general framework for thinking about how such provocation proceeds. Darwin's less heralded mechanism of selection—sexual selection—might connect representing life for a living to the work of rendering life. That is, given that "life" and "representation" are different effects whose ecologies do not necessarily overlap, how do Alife creatures make a living, representing life for a living?

In both The Origin of Species and The Descent of Man, Darwin reminds us over and over that it is not only through war and struggle that evolutionary fitness emerges. Elaborately choreographed seduction is, for example, an important vector of ornithological sexual selection, or the struggle for mates among birds of the same species:

> All those who have attended to the subject, believe that there is the severest rivalry between the males of many species to attract, by singing, the females. The rock-thrush of Guiana, birds of paradise, and some others, congregate; and successive males display with the most elaborate care, and show off in the best manner, their gorgeous plumage; they likewise perform strange antics before the females, which, standing by as spectators, at last choose the most attractive partner.[30]

While the ascription of "choice" to the female bird here clearly troubled Darwin with its implications for female agency, he was perhaps more bothered by the sheer exuberance of plumage, seeing in it a monstrosity exceeding any usual notion of fitness. "The tuft of hair on the breast of the wild turkey-cock cannot be of any use, and it is doubtful whether it can be ornamental in the eyes of the female bird; indeed, had the tuft appeared under domestication, it would have been called a monstrosity."[31] Indeed, Darwin's remark here seems to indicate that "choice" is an inadequate model for thinking these scenes of seduction and their feedback onto ornament, song, speech. Darwin's doubt concerning the possible excitement of a female bird by the tom's waddle speaks less to Darwin's impoverished ornithological imagination than it does to the weakness of the choice model. Indeed, rather than resulting from a site of careful evaluation and control (such as domestication) the evolution of such spectacularly useless ornaments speaks to the operation not of decision but of emergence: the monstrousness of the tuft sprouts out of a feedback loop whose origin is neither male nor female, passive nor active: interaction. In this sense, Alife organisms and humans form an extended, interactive, phenotype of each other, with rhetorical softwares serving as the ribotypic translation apparatus that enables Alife code to operate on human bodies and vice versa, the becoming-silicon of flesh, the becoming-flesh of silicon. On Darwin's model, perhaps artificial selection is an instance of transpecies sexual selection.[32]

As with the other levels of translation, one simply cannot tell in advance if a given interaction between ribotype (RTYPE) and phenotype will succeed in yielding organisms that will achieve actualized success as lively. Each rhetorical software or RTYPE must be run—and the Alife organisms' "lives," as actualized lives, are at stake. It is only in practice—what befalls the GTYPE and its contingent virtual vitality—that the actualized vitality of the Alife organism can emerge. Indeed, in some sense such liveliness can occur only in the future, for each practice encounters the news that any formal definition of life is yet to come.

Perhaps there is good reason for the apparently irreducible contingency of the GTYPE–RTYPE–PTYPE interaction. John Von Neumann, polymath propagator of the theory of self-reproducing automata, describes complexity as being more difficult to describe than to practice: "There is a good deal in formal logics to indicate that the description of the functions of an automaton is simpler than the automaton itself, as long as the automaton is not very complicated, but that when you get to high complications, *the actual object is simpler than the literary description*" (italics mine).[33] However one characterizes the life effect, it is certainly "complicated"—it involves passages over borders, the organization of a "complex" or an ecology. Describing or translating the liveliness of artificial life is perhaps more complex than the automata of Alife themselves. As in the joke I mentioned earlier, it is perhaps simpler to practice or "grow" Alife than it is to describe it. Truly an abductive enterprise, Alife continually seeks confirmation in a practice that is yet to come—the translation of Alife as lively, an understanding of the formal nature of life, life-as-we-know-it within the context of life-as-it-could-be.

More than a logical tangle, though, the tactic of interrogating the comparative probability of artificial life and alien life provokes an experience of life's ubiquity—no longer confined to either the planet or to flesh, it finds itself distributed across the desktop and the universe.

Literary ribotypes

Not all creatures can be rendered equally visible, narratable, and therefore abductable, so some rhetorical tactics would seem to be more successful than others in Alife's evolving ribotype and its operations of transhuman sexual selection. In the literary phylum, authors such as Philip K. Dick and William S. Burroughs have generated remarkable rhetorical effects of vitality in diverse and divergent contexts. Dick, the speed-typing author of over thirty-six novels and several volumes of short stories, describes "vugs," a Titanian, silicon-based life-form that inhabits his 1963 novel Game-Players of Titan.

> They were a silicon-based life form, rather than carbon-based; their cycle was slow, and involved methane rather than oxygen as the metabolic catalyst. And they were bisexual. . . ."Poke it," Bill Calumine said to Jack Blau. With the vug-stick, Jack prodded the jelly-like cytoplasm of the vug. "Go home," he told it sharply.[34]

Dick's figuration of the in silico creature quickly overtakes the comfortable distance established between the stick-wielding humans and the amorphous blobs from Titan. Telepathic, vugs render the distinction between the interiority of a human and its "outside" indeterminable, as the boundaries of human identity become as fluid as the physical outline of the vug. Before long, both the characters in and the readers of Dick's novel find themselves "surrounded." "As he sat on the edge of the bed removing his clothes he found something, a match folder under the lamp by the bed and examined it. . . . On the match folder, in his own hand penciled words: WE ARE ENTIRELY SURROUNDED BY BUGS RUGS VUGS."[35] Crucial to the effect garnered here—one might call it panic—is the vehicle of vug knowledge. Writing, that allegedly stable reservoir of memory, becomes the vector not of certainty but of potential. The reception of a message to oneself becomes the occasion not just for recall but for the disturbance of recollection. Dick's character, Pete Garden, could not remember writing such a note: "I wonder when I wrote that? In the bar? On the way home? Probably when I first figured it out, when I was talking with Dr. Philipson."[36] By disturbing both the interiority of his characters—through telepathy, you're thinking—and the interiority of his reader—through the suggestion that Dick's writing, too, may contain a strange message, one so inarticulable that one must try out different bonsonants, ronsanants, consanants—Dick dislocates the vug's vitality and distributes it across other substrates. The very existence of a silicon life-form immediately leaves us, possibly, "surrounded," but Dick's play on the simulatable character of life and the incessant movement of writing implicates both his characters and his readers in a strangely interactive paranoid world where anything, even a book, could be alive.

This dislocation of vitality from its "home," carbon, instills many of Dick's novels with this uncanny sense of being surrounded by vitality. In *Radio Free Albemuth*, a novel found with Dick's papers after his death in 1982, Nicholas Brady receives a visit from his future self: "He had the impression that the figure, himself, had come back from the future, perhaps from a point vastly far ahead, to make sure that he, his prior self was doing okay at a critical time in his life. The impression was distinct and strong and he could not rid himself of it."[37] This unforgettable memory of the future seems primarily concerned with one thing: that the universe is itself alive. Having encountered VALIS—the Vast Living Intelligent System—Brady is now plugged in to the enormous vitality of the "void." "By now I knew what had happened to me; for reasons I did not understand, I had become plugged into an intergalactic communications network, and I gazed up trying to locate it, although most likely locating it was impossible."[38] Crucial to Dick's formulation is the notion that such vitality cannot be "located." Beyond the boundaries of any given organism—whether human, vug, or the universe itself—vitality is characterized by its excess, a surplus that renders the desire to locate the territory of vitality difficult if not impossible. Only violence—the repetitive prodding of the vug stick—connects the flowing turbulence of life to its alleged container, "home," and the network can be "plugged in to" but not located, as each node leads to another in the distributed effect Dick renders as VALIS, an artificial, computational but living god.

So too does much of Alife create this sense of dislocation, as the incessant vitality, like the Eveready Energizer Bunny, goes "on and on," everywhere. Indeed, as I have noted elsewhere, the practice of Alife seems to be involved in the desire for transcendence, the desire to be "above everywhere," a position from which one can articulate, finally, the formal characteristics of life in this moment of its dislocation. But the excessive, transformative character of Alife is effected in more than a simply transcendental fashion; the pesky vitality periodically invades the identity of the user, that wetware sufficiently charmed by the Tamagotchi to bury it.[39] As a classic inoculation of subjectivity—the very insemination of the name—

nothing would seem more masterful and transcendental than this act of self-replicating memorialization. Some scholars argue that it was indeed the tombstone that first provided the ecology for the emergence of writing itself, an old but erratic technology of immortality.

Still, as with Dick's novels, such vitality exists within a frame: the virtual ecology of the computer. If, after a time, the reader of Philip K. Dick's novels slowly remembers the memorializing capacities of writing, so too does silicon's vitality remain confined, primarily to the postvital window through which it emerges. If the operation of abduction—the encounter with a fundamentally alien, unprecedented future—makes the appearance of *in silico* organisms possibly ubiquitous in the distributed network of contemporary life, what ensures the autonomy and isolation of the vital silicon creature in the context of its leaky legacy of excess?

Life is for fetuses, or insane in the membrane

"A universe comes into being when space is severed into two," say Humberto Maturana and Francisco Varela. "A unity is defined. The description, invention and manipulation of unities is at the base of all scientific inquiry."[40] In a different register, Donna Haraway remarks, "Life itself, a kind of technoscientific deity, may be what is virtually pregnant."[41] My aim here is to put these two ideas together.

Humberto Maturana and Francisco Varela, in their 1968 text "Autopoiesis: The Organization of the Living," offer a general theory of living systems that would characterize vital systems in terms of their autonomy:[42] In this respect, Maturana and Varela are placed firmly within an Aristotelian tradition that saw organisms as wholes mobilized by their purpose or telos, but with a difference: the purpose of an organism is autonomy itself.

This theory of biological systems allows one to dispense with the classical category of teleology and its shadow of God, but perhaps crucially it also evaporates the distinction between living systems and machines. In the place of this distinction—one based, perhaps, on some vitalist trace in biology—Maturana and Varela generate the difference between "autopoietic" and "allopoietic" machines. Autopoietic machines work on themselves, as it were, generating their identity as an effect of their ongoing self-organization:

> An autopoietic machine is a machine organized (defined as a unity) as a network of processes of production (transformation and destruction) of components that produce the components which: (i) through their interactions and transformations continuously generate and realize the network of processes (relations) that produced them; and

(ii) constitute it as a concrete unity in the space in which they (the components) exist by specifying the topological domain of its realization of such a network.[43]

Such a lengthy and tangled definition reminds us of Von Neumann's observation cited earlier: that an event as complex as a living system is easier to achieve than to describe. Still, Maturana and Varela's definition has serious influence within certain strands of artificial life, so its rhetorical management of this complex problem—the very definition and borders of living systems—is crucial to understanding how the interior of autonomous Alife organisms are invested with "life."

On the one hand, it would seem obvious that Maturana and Varela's arguments enable artificial life. By decoupling life from its usual lodging, organisms, autopoiesis makes possible the dislodging of life from any organic location whatsoever. It is in this sense that Maturana and Varela's work resonates with the news that living systems are primarily informatic. They seek to situate the autopoietic effect within a larger integrated *process* rather than confining it to its network of molecular effects, even as they refuse the claim that living systems can be characterized as operations of coding.[44]

This deterritorialization of life via autopoiesis, then, continued the erosion of the distinction between machines and organisms wrought by the postvital understanding of living systems. Unlike the distributed character of networked life, whose vitality is tied to the possibility of a contingent outside with which each component could connect, Maturana and Varela's vision emphasizes the autonomy and closure of the autopoietic system. If the postvital understanding of life emphasizes connection—as in Stuart Kauffman's Boolean nets where nodes garner vitality and order through relation to multiple, other nodes—then the cybernetic argument of Maturana and Varela renders life as an interiority, one constantly making itself *as a self.*

This notion of the *interiority* that inheres in autopoietic systems maps logically onto Maturana and Varela's claim that autopoiesis is "necessary and sufficient for the occurrence of all biological phenomena."[45] This necessary and sufficient status accorded autopoiesis retains the historical sovereignty and interiority of life—the agency of an organism consists in its construction only of itself—even as Maturana and Varela seek to offer a theory of living *organizations,* a theory that stresses the relentlessly relational character of living systems.

This self-contained, logical character of the autopoietic system—as necessary and sufficient—marks the topological map of the living system as well. "In the beginning," to paraphrase the quote with which I began this segment, "was the inside and the outside." The authors begin with this distinction, warranted by their emphasis on autonomy, what they deem to be "so obviously an essential feature of living systems."[46] For a distinction that possesses so much self-evidence, the claim for the autonomy of the autopoietic system—that process of self-organization that emerges between the inside and the outside—poses many problems. Even as Maturana and Varela attempt to demarcate living systems' distinctive qualities, they find themselves unable either to confirm or deny the difference between social organizations and biological ones. Faced with what they see as the ethical problems that inhere in the answer to such a question, problems they deem to be the problems of the future, Maturana and Varela defer the answer to this question to the future itself: "In fact no position or view that has any relevance in the domain of human relations can be deemed free from ethical and political implications. . . . This responsibility we are ready to take, yet since we—Maturana and Varela—do not fully agree on an answer to the question . . . we have decided to postpone this discussion."[47] That is, Maturana and Varela cannot agree on the status of the following question: Are social organizations inside or outside the purview of biological laws? The decision not to decide, to postpone or defer, allegorizes the futural character of the very membrane between inside and outside, the autonomy machine, that Maturana and Varela deploy. Only retroactively— in the future—is the distinction between inside and outside self-evident. Far from obvious, the mobius space of inside and outside are sites of indeterminacy and undecidability that emerge in the abductive process of living systems. Maturana and Varela's very autonomy is threatened by the force of the problem: they cannot choose not to decide about the future, ethical problems posed by the theory of autopoiesis they offer.[48] Even on their own terms, such a topological distinction poses a cognitive problem that "has to do with the capacity of the observer to recognize the relations that define the system as a unity, and with his capacity to distinguish the boundaries which delimit the unity in the space in which it is realized."[49]

Despite its conceptual trouble, the localization of life rendered by the theory of autopoiesis does much to ensure that artificial life remains confined in its window. With their clear, if tangled, exposition of the claim that autonomy is the fundamental life behavior, they buttress the classical claim for the interiority of organisms even as the life effect is distributed across the multiple links of networks.

By propagating such a clear demarcation between the inside and outside of the living system, Maturana and Varela's argument replicates the historical comportment of life as an entity distinguishable from its environment. The contemporary localization of life extends this encapsulation and treats the fetus as an entity distinct from its mother's body. As scholars such as Susan Squier, Barbara Duden, Karen Newman, Val Hartouni, and Donna

Haraway have argued, the fetus was "born" as a distinct entity through rhetorical and visual techniques that severed it from the maternal body and invested it with life and subjectivity. Both the rhetorics of "choice" and "pro-life," Newman argues, emerge out of a discourse full of rights-laden bodies, individuals in direct conflict that paradoxically dwell in the same body. While Newman overlooks the historical transformations of life in the period that she analyzes, she carefully documents the persistent, historical occlusion of the maternal body and the emergence of the fetus, an emergence that functions through the attribution of distinct interiority to an entity which is, paradoxically, *inside* the invisible maternal body.

In "The Virtual Speculum in the New World Order," Donna Haraway highlights the particularly odd status of such a distinct fetus at a moment when life has been dislocated. Haraway offers multiple readings of a cartoon that she dubs "Virtual Speculum," which features "a female nude . . . in the position of Adam, whose hand is extended to the creative interface with not God the Father but a keyboard for a computer whose display screen shows the global digital fetus in its amniotic sac."[50] Among Haraway's proliferating readings, she argues that the digital fetus is "literally . . . somehow in the computer" and thus "more connected to downloading than birth or abortion . . . the on-screen fetus is an artificial life form."[51]

In Haraway's formulation of the "fetus in cyberspace," the topological comportment of life as outlined by Maturana and Varela returns; life is conceptually or "virtually pregnant"—disturbed at its border between inside and outside, a fetus "in" a nonspace, life has missed its (historical) period. The conundrum posed by "Virtual Speculum" is literally, where is life? Its "source" appears to be the gleaming screen of the workstation, as capital transforms more than the global markets via the new technologies of pixel, keyboard, and perhaps, network. But such a reading immediately overlooks the corporeal connection of the female nude, whose hand touches the keypad, whose digits, perhaps, experience the pain of labor via carpal tunnel syndrome. Thus the digital fetus, awash in amniotic and semiotic fluid, exemplifies the persistent exteriority of living systems. Even under its greatest ideological pressure to be an interiority—the fetus in a box—life resists attributions of autonomy as effectively as genetically modified corn finds its way into a taco shell.[52] Impossible to locate, the life effect occupies a Möbius body, a rhizome that traverses the interiority of the screen and its outside. As a membrane, the screen marks less a clean boundary than a multiplicity. Difficult to narrate—which is before, which after?—such a multiplicity fosters the implosion of the virtual and the actual even as it highlights an odd morphology of life whose representation is a fluctuating network rather than an organism.

The sheer strangeness of this entanglement—the becoming flesh of silicon, the becoming silicon of flesh—seems to foster an abduction, one that *forestalls* the difference of the future by substituting "for a complicated tangle of predicates attached to one subject, a single conception," that is, life. Newman describes this demand as a "referential panic, a need for realist images" that would render the strange new tangles of technoscience and life consumable and narratable.

And yet this panic is not confined to the work of representation—Alife's power emerges not out of the barrel of a gun but from the gestures of mouse and pixel, signifying and asignifying grapple with the machinic phylum. To be sure, the familiar seductions of Alife are rhetorical, but they involve the sculpted, implicitly choreographed movements of bodies as well as the affects provoked by the encounter with Alife creatures. Strange antics indeed.

Notes

1 James Watson, *The Double Helix* (New York: Signet Books, 1968), 18.

2 Here I am borrowing from Stephen Wolfram's understanding of irreducibility—"many times there is no better way of predicting a computation than to actually run it, or to simulate it in some way"; see Rudy Rucker, *Mind Tools: The Five Levels of Mathematical Reality* (Boston: Houghton Mifflin, 1987), 313.

3 François Jacob, *The Logic of Life: A History of Heredity*, trans. Betty E. Spillmann (New York: Pantheon, 1973), 299.

4 Stuart Kauffman, *At Home in The Universe* (New York: Oxford University Press, 1995), 56.

5 Ibid., 57.

6 Ibid., 77.

7 Elias Canneti, *Crowds and Power*, trans. Carol Stewart (New York: Noonday Press, 1962), offers remarkable insight into the different forms—crowds, packs, mobs—such multiplicities can become. See also Gilles Deleuze and Félix Guattarl's treatment of centered versus acentered networks in "Rhizome," in *A Thousand Plateaus: Capitalism and Schizophrenia*, trans. Brian Massumi (Minneapolis: University of Minnesota Press, 1987).

8 Bretton Woods, the agreement ending the orientation of the world currency markets to the gold standard, would mark a comparable dislocation of economic value. For a discussion of the loss of reference associated with this shift, see Brian Rotman, *Signifying Nothing* (Stanford, Calif.: Stanford University Press, 1993), 87–107.

9 My argument that scientists have recently begun interfacing with organisms as informational constructs, of course, does not rule out the possibility and the probability that organisms have always already been, in part, such constructs. My point is, rather, that such an interface has only been made visible and actual via the rhetorics and practices of late-twentieth-century biology. These rhetorics and practices, therefore, disciplined and comported organisms in a way that made the informatic character of life available to the scientific register.

10 See for, example, the description of life as a "duplex" system by Alife researcher Fred Cohen, in Steven Levy, *Artificial Life: The Quest for a New Creation* (New York: Pantheon Books, 1992), 127. While Cohen's insight concerning computer viruses treats artificial life as an ecological event—"a recipe for interactions capable of cooking up the me-

chanics of life required the settting that allowed its potential to be fulfilled"—it also thoroughly dislocates the emergence of life, distributing it to the (at least) parallel processes of program and instantiation, instruction and construction. On the conflation of construction and instruction, see Richard Doyle, *On Beyond Living: Rhetorical Transformations of the Life Sciences* (Stanford, Calif.: Stanford University Press, 1997).

11 Computer scientist Pierre Lévy characterizes the positively affective character of this excited uncertainty: "it is the vertiginous sensation of plunging into the communal brain that explains the current enthusiasm for the Internet"; see Lévy, *Becoming Virtual: Reality in the Digital Age*, trans. Robert Bononno (New York : Plenum, 1998), 145. Crucial to this analysis is the notion that something besides representation fuels the encounter with the machine—a practiced vertigo.

12 Ibid., 29.

13 Charles Sanders Peirce, *Collected Papers* (Cambridge: Belknap Press of the Harvard University Press, 1932), 2:270.

14 An iteration of the most famous example of the syllogism will, I hope, help explicate Peirce's discussion.
    All men are mortal. (Major Premise)
    Socrates is a man. (Minor Premise)
    Therefore, Socrates is mortal. (Conclusion)
    The validity of the conclusion in syllogistic reasoning depends on the validity of the premises. In this instance, the major premise emerges out of the persistent and habitual encounter with death, while the minor premise becomes debatable in the light of Socrates's behavior as a "gadfly."

15 Peirce, *Collected Papers*, 2:643.

16 The exclamation mark indicates my astonishment that the enormous and not yet calculated biodiversity can be represented as "narrow."

17 Chris Langton, "Artificial Life," in *Artificial Life* (Redwood City, Calif.: Addison-Wesley, 1989), 33.

18 Peirce, *Collected Papers*, 2:642.

19 Langton, "Artificial Life," 2.

20 The distributed computing solution Seti@home ⟨http://setiathome.ssl.berkeley.edu⟩ allows users to become part of a network that trolls data for alien signals, right from your computer desktop. The problem of authentication is articulated in the program's software license agreement, where a strange promise is acceded to—the impossible promise *not to become alarmed:* "I understand that strong signals will occasionally be detected and displayed on the screensaver during the course of data analysis. I will not get alarmed and call the press when such signals appear."

21 Langton, "Artificial Life," 24.

22 Ibid., 23.

23 Lynn Margulis, *Early Life* (Boston: Science Books International, 1982), 85.

24 Even within the nucleus of eukaryotes, there are differentials of blockage: "Chromatin is a complex of DNA and protein diffused throughout the cell nucleus most of the time. As a eukaryotic cell prepares to divide, the chromatin condenses into rod-shaped bodies—the chromosomes—and in many the nucleolus disappears"; see Margulis, *Early Life*, 3.

25 Gilles Deleuze and Félix Guattari have described the orchid's "imaging" of the wasp as an example of a rhizomatic relation. The wasp, they argue, is an element in the reproductive system of the orchid, and as such is *part of the orchid's morphology*, its becoming-wasp, the wasp's becoming-orchid. "The orchid deterritorializes by forming an image, a tracing of the wasp; but the wasp reterritorializes on that image. The wasp is nevertheless

deterritorialized, becoming a piece in the orchid's reproductive apparatus. But it reter-retorializes the orchid by transporting its pollen. Wasp and orchid, as heterogeneous elements, form a rhizome"; see *Thousand Plateaus*, 10.

Darwin, by contrast, was convinced that insects must, as autonomous organisms, de-rive some material benefit from the pollination practice, such as nectar, while Sprengel posited the existence of what he called *Scheinsaftblumen* "or sham-nector producers; he believed that these plants exist by an organized system of deception, for he well knew that the visits of insects were indispensable for their fertilization. But when we reflect on the incalculable number of the pollen-masses attached to their proboscides, that the same insects visit a large number of flowers, we can hardly believe in so gigantic an imposture"; see Darwin, *Various Contrivances by which British and Foreign Orchids are Fertilized by Insects* (London: John Murray, Albemarle Street, 1862; CD-Rom, Lightbinders, 1997), 47.

26 Lévy, *Becoming Virtual*, 171 (emphasis in original).

27 Elizabeth Grosz, "Thinking the New," *Symploke* 65 (1998): 38–55.

28 Compare, for example, the status of Alife organisms with other boundary-troubling life-forms. Virus's status as a life-form—is it alive or isn't it?—has no effect on its ability to propagate. By contrast, the success of Alife organisms is tied to the replicability, if not description, of the intensely affective responses to Alife creatures. These affects are replicated, in part, by rhetorical softwares that comport Alife *as* life, and which narrate the liveliness of the creatures to ourselves or others. These narratives render us as what Simon Schaffer and Steven Shapin might characterize as "virtual witnesses" to actu-alized life; see *Leviathan and the Air-Pump: Hobbes, Boyle, and the Experimental Life* (Princeton: Princeton University Press, 1985).

29 Consider, for example, Alife researcher Chris Langton's close encounter of a silicon kind, as told by new media prophet Kevin Kelley:

> Langton remembered working alone late one night and suddenly feeling the presence of someone, something alive in the room, staring at him. He looked up and on the screen of Life he saw an amazing pattern of self-replicating cells. A few minutes later he felt the presence again. He looked up again and saw that the pattern had died. He suddenly felt that the pattern had been alive—alive and as real as mold on an agar plate—but on a computer screen instead. The bombastic idea that perhaps a computer program could capture life sprouted in Langton's mind.

See Kevin Kelley, *Out of Control* (New York: Addison Wesley, 1994), 344. Note, of course, the creepy vitality of ideas of life that themselves "sprout." See also Pier Luigi Luisi, "Defining the Transition to Life," in Francesco Varela and Wilfred Stein (eds.), *Thinking About Biology* (Redwood City, Calif.: Addison, 1992), 35. Luisi writes of the affective charge carried by the claim for the synthesis of vitality: "the self-replicating bounded structures . . . should be considered as minimal synthetic life. Such a statement may possess an unappealing flavor, but I believe one should not be afraid of it. This feeling of unappealingness probably arises for pyschological reasons, but it should not cloud the scientific issue."

30 Charles Darwin, *The Origin of Species* (New York: Modern Library Edition, 1962), 69.

31 Ibid., 70.

32 Richard Dawkins defines his notion of the extended phenotype, where gene action is not confined to the interior of sovereign bodies, in terms of the following "central theorem": "An animal's behavior tends to maximize the survival of the genes 'for' that behavior, whether or not those genes happen to be in the body of the particular animal performing it"; see *The Extended Phenotype* (Oxford: Oxford University Press, 1982), 233. In the case of

Alife creatures, then, the behavior in question is representational "life," and the maximization of the survival and propagation of this behavior is carried out, actualized, by humans. And just as the reproduction of flowering plants depends on imaging insect pollinators, so too does the actualization of Alife depend on imaging vitality for its human propagators. Dawkins discusses a fascinating example of anelid worms that simulated, as a group, anemones. My example here takes the notion one step further, to where the very life of the organism is itself simulated as a seductive tactic.

33 John von Neumann, *Theory of Self-Reproducing Automata*, ed. Arthur W. Burks (Urbana: University of Illinois Press, 1966), p. 47. Remarkably, this treatise was not produced solely by von Neumann himself. Arthur Burks—who also edited several volumes of Charles Sanders Peirce's selected papers—worked with the wire recorders that had preserved von Neumann's speech and rendered it into text.

34 Philip K. Dick, *Game-Players of Titan* (New York: Vintage, 1992), 6

35 Ibid., 116.

36 Ibid.

37 Philip K. Dick, *Radio Free Albemuth* (New York: Vintage, 1985), 10.

38 Ibid., 110–111.

39 Numerous burial services—both of the electronic and the earthly kind—have been offered worldwide for artificial life pets. See Caroline Wyatt, "Despatches," BBC News, November 20, 1997, ⟨http://news6.thdo.bbc.co.uk/hi/english/despatches/newsid__ 32000/32630.stm.⟩

40 Humberto R. Maturana and Francisco J. Varela, *Autopoiesis and Cognition: The Realization of the Living* (1968; Dordrecht: Reidel, 1980), 101.

41 Donna Haraway, *Modest__Witness@Second__Millennium.FemaleMan@___Meets__Onco Mouse*™: *Feminism and Technoscience* (New York: Routledge, 1997), 186.

42 Maturana and Varela, *Autopoiesis and Cognition*, 101.

43 Ibid., 78–79.

44 Ibid., 102.

45 Ibid.

46 Ibid., 73.

47 Ibid., 118.

48 Instructive in this regard is the ethics of Emmanuel Levinas, for whom the autonomous subject is constituted as the continual encounter with alterity, an encounter that renders it topologically not as a unity but as a Klein bottle: "Subjectivity realizes these impossible exigencies—the astonishing feat of containing more than it is possible to contain"; see Levinas, *Totality and Infinity: An Essay on Exteriority*, trans. A. Lingis (Pittsburgh: Duquesne University Press, 1969), 27.

49 Maturana and Varela, *Autopoiesis and Cognition*, 108.

50 Haraway, *Modest__Witness@Second__Millennium*, 176.

51 Ibid., 186.

52 Indeed, if life has the attribute of autonomy, it has an equal measure of deterritorialization; see Nell Boyce, "Taco Trouble," *The New Scientist*, 7 October 2000.

The word for world is computer: simulating second natures
in artificial life

Stefan Helmreich

In a video promoting the Santa Fe Institute, biologist Thomas Ray introduces us to his computer simulation of evolution, Tierra. Situated in the
midst of a *Star Trek*–like set, he narrates us into the world shimmering
beyond his computer screen. Our gaze, guided by the camera's eye, passes
through the cathode ray looking glass and is catapulted into a dreamlike
fantasy flight over the rich "ecosystems" of an "artificial world," where
"populations" of "digital organisms" vie to "self-replicate." Ray provides a
voice-over in the style of a nature show host:

> Now we're going into the computer which is where the creatures
> actually live, and we can see the environment that they inhabit. This is
> the central processing unit that provides their energy, the CPU, like the
> sun. And now we're going to fly over the memory which provides the
> space that the creatures inhabit. . . . . In the real world, plants cover
> the surface of the land. In the digital world, digital organisms fill the
> memory space.[1]

For Ray and many others working in the field of Artificial Life, computer
programs that self-replicate can be seen as new forms of life, forms that
can be sparked into existence by scientists who view the computer as an
alternative universe. As Ray tells us in his video narration, "Tierra is Spanish for Earth. I called it that because I thought of it as a different world that
I could inoculate with life. It's not the same world we live in. It's not the
world of the chemistry of carbon and hydrogen. It's a different universe.
It's the chemistry of bits and bytes."[2]

How have computers become different universes? How have practitioners of Artificial Life, and of the sciences of complexity more generally,
managed to transubstantiate computer simulations into epistemological
and ontological spaces capable of harboring "life"? There are many answers. Some come from historians and philosophers of science, who have

alerted us to how theory and experiment fold into one another in the era of computer modeling, positioning simulations as halfway houses between the real and the imaginary.³ Others come from simulation scientists themselves, who maintain that the universe is a mammoth computer, a conviction that underwrites the idea that computers are kinds of universes.⁴ In an age when the essence of vitality is often compressed into "genes" figured as equivalent to "programs,"⁵ it is not surprising that computational universes can support "life."

But these historical, philosophical, and internalist accounts of simulation science must be supplemented by attention to the cultural contexts within which scientists have worked. Drawing on anthropological fieldwork I conducted among Artificial Life scientists at the Santa Fe Institute for the Sciences of Complexity in Santa Fe, New Mexico, from 1993 to 1994, I maintain that simulations "come to life" when elements from complex systems theory and computer science combine with elements from Judeo-Christian cosmology, science fiction, an American frontier imagination, a calculative rationality that is often coded masculine, and a visual common sense committed to one-to-one mappings between representation and reality. Welded together, these elements persuade Artificial Life scientists that computers can be considered their own worlds, a belief that may foreshadow new ways of thinking about nature in the age of simulation. Downloading nature into computers may offer a new court of appeal for those who would point to nature as a source of lessons about what is inevitable, given, and even rationally designed.⁶ On the other hand, bit-mapping nature in silico may make obvious the role of human interpretation in constructing any nature.

The title of this essay puns on science fiction author Ursula Le Guin's The Word for World Is Forest, a tale about a forest-dwelling people who use their dreams as resources for guidance in the waking world.⁷ The Le Guin story is a romantic one that imagines a people existing outside the stream of history, so it may seem odd to tweak its title to speak of late-twentieth-century scientists who exist largely within a post–Judeo Christian, European-American cultural context. There is, however, a romantic incandescence to the project of Artificial Life, to practitioners' claims that they can mime nature technologically, that leads me to discern similarities between Artificial Life researchers and Le Guin's mythical Athsheans. Artificial Life workers, like Athsheans, resource a set of culturally specific dreams— of transcendence, discovery, creation, colonization—to manufacture the worlds, the second natures, they interpret. I take the concept of "second nature" from a lineage that begins with Hegel, who "taught us to see a difference between 'first nature'—the given, pristine, edenic nature of physical and biotic processes, laws, and forms—and 'second nature.' Second nature comprises the rule-driven social world of society and the market, culture and the city, in which social change is driven by a parallel set of socially imposed laws."[8] Computer worlds are second natures in the sense that they are human constructions but also in that they are modeled after first nature. The artificial worlds of Artificial Life are "second natures" in still another sense: they not only ape first nature but seek to replace it, to succeed it as a resource for scientific knowledge. Before traveling into these second natures, background on Artificial Life will be helpful, as will a soirée into Thomas Ray's Tierra, a popular artificial life world.

Artificial life and Tierra

Artificial Life is dedicated to the computer simulation—or, more grandly, synthesis—of biological systems. It emerged in the late 1980s, growing out of conversations among computer scientists, biologists, and physicists. Artificial Life researchers envision their project as a reinvigorated theoretical biology and as an initially more modest but eventually more ambitious enterprise than Artificial Intelligence. Where Artificial Intelligence attempted to model the mind, Artificial Life hopes to capture on computers the formal properties of organisms, populations, and ecosystems—the life processes that support the evolution of such things as minds. The intellectual charter for this practice is summarized in Artificial Life scientist Christopher Langton's declaration that life "is a property of the organization of matter, rather than a property of matter itself."[9] Some have found this claim so compelling that they believe real artificial life-forms can exist in computers, and they hope the creation of computer life-forms will expand biology's purview to include not just life-as-we-know-it, but also life-as-it-could-be.[10] Though Artificial Life research is carried out at many places, it has been most popularly associated with the Santa Fe Institute (SFI), a gathering ground for an international community of scientists interested in computer modeling of nonlinear dynamics in physical, biological, and social systems.

Ray's Tierra exemplifies the approach of the sciences of complexity, setting computational entities into interactions which result in nonlinear, emergent dynamics. It also offers up the most ambitious claim of Artificial Life, namely that propagating information structures can count as real life in a virtual world. Tierra is a computer program that serves as an environment within which short assembly language programs "resident" in random-access memory (RAM) can replicate based on how efficiently they make use of central processing unit (CPU) time and memory space. When the program is set in motion, an array of "digital organisms" emerge, descending in a cascade from an ancestral self-replicating program inoculated into the system by the programmer at the beginning. Only the "fittest"—those which replicate most quickly, those which pirate the replicative subroutines of other programs—survive. According to Ray, Tierra is not just a *model* of evolution. It is "an *instantiation* of evolution by natural selection in the computational medium."[11] Ray's definition of evolution allows him to enfranchise Tierran organisms into the dominion of life: "I would consider a system to be living if it is self-replicating, and capable of open-ended evolution."[12] Indeed, for Ray, the parallels are many between organic and computer life:

> Organic life is viewed as utilizing energy, mostly derived from the sun, to organize matter. By analogy, digital life can be viewed as using CPU (central processing unit) time, to organize memory. Organic life evolves through natural selection as individuals compete for resources (light, food, space, etc.) such that genotypes which leave the most descendants increase in frequency. Digital life evolves through the same process, as replicating algorithms compete for CPU time and memory space, and organisms evolve strategies to exploit one another.[13]

The digital life in Tierra consists of assembler language programs. The code of each creature occupies some block of memory and reproduces by copying its code into another block. Tierran creatures are nothing more than short programs that code for self-replication.

Ray has claimed that in his observations of Tierra, he has seen the emergence of evolutionary dynamics, including the rise of digital organisms that parasitize others. Ray argues that getting to know the alternative biology in Tierra can help biologists expand their theories about what is necessary and what contingent in biological systems. There is of course a paradox here, one Ray recognizes but finesses.[14] This is that the system must be engineered with basic evolutionary ideas in mind (scarcity of resources, the idea that organisms can only be one place at one time, etc.), but must *also* be general enough to allow new, alternative evolutionary phenomena to emerge. One researcher I interviewed said that deciding which lifelike notions to include as basic was "where the 'art' in Artificial Life comes in." The programmer-experimenter must mix intuitions about where science is already correct with programming decisions general enough to allow new phenomena to grow. Ray describes the process this way: "We must derive inspiration from observations of organic life, but we must never lose sight of the fact that the new instantiation is not organic and may differ in many fundamental ways."[15] Ray sees a new nature inside the computer but has in fact already engineered this nature to have features with which he is familiar. It is these features that render the program legible as a simulation of evolution at all. As N. Katherine Hayles writes of Tierra, "The interpretation of the program through biological analogies is so deeply bound up with its logic and structure that it would be virtually impossible to understand the program without them. These analogies are not like icing on a cake, which you can scrape off and still have the cake. . . . . The biological analogies do not embellish the story; in an important sense, they constitute it."[16] I have explored these biological analogies in depth elsewhere.[17] Here, I focus on scientists' foundational belief that computers count as alternative worlds or universes.

Computers as worlds or universes

For many people I interviewed, a *world* or *universe* could be understood as a self-consistent, complete, and closed system governed by low-level laws that supported higher-level phenomena which, while dependent on these elementary laws, could not be simply derived from them (here *low* refers to physics and *high* to chemical, biological, and social phenomena). Put into the language of the sciences of complexity, a world or universe is a dynamical system capable of generating surprising emergent properties. If we accept this definition, computational systems reasonably count as worlds or universes. This is the view taken by scientist David Hiebeler in an article titled "Implications of Creation," where he writes, "Computers provide the novel idea of simple, self-contained 'artificial universes,' where we can create systems containing a large number of simple, interacting components. Each system has its own dynamics or 'laws of physics'; after specifying those laws, we set the system into motion and observe its behavior."[18] One researcher drew the similarity between worlds and universes and computer simulations for me rather tightly: "Worlds and universes are complex processes, based on fixed, low-level principles. Computer simulations are complex processes, based on fixed, low-level principles." Computer simulations, he concluded, can therefore be worlds or universes. How have pronouncements like these become possible? A glance at the history of computer modeling offers preliminary answers.

Computers have been used for quite a while to simulate aspects of the world we inhabit. Los Alamos scientists used them early on to simulate nuclear reactions, and the American military has also used them to model nuclear war scenarios. Simplified computer models of the world became a component in Artificial Intelligence research in the early 1970s, as scientists attempted to provide programs with representations of the world about which the programs were to make reasoned decisions.[19] These representations of aspects of the world came to be called microworlds. Paul Edwards has argued that computer microworlds were also "closed worlds," and were manufactured as such by the military, which favored computer simulations of conflict that simplified international relations into formal game scenarios. Edwards has argued that the military origins of computer modeling set the stage for understanding computers as surrogate worlds, and has written that the image of computer as world has served as a powerful attraction for rationalist epistemologies:

> Every microworld has a unique ontological and epistemological structure, simpler than those of the world it represents. Computer programs are thus intellectually useful and emotionally appealing for the same reason: they create worlds without irrelevant or unwanted complexity. . . . The programmer is omnipotent but not necessarily omniscient, since highly complex programs can generate totally unanticipated results. . . . This makes the microworld exceptionally interesting as an imaginative domain, a make-believe world with powers of its own.[20]

Edwards captures the textures of computing that have attracted many people to fields like Artificial Life. He maintains that the microworlds of computers have been appealing particularly to many men, to whom "power is an icon of identity and an index of success. . . . With a 'hard' formalized system of known rules, operating within the separate reality of a microworld, one can have complexity and security at once: the score can always be calculated; sudden changes of emotional origin do not occur. Things make sense in a way human intersubjectivity cannot."[21] Following Edwards, I suggest that rational, calculative masculinity has been one resource that has helped transform computers into worlds for many Artificial Life scientists. In the late seventies and early eighties, many of those men who came to be involved in Artificial Life purchased PCs as personal rites of passage, rites they narrated to me as accounts of mastery and self-actualization through the taming of machinery (some 90 percent of scientists in Artificial Life are men). Several told me their wives would complain when they stayed up all night programming, "lost in their own worlds." One said of his spouse, "I think she is jealous of the computer." Another self-consciously reflected that he built simulations because he wanted to get away from a world that felt difficult to control. He told me, "I didn't like chemistry because it involved measuring the real world, and I've never been a big fan of the real world. I'd much rather make my own world on the computer and then measure that." Of course, to boil down to masculinity people's perceptions of computers as worlds is overly simple. Masculinity is by no means monolithic, and gender is a structure of dispositions that, while often bundled together in people referred to as men or women, are not all or ever the exclusive property of such persons. One woman researcher I spoke with well expressed what some might see as a typically masculine motivation for using computers: "Having the world at your control is something that has a lot of appeal. You can figure the whole thing, with the parameters you want. It gives you this sense that everything's fine and under control."

Of course, in many Artificial Life worlds, everything is not under control. Many researchers hope that their worlds will surprise them. In Sherry Turkle's terms, an aesthetic of "hard mastery," in which programmers seek to comprehend and structure programs, is giving way to one of "soft mastery," in which people encounter and appreciate computers as dynamically changing systems only partially available to complete understanding.[22] The masculinist imperatives of standard AI—rationality, objectivity, disinterestedness, and control—have yielded to what are commonly characterized as more "ecological" programming techniques. Turkle argues that new modes of computing may be more appealing to women, who have stereotypically been trained to value negotiation, relationship, and attachment. But this is only part of the story. Many men in Artificial Life see themselves refiguring their masculinity as they work with modes of computation that mimic nature. In a conversation with two younger heterosexual men about how their gendered subjectivity might be implicated in their science, they told me Artificial Life allowed them to express and work with a side of themselves that was more intuitive, perhaps more stereotypically "feminine," even "Gaian." This way of putting things shores up the category of gender even as it purports to erode it, keeping "femininity" stable as a resource that men might mine to broaden their intellectual work. As Judith Genova notes, "Perhaps it is no accident that men become intuitive and holistic just at the time when computers can successfully simulate logical, analytic thought."[23] The ways gender is recoded in Artificial Life scientists' subjectivities troubles any simple attachment of Artificial Life aesthetics to masculinity, old or new, but it is nonetheless important to keep masculinity in view, especially when male researchers speak of themselves, as I will discuss later, as masculine gods "inoculating" computational matrices (often, "soups") with self-replicating "seed" programs.

The notion of microworld has largely been superseded by the "artificial world," signaling a shift from understanding computer worlds as toy worlds to seeing them as realities in their own right. How do people articulate this belief that computers contain real worlds? Examining what I will tag "scientific warrants" provides a view of how many Artificial Life researchers I interviewed reflected on this question.

Scientific warrants for construing computers as worlds

I want to back up a bit more into history to revisit high speed computing and simulation as they developed at Los Alamos during and after World War II. Los Alamos is important for the Artificial Life tale because senior fellows at the lab created SFI, and much common sense about computers comes directly from this institution.

Peter Galison has discussed how computer simulations were employed by nuclear weapons researchers at Los Alamos and elsewhere for problems too complex to solve analytically and impossible to investigate experimentally. Computer simulations of microphysical events like nuclear reactions came to occupy a liminal place between theory and experiment. On the one hand, they resembled theory because they set into motion processes of symbolic manipulation. On the other hand, they resembled experiments because they exhibited stable results, replicability, and amenability to error-analysis procedures. "Data" generated by simulations could be given the same epistemic status as data from "real" experiments.[24] On a deeper level, the simulations in question—those using pseudorandom numbers as starting points for the emulation of physical processes—were seen to share a "fundamental affinity" with "the statistical underpinnings of the world itself."[25] As Galison notes, "The computer began as a 'tool'— an object for the manipulation of machines, objects, and equations. But bit by bit (byte by byte), computer designers deconstructed the notion of a tool itself as the computer came to stand not for a tool, but for nature itself."[26] Galison claims that Monte Carlo simulations, as these were called, were assimilated to experiment in part because the stochasticity embedded in them was seen as directly analogous to the stochastic processes that characterized microphysical nature. Theories could be tested with reference to an artificial reality that was just as good as the real thing, that was in fact itself a sort of alternative, artificial reality. Simulations could be stand-ins for experiment, but more boldly they could be seen as understudies for nature itself.

Computers as nonlinear and complex systems

Researchers in Artificial Life inherited this tradition of simulation. In my interviews, people gave me a more current take on the ontological similarities between worlds and computers: both worlds and computers are nonlinear dynamical systems. More, they imagined computer and natural systems as transforming information. Many complexity scientists hold that the universe is just a computer transforming information, and that computers, which do this so well, should on this definition be considered universes or worlds. This belief allows Thomas Ray to write, "The computational medium of the digital computer is an informational universe of boolean logic, not a material one."[27] But matter need not be factored out entirely. Steen Rasmussen and others have insisted that matter is self-programmable, that the world we inhabit is really a computation instantiated in the pattern of matter that composes our universe.[28]

Cellular automata

Many in Artificial Life have been enamored of a mathematical formalism known as the cellular automaton (CA), which allows a programmer to specify rules for local interaction between "cells" on a lattice-like grid, specifications which then produce a variety of emergent patterns. States of cells change according to the states of their neighbors, and using CAs, a variety of simple systems can be simulated, including self-reproducing automata.[29] Because of the CA formulation's generality, virtually any process that can be algorithmically specified can be modeled. CAs thus support a sort of alternative universe capable of sustaining alternative realities. Langton writes that "the transition function for the automata constitutes a local *physics* for a simple, discrete space/time universe. The universe is updated by applying the local physics to each 'cell' of its structure over and over again."[30] Artificial Life researcher Andrew Wuensche writes that a CA may "be viewed as a logical universe with its own local physics, with [emergent structures] as artificial molecules, from which more complex [emergent structures] with the capacity for self-reproduction and other essential functions of biomolecules might emerge, leading to the possibility of life-like behaviour."[31] Some people I interviewed claimed that our universe might be thought of as a giant cellular automaton.[32] While I was at SFI, I overheard many jokes and serious comments about how this could be so. In 1994 one researcher wrote me in an e-mail: "Many years before the term 'Artificial Life' was coined by Langton I learned about cellular automata and von Neumann's research. . . . From that point on I tended to think of the physics of our universe as a 3-D cellular automaton (even though I know such models tend not to have gravity, other fields, distortions of space, quantum effects, etc.). I view myself as a pattern in a CA world—one in which motion is just an illusion."[33]

Many people studying Artificial Life at Santa Fe began experiments at home with CAs, and some were fascinated by the most popular of CA formulations, Conway's "Game of Life," in which the switching off and on of cells results in patterns that can look like shapes growing or traveling across the screen. One researcher told me in an e-mail interview, "By chance, my work in computers got me exposed, in a peripheral way, to

Cellular Automata (CA). What I like about CA is the very thing that makes them so unusual in the clockwork world of computers: they harbor the unexpected; emergent behavior that's not designed into them." CAS are compelling metaphors for thinking about complex emergent phenomena in part because they are visually surprising, and the consequences of low level rules are quite unpredictable.

The physical Church-Turing thesis

Because SFI is so focused on computation, many SFI scientists subscribe to versions of the physical Church-Turing thesis: the notion that any physical process can be thought of as a computation, and that therefore any physical process can be recreated in a computational medium. In the 1930s, mathematician Alan Turing proposed that any human cognitive process that could be described algorithmically could be translated into a sequence of zeroes and ones and could therefore be implemented on a computer. After it was shown that the lambda calculus (a recursive mathematical system that allows functions to act as objects of other functions) of logician Alonzo Church was equivalent in power to the Turing machine (itself equivalent to CAS), the Church-Turing thesis was formulated, stating that all reasonable computational processes are equivalent to discrete, digital models of computation. The physical Church-Turing thesis figures the universe as such a computational process, as a physical system that converts inputs, or initial conditions, into outputs, the system's final state.

All of these elements—ideas about computers as stochastic and nonlinear systems, formulas drawn from cellular automata theory, and the physical Church-Turing thesis—are scientific resources that Artificial Life researchers use to think of computers as ontologically like nature and hence capable of being worlds. The idea that the physical world is a computation is so prevalent that it affects the ways some people construct their intuitions about the physical world. As one Artificial Life researcher told me, "The universe is a computation of which I am a part. . . . It's the information patterns of things that make them important rather than anything else. . . . I'm perfectly comfortable . . . with the notion that we're running on some big simulator out there. It seems to me as good as anything."

Philosophical positions

While there are many scientific warrants for construing computers as worlds, the equation remains controversial. Some philosophers of computing have argued that computers are tools for manipulating symbols; the fact that they can be systematically interpreted as worlds does not make them worlds. Langton told me of a conversation he had with philosopher Stevan Harnad on this topic. Harnad said he could imagine an artificial system that produced signs systematically interpretable as a world. As Harnad reconstructs his position in the journal *Artificial Life*,

> The virtual system could not capture the critical (indeed the essential) difference between real and virtual life, which is that the virtual system is and always will be just a dynamical implementation of an implementation-independent symbol system that is systematically interpretable as if it were alive. Like a highly realistic, indeed oracular book, but a book nonetheless, it consists only of symbols that are systematically construable (by us) as meaning a lot of true and accurate things, but without those meanings actually being in the symbol system.[34]

Harnad argues that virtual worlds and virtual life can no more be real worlds and life than simulated fires can make things burn.

Computationally inclined Artificial Life researchers have a response: things *can* be alive or on fire *with respect to* computer worlds. One researcher told me: " 'Life' can only be defined WITH RESPECT TO A PARTICULAR PHYSICS. A computer virus is almost as 'alive' as a real virus (not yet, but close) but only within the physics of the computer memory." This notion that computers can contain separate, closed worlds came up again and again, and was one of the rhetorical moves people used to grant simulations an ontology usually reserved for universes. One person told me that, with computer simulations, "we have something totally self-contained in the computer." Another argued for the distinctness of computational realities by saying that self-reproducing programs were "alive in there," but only "a model out here." He continued, "They're alive with respect to their own universe, their own rules. Wimpy, pitiful little life, but I can't rule it out."

But researchers in computational Artificial Life are not content simply claiming that computers contain symbolic or informational worlds. Some have been adamant about asserting the physicality or materiality of computers and have used this as a lever to claim that real artificial worlds and life can exist in the computer. Bruce MacLennan makes an argument that many Artificial Life workers have found compelling:

> I want to suggest that we think of computers as programmable mass-energy manipulators. The point is that the state of the computer is embodied in the distribution of real matter and energy, and that this matter and energy is redistributed under the control of the program.

In effect, the program defines the laws of nature that hold within the computer. Suppose a program defines laws that permit (real!) mass-energy structures to form, stabilize, reproduce, and evolve in the computer. If these structures satisfy the formal conditions of life, then they are real life, not simulated life, since they are composed of real matter and energy. Thus the computer may be a real niche for real artificial life—not carbon based, but electron-based.[35]

This interpretation has pleased people fussy about the distinction between worlds and universes; here computers are figured as worlds that exist in the same universe as the familiar organic world we know. A computer simulation manipulates structures and patterns of real voltages and so is not purely "symbolic," though its symbolic character is important for how those real events take place.

But the fact of materiality does not solve the question of symbols. When MacLennan writes, "If these structures satisfy the formal conditions of life, then they are real life, not simulated life, since they are composed of real matter and energy," he not only retreats from the materiality he is trying to assert (by stating that what matters is form), but also ignores the fact that whatever the "formal conditions of life" are, they will be defined in language. Brian Cantwell Smith has argued that computers' very physical existence is completely enmeshed in our social and linguistic world, that the way disc drives, windows programs, file systems, and RAM caches work results from many scientific, cultural, and economic decisions.[36] To treat computers as entities that come to us straight from nature ignores the decisions that have produced them, that allow us to contemplate them as worlds. Material practices—of which language is one—have made and will make or unmake computers as worlds or as niches for artificial life. Langton's argument that computer programs that cannot be linguistically distinguished from known life-forms should be considered life—the position he took against philosopher Stevan Harnad—recognizes that language is an essential technology for materializing vitality. I would side with Langton against Harnad, though argue that Langton does not adequately recognize the specificity of his own use of a language that names life as a process that haunts both organic and electronic entities. As Richard Doyle points out in On Beyond Living, there is nothing about computers in themselves (their speed, capacity to set math in motion, informatic logic) that forces us to see them as homes for artificial life. Rather, Doyle notes, our ways of seeing and working with them are structured by a kind of "rhetorical software" that allows us to enliven them with narratives fished from the reservoirs of our culture. In the next sections, I write about the extrascientific or "cultural" resources researchers employ to animate artificial worlds, resources I rhetorically separate from "science" because they are not called upon in official discourse about why computers are artificial worlds.

Cultural resources for constructing computers as worlds

Many Artificial Life researchers use *world* or *universe* as synonymous with *nature*, and Western science usually constructs *nature* as a system ordered by physical and chemical laws. Thomas Ray says that Tierra can be considered an artificial world because it has its own physics and chemistry whose rules produce a nature in the computer.[37] Ray writes that we must "understand and respect the *natural* form of the digital computer, to facilitate the process of evolution in generating forms that are adapted to the computational medium, and to let evolution find forms and processes that *naturally* exploit the possibilities inherent in the medium."[38] Following the logic of Tierra as alternative nature, a supporter of Ray said at one workshop, "I would argue that your work is empirical, not theoretical. You've built a new world and are doing empirical work in it." The creation of laws in the universe of the computer, Artificial Life researchers believe, is an important step toward evolving real virtual life. For Ray, the future for artificial life looks bright once we recognize that computers can be full-blown worlds:

> Until recently, life has been known as a state of matter, particularly combinations of the elements carbon, hydrogen, oxygen, nitrogen, and smaller quantities of many others. However, recent work in the field of AL has shown that the natural evolutionary process can proceed with great efficacy in other media, such as the informational medium of the digital computer. These new natural evolutions in artificial media are beginning to explore the possibilities inherent in the "physics and chemistry" of those media. They are organizing themselves and constructing self-generating complex systems. While these new living systems are still so young that they remain in their primordial state, it appears that they have embarked on the same kind of journey taken by life on earth and presumably have the potential to evolve levels of complexity that could lead to sentient and eventually intelligent beings.[39]

I've been following Paul Edwards in understanding computers as closed worlds, ordered and technically created as bounded conceptual spaces. Edwards writes that the alternative to the closed world has been not the open world, but the "green world": "The green world is an unbounded natural setting, such as a forest, meadow, or a glade. . . . . Green world drama thematizes the restoration of community and cosmic order through the transcendence of rationality, authority, convention, and technology."[40]

In the imagination and practice of Artificial Life scientists, the closed world comes together with the green world; computers are cultivated to contain worlds open with possibility, resistant to total rational explanation, and full of surprising (but law-governed) potential. Claus Emmeche's book *The Garden in the Machine* diagnoses this fusion of closed and green worlds in Artificial Life, this hoped-for coming together of the planned and the possible in the sphere of simulation.

Western creation stories

Definitions of the universe, worlds, or nature as law-governed resonate with conceptions specific to Western Judeo-Christian cosmology as it has been shaped in the wake of the scientific revolution; these definitions summon up images of a law-giving Creator, and recall pictures of the universe as a giant clockwork or as a book in need of careful reading and deciphering. Many of the people I interviewed were raised as Jews or Christians, but virtually all now count themselves as atheists, a belief system that did not prevent them, when speaking of artificial worlds, from relying on a host of Creationist mythologies, sometimes directly, sometimes channeled through the medium of science fiction. In one conversation, a researcher explicitly asked me "to think theologically for a moment" when trying to understand artificial worlds. He told me that there is a moment of creation when the programmer writes the formal rules that will govern the system. To buttress his arguments about why simulations might be considered worlds, one man appealed to the possibility that our universe might be just a cosmic simulation. He said, "If God up there turned off the simulator and then turned it back on again, we wouldn't know. So, that puts us in some kind of epistemologically inferior position, a lesser degree of reality than Him, since we can't do the reverse. And that would be true of the guys inside [the computer]."

Describing programmers as a genus of god was a frequent strategy among people with whom I spoke, and this was not just a playful way of speaking, but a move that granted programmers the epistemological authority to erase their own presence as the beings who gave their simulations meanings as worlds. This permitted programmers to have the same relation to their simulations as the god of monotheism has to His creation (most vividly in his Deist incarnation as a divine watchmaker). Both occupy a transcendent position, the position of the "unmoved mover." When I asked one researcher how he felt when he built simulations, he replied bluntly, "I feel like God. In fact, I am God to the universes I create. I am outside of the time/space in which those entities are embedded. I sustain their physics [through the use of the computer]." MacLennan's discussion of the experimental use of artificial worlds relies on images of a god who creates and then dispassionately observes the world: "Because synthetic ethology creates the worlds it studies, every variable is under the control of the investigator. Further, the speed of the computer allows evolution to be observed across thousands of generations; we may create worlds, observe their evolution, and destroy them at will."[41] The use of this kind of God imagery allows Artificial Life researchers to disappear themselves from the scene and give their simulations what Hayles has called "ontological closure."[42] This desire to push to one side the human activity involved in making artificial worlds was in evidence at one Artificial Life conference, when Langton contended that some simulations have less human agency embedded in them than others. Many allow the programmer to be the agent of natural selection, but it is Ray's Tierra that Langton describes as "the system that went all the way," the system in which "we've removed the hand of God."[43] The God imagery allows programmers to alternately shape and observe their worlds, to minimize the importance of their own post-creation interventions, and to draw the boundaries between creation and tinkering strategically. God imagery also secures a sort of ultimate objectivity. As Ray puts it in one interview, "Even if my world gets as complex as the real world, I'm god. I'm omniscient."[44]

God imagery is ubiquitous in popular treatments of Artificial Life. An advertisement for SimEarth™, computer software that simulates global ecological dynamics, reads: "More than just a home computer game, Sim-Earth™ was developed with Professor James Lovelock, originator of the Gaia hypothesis, to give us all an opportunity to play god—from the safety and security of our own homes." Ad copy for SimLife™, a computer toy for experimenting with evolution of artificial organisms, reads: "Build your very own ecosystem from the ground up, and give life to creatures that defy the wildest of imaginations." The theological elements that make Sim-Life™ thinkable are apparent in a review of the product written by Chris Langton: "The role of the user in these games is not so much participant in the action, as is the case with most computer games, but rather as the reigning 'God' who designs the universe from the bottom up. . . . In SimLife, Maxis has essentially created a flight simulator that gives one a taste of what it would be like to be in the pilot's seat occupied by God."[45] Note the identification of God with a pilot in a flight simulator; this reminds us that the god we are supposed to have in mind is a god in heaven. This is a god who uses the tools of extraterrestrial technology to examine and create the world, and the satellitelike map given to the user of Sim-Life™ is a perfect tool for this floating panoptic entity.

The god after which Artificial Life researchers are imagined is a biblical god. And more, a masculine god. Ray names himself a god who "inoculates" the "soup" of Tierra with a single self-replicating "seed" "ancestor" program. Carol Delaney has argued that in cultures influenced by Judeo-Christian narratives of creation and procreation, to use the word "seed" to speak of the impetus of creation summons forth gendered images. In the creation tales of these traditions, God, imagined as masculine, sparks the formless matter of earth to life with a word, a kind of divine seed, the *logos spermatikos*.[46] Creation and procreation in these narratives is monogenetic, generated from one source, symbolically masculine. "Man" and "God" take after one another. The creation in Tierra—and note that *Tierra* means *soil* as well as *Earth* in Spanish—symbolically mimics the biblical story of creation. We might see in Tierra images of a symbolically "male programmer mating with a female program to create progeny whose biomorphic diversity surpasses the father's imagination."[47] The programmer in Artificial Life becomes God the Father, not surprising given the gendered character of the Judeo-Christian deity after which researchers model themselves.

Science fiction

Science fiction was a rich resource for my informants' imaginings of computers as worlds. Several referred me to stories in which artificial worlds were created by entities who imagined themselves gods. Stanislaw Lem was mentioned frequently, particularly a story in his book, *The Cyberiad*.[48] I reproduce a fragment of the story here because for Artificial Life it has become almost scripture. The discussion between the two characters anticipates almost exactly the conversation between Langton and Harnad reported earlier—and anticipates it so exactly that one might conclude that science fiction informs science as much as the reverse. Trurl the constructor has just returned home to explain to his friend Klapaucius that he has fashioned a model of a kingdom for a despotic king. He feels he has given this king a toy that will keep him from oppressing real creatures.

> "Have I understood you correctly?" [Klapaucius] said at last. "You gave that brutal despot, that born slave master, that slavering sadist of a painmonger, you gave him a whole civilization to rule and have dominion over forever? . . . Trurl how could you have done such a thing?!"
>
> "You must be joking!" Trurl exclaimed. "Really, the whole kingdom fits into a box three feet by two by two and a half . . . it's only a model . . ."
>
> "A model of what?"
>
> "What do you mean of what? Of a civilization, obviously, except that it's a hundred million times smaller."

"And how do you know there aren't civilizations a hundred million times larger than our own? And if there were, would ours then be a model? And what importance do dimensions have anyway? In that box kingdom, doesn't a journey from the capital to one of the corners take months—for those inhabitants? And don't they suffer, don't they know the burden of labor, don't they die?"

"Now wait just a minute, you know yourself that all these processes take place only because I programmed them, and so they aren't genuine. . . ."

"Aren't genuine? You mean to say the box is empty, and the parades, tortures and beheadings are merely an illusion?"

"Not an illusion, no, since they have reality, though purely as certain microscopic phenomena, which I produced by manipulating atoms," said Trurl. "The point is, these births, loves, acts of heroism and denunciations are nothing but the minuscule capering of electrons in space, precisely arranged by the skill of my nonlinear craft, which—"

"Enough of your boasting, not another word!" Klapaucius snapped. "Are these processes self-organizing or not?"

"Of course they are!"

"And they occur among infinitesimal clouds of electrical charge?"

"You know they do."

"And the phenomenological events of dawns, sunsets and bloody battles are generated by the concatenation of real variables?"

"Certainly."

"And are we not as well, if you examine us physically, mechanistically, statistically and meticulously, nothing but the minuscule capering of electron clouds? Positive and negative charges arranged in space? And is our existence not the result of subatomic collisions and the interplay of particles, though we ourselves perceive those molecular cartwheels as fear, longing, or meditation? And when you daydream, what transpires within your brain but the binary algebra of connecting and disconnecting circuits, the continual meandering of electrons?"[49]

Trurl is being accused of playing a god fully outside creation. In this story, we reread a familiar theology, translated into cybernetic jargon.

Science fictional common sense encodes a number of cultural themes aside from those around creation. The most salient is of exploring and colonizing the universe. Several researchers told me that creating artificial worlds in computers is necessary for a universal biology since, at the moment, we are unable to do natural history on other planets. Since we don't have Star Trek's USS Enterprise at our disposal, we must construct the worlds we would explore. This imagining of exploring the universe is warranted in the Artificial Life's official ideology. As Langton puts it in a promotional video, "It's very difficult to build general theories about what life would be like anywhere in the universe and whatever it was made out of, when all we have to study is the unique example of life that exists here on Earth. So, what we have to do—perhaps—is the next best thing, which is to create far simpler systems in our computers."[50] Doyle has commented that the idea that theoretical biology is hamstrung by its imprisonment on earth derives from a desire to occupy a transcendent position from which to scan the universe. It is this position which Artificial Life scientists strive to create for themselves by manufacturing their own galaxies of artificial worlds.

There is also another figure at work here: the future. Like people who have their bodies or heads put in cryonic suspension, Artificial Life researchers are impatient for "the future" to arrive, impatient enough to want to reel it into the present. One person said in conversation with me that he was frustrated by his finitude and mortality. He wished he could live long enough to see "all the cool things that would happen in the future." He wished he could be around when and if humans eventually contacted life on other planets, and his desire to create life in computers had to do with this curiosity. In this vision, cyberspace is figured as the new outer space.

Perhaps the impulse to think of computers as worlds results in part from a desire to author a reality, just as a science fiction writer does. One person told me that he thought of himself as a storyteller, and programming, like the Dungeons and Dragons designing he used to do, was one way of telling stories. Ursula Le Guin's essay "Do-It-Yourself Cosmology" provides a useful account of how science fiction writers and scientists create speculative alternative worlds:

> Scientist and science-fictioneer invent worlds in order to reflect and so to clarify, perhaps to glorify, the "real world," the objective Creation. The more closely their work resembles and so illuminates the solidity, complexity, amazingness and coherence of the original, the happier they are."[51]

This is clear in Langton's review of SimLife™:

> SimLife allows the construction and study of simple, artificial ecologies, and the surprising complexity and richness of behavior that emerges in even these extremely simple "artificial natures" has already given me a greater appreciation of the real thing—Nature in all her real glory—writ with a capital "N." That glory shines all the more

brightly even from the meagre illumination that this simple "Software Toy" is able to shed upon Her.[52]

Science fiction–fueled simulation has become a new tool for thinking natural theology, a new tool for revealing and reproducing the plans of an omnipotent creator.

Cyberspace as a new creation and colonial space

The idea that computational processes can be thought of as existing in a kind of territory is supported by discourse about computer networks as "cyberspace." David Noble summarizes the often millenarian, Christian language surrounding cyberspace:

> The religious rapture of cyberspace was perhaps best conveyed by Michael Benedikt. . . . Editor of an influential anthology on cyberspace, Benedikt argued that cyberspace is the electronic equivalent of the imagined spiritual realms of religion. The "almost irrational enthusiasm" for virtual reality, he observed, fulfills the need "to dwell empowered or enlightened on other, mythic, planes." Religions are fueled by the "resentment we feel for our bodies' cloddishness, limitations, and final treachery, their mortality. Reality is death. If only we could, we would wander the earth and never leave home; we would enjoy triumphs without risks and eat of the Tree and not be punished, consort daily with angels, enter heaven now and not die." Cyberspace, wrote Benedikt, is the dimension where "floats the image of the Heavenly City, the New Jerusalem of the Book of Revelation. Like a bejeweled, weightless palace it comes out of heaven itself . . . a place where we might re-enter God's graces . . . laid out like a beautiful equation."[53]

Computers are figured as a place to begin again. Artificial Life participates in this imaginary, maintaining that simulated worlds might be places to see possible ways life could have evolved. At one conference, I saw a simulation called Aleph, a nice pun on Alife, but also a name that summoned up, with its Hebrew letter name, the beginning of things. The subtitle of Steven Levy's popular book, *Artificial Life: The Quest for a New Creation*, captures the imagination at work here.

This image of computers as new worlds dovetails with another image: the electronic frontier. As Robert N. Bellah has argued, notions of the new holy land and of the frontier are often mutually constitutive.[54] People imagine new frontiers as allowing them to start anew from an Edenic state: early English colonists saw in America a land of milk and honey, pilgrim John

Winthrop saw in America a "City upon a Hill" (a reference to a new Jerusalem), and Mormons saw in Utah a new Zion. The Americas, in their guise as "the New World" were often understood as a place for (European) humanity to begin again. In the frontier territories of Artificial Life, we find that Man will be as God again, having regained a creative capacity modeled after God. He will create creatures and give them their rightful names, just as Adam did, and just as Linnaeus, the eighteenth-century taxonomist who called himself a "second Adam," did at the moment modern biology was born. The image of a Garden of Eden in cyberspace fuses nicely with the spatialized metaphor of the electronic frontier and allows Artificial Life researchers to think of themselves as creating, populating, and exploring new lands.

This notion of colonization and taming underwrites Thomas Ray's plan to have a global version of his Tierra system. Ray hopes that Tierran organisms can "run wild" in an Internet reserve, traveling around the globe in search of spare CPU cycles, terraforming cyberspace in the process. To hear Ray speak of global Tierra is to hear him speak of an empty world he plans to colonize with his digital organisms. Cyberspace becomes a new frontier, a second nature ready for colonization by the first-world imagination.[55]

In a paper titled "Visible Characteristics of Living Systems: Esthetics and Artificial Life," a company of philosophers and artists articulate—apparently without irony—the idea that Artificial Life researchers should think of themselves as colonizing new spaces:

> Artificial Life researchers habitually give names to their creations, build up complicated typologies for them and come up with bold new syntactic designations for areas that are still terra incognita. They are the true successors of the Conquistadors of the 15th and 16th centuries, of the explorers of the 18th and 19th. . . . Having created his creature in real or virtual space, the researcher or artist-researcher tries to capture it with language and to master its imaginative dimension in such a way as to turn it into a usable object, one that can be proposed for comparison, criticism, and reconstruction.[56]

The rhetorics of exploration, colonization, and conquest in Artificial Life are intensely masculine. Imagining programmers as gods, as single-handed creators of life, as objective, transcendent observers, and as intrepid explorers of the final frontiers of cyberspace all invoke masculine imagery and not just this, but imagery of a kind of white masculinity, of a white man who hunts, explores, and goes on adventures in undiscovered lands, and feels at ease and assured in his power in naming and conquering. The fact that the majority of Artificial Life practitioners are white men who grew up reading cowboy science fiction is not trivial. The masculine imagery of exploration also reinforces a gendering of the "lands" of cyberspace as feminine, waiting to be penetrated or unveiled, continuing a Western tradition of seeing Nature as a woman (recall Langton's pronouncements about "Nature in all her real glory," coupled with his images of a masculine god using SimLife™ to create the world). Nature, a living—sometimes nurturing, sometimes wild—being in seventeenth-century European cosmology was devivified and mechanized in the eighteenth and nineteenth centuries and has now been resuscitated as an enormous computer.[57]

Technologies of vision

The data researchers garner from their simulations are usually presented visually, and in Artificial Life simulations, visuals almost always afford a god's eye view, allowing experimenters to survey an entire world at once. In many systems, we are provided with a panoptic picture of a world, over which we can witness the careenings of computational creatures. In Tierra, we are presented with graphs that summarize evolutionary activity and chart the rise and fall of program lineages. The visual access we are granted to artificial worlds positions us as "objective" observers, located everywhere and nowhere at once. Artificial Life programs instantiate the dream of objectivist science. Donna Haraway has written that in Western thought, "the eyes have been used to signify a perverse capacity—honed to perfection in the history of science tied to militarism, capitalism, colonialism, and male supremacy—to distance the knowing subject from everybody and everything in the interests of unfettered power. . . . [The] view of infinite vision is an illusion, a god-trick."[58] This illusion is manufactured as reality in computational Artificial Life.

Vision is a sense researchers invoke to speak about unmediated pictures of the empirical world. They speak of "seeing" phenomena emerge in their simulations, and when speaking this way they take themselves to be reporting—rather than, say, hallucinating—real results. At many conferences I attended, people referred to the "real world" or "nature" by pointing out the window, while they indicated artificial worlds by pointing at images on computer screens. At an SFI workshop on artificial worlds, one researcher compared the technology of computer simulation to telescopes and microscopes, saying that these technologies allow us to peer into whole new worlds.

The idea that life only evolved once on Earth, and that it would be interesting to see what else might have happened, is one to which Artificial Life scientists frequently refer. The world is often compared to a videotape, one that we might rewind, fiddle with, and watch for different story lines. Langton writes, "Although studying computer models of evolution is not the same as studying the 'real thing,' the ability to freely manipulate computer experiments, to 'rewind the tape,' perturb the initial conditions, and so forth, can more than make up for their 'lack' of reality."[59]

This image of the world as videotape fast-forwards me to a description of a compelling video I saw at one Artificial Life conference. Accompanying a talk entitled "Artificial Fishes with Autonomous Locomotion, Perception, Behavior, and Learning in a Simulated World," the video showed simulations of swimming fish.[60] The audience was enraptured as simulated fishes acted out their artificially evolved capacities to swim and hunt. The presentation ended with simulations strung together in an extended parody of a Jacques Cousteau documentary. As the audience laughed at the video and at the movements of the fishes on the screen, it became clear that the lifelike quality of these simulations produced an unease and sense of the wonder that was itself precisely the cultural resource that made these creatures seem lifelike. The laughter bespoke a set of intuitions and untheorized thoughts about autonomy and agency. This reference to Cousteau invokes another ocular technology: the aquarium, which also interposes a glass barrier between an observer and a bundle of phenomena. A few Artificial Life researchers call upon people's familiarity with aquariums or terrariums to rhetorically structure their simulations as closed worlds. Some systems show us fishes or birds moving around as though in an enclosed space.

Artificial Life worlds, visualized on computer screens, depend on a complex of viewing habits developed in such diverse activities as watching documentary video, looking out windows, and gazing into aquariums. What all of these practices have in common is a commitment to the notion that representation *represents*, that visual accounts of the world can be isomorphic with the world itself.[61]

Exit

At an SFI workshop on artificial worlds, linguist George Lakoff argued that seeing computers as worlds is enabled by metaphors that allow scientists to import everyday language into the space of simulation. But while SFI and practitioners of Artificial Life may come up with a stable practice for interpreting computer simulations as model worlds, or even as real, independent worlds, this does not mean that they become such things for people who do not share the enabling metaphors. Lakoff maintained that what models do is "create an expert priesthood, and what SFI is doing is setting up a new kind of priesthood."[62] Following this pronouncement, Lars Risan has suggested that just as the authority of Catholic priests empowers them to convince people that wine can literally, not just metaphorically, be turned into blood, so it might be that the expert priesthood of Artificial Life, through skilled interpretation of computer simulation, can convince people that computer simulations can literally be turned into worlds and life. Artificial Life techniques have already made their way into everyday common sense through such programs as SimLife™, extending the network of priestly training in a kind of Protestant dispersal of Artificial Life.[63]

The transubstantiation work of Artificial Life researchers signals new ways of understanding nature. For these scientists, worlds generated in simulation are congealing into new epistemological and ontological territories. As I've argued here, this is the result of a complex commerce of language and sensibility between silicon second natures and the second natures represented by the cultural worlds of Artificial Life researchers. As Evelyn Fox Keller and Elisabeth Lloyd write about scientific meanings of words, "By virtue of their dependence on ordinary language counterparts, technical terms carry, along with their ties to the natural world of inanimate and animate objects, indissoluble ties to the social world of ordinary language speakers. . . . [They] have insidious ways of traversing the boundaries of particular theories, of historical periods, and of disciplines. . . . They serve as conduits for unacknowledged, unbidden, and often unwelcome traffic between worlds."[64] This semiotic traffic is not only at high density between social and simulated worlds, but has become knotty enough to produce simulations *as* worlds, as new kinds of spaces, as Gardens of Eden where explanations grown in silicon can supplement or supplant the very natures they seek to illuminate. But simulations can only crystallize as worlds in complex webs of hardware, software, and human wetware. In this essay, I have fixed attention on semiotic traffics that many Artificial Life practitioners would disavow, but I maintain that it is precisely at the intersections of domains coded as *science* and *culture* that Artificial Life worlds come to life for the people who craft them. It is at these junctures that scientists of the artificial solidify simulations as sites for growing explanations.

Notes

1   Thomas Ray on Tierra, in Santa Fe Institute, *Simple Rules . . . Complex Behavior* (Santa Fe Institute video, 1993).
2   Santa Fe Institute, *Simple Rules.*
3   Claus Emmeche, *The Garden in the Machine: The Emerging Science of Artificial Life*, trans. Steven Sampson (Princeton, N.J.: Princeton University Press, 1994); also Peter Galison, "Computer Simulations and the Trading Zone," in Galison and D. J. Stump (eds.), *The Disunity of Science: Boundaries, Contexts, and Power* (Stanford, Calif.: Stanford University Press, 1996), 118–157.
4   Steen Rasmussen, Rasmus Feldberg, and Carsten Knudsen, "Self-Programming of Mat-

ter and the Evolution of Proto-Biological Organizations" (SFI preprint 92–07–035, 1992).

5 Evelyn Fox Keller, *Refiguring Life: Changing Metaphors of Twentieth Century Biology* (New York: Columbia University Press, 1995); also Richard Doyle, *On Beyond Living: Rhetorical Transformations of the Life Sciences* (Stanford, Calif.: Stanford University Press, 1997).

6 Sylvia Yanagisako and Carol Delaney, "Naturalizing Power," in Yanagisako and Delaney (eds.), *Naturalizing Power: Essays in Feminist Cultural Analysis* (New York: Routledge, 1995), 1–22.

7 Ursula K. Le Guin, *The Word for World Is Forest* (New York: Berkeley, 1976); and Le Guin, "Do-It-Yourself Cosmology," in *The Language of the Night: Essays on Fantasy and Science Fiction* (New York: HarperCollins, 1989), 118–122.

8 Neil Smith, "The Production of Nature," in G. Robertson, M. Mash, L. Tucker, J. Bird, B. Curtis, and T. Putnam (eds.), *Future Natural: Nature, Science, Culture* (London: Routledge, 1996), 35–54.

9 Christopher G Langton, "Toward Artificial Life," *Whole Earth Review* 58 (1988): 74.

10 Christopher G. Langton, "Artificial Life," in Langton (ed.), *Artificial Life* (Redwood City, Calif.: Addison-Wesley, 1989), 1.

11 Thomas Ray, "An Evolutionary Approach to Synthetic Biology: Zen and the Art of Creating Life," *Artificial Life* 1 (1994): 183.

12 Thomas Ray, "An Approach to the Synthesis of Life," in C. Langton, C. Taylor, D. Farmer, and S. Rasmussen (eds.), *Artificial Life II* (Redwood City, Calif. Addison-Wesley, 1992), 372.

13 Ibid., 373–374.

14 Ray, "Evolutionary Approach," 183.

15 Ibid.

16 N. Katherine Hayles, "Simulated Nature and Natural Simulations: Rethinking the Relation between the Beholder and the World," in W. Cronin (ed.), *Uncommon Ground: Toward the Reinvention of Nature* (New York: W. W. Norton, 1995), 421.

17 Stefan Helmreich, *Silicon Second Nature: Culturing Artificial Life in a Digital World* (Berkeley: University of California Press, 1998).

18 David Hiebeler, "Implications of Creation" (SFI preprint 93–05–025, 1992), 2.

19 Terry Winograd, *Understanding Natural Language* (New York: Academic Press, 1972); also Hubert L. Dreyfus, *What Computers Can't Do: A Critique of Artificial Reason*, 2d ed. (New York: Harper and Row, 1979); Paul Edwards, *The Closed World: Computers and the Politics of Discourse in Cold War America* (Cambridge: MIT Press, 1996).

20 Edwards, *Closed World*, 171–172.

21 Ibid., 172.

22 Sherry Turkle, *Life on the Screen: Identity in the Age of the Internet* (New York: Simon and Schuster, 1995).

23 Judith Genova, "Women and the Mismeasure of Thought," in N. Tuana (ed.), *Feminism and Science* (Bloomington: Indiana University Press, 1989), 212.

24 Galison, "Computer Simulations," 142–143.

25 Ibid., 144.

26 Ibid., 156–157.

27 Ray, "Evolutionary Approach," 184

28 See Rasmussen, Feldberg, and Knudson, "Self-Programming of Matter."

29 Christopher G. Langton, "Self-Reproduction in Cellular Automata," *Physica D* 10 (1984): 135–144.

30 Langton, "Artificial Life," 28.

31 Andrew Wuensche, "Complexity in One-D Cellular Automata: Gliders, Basins of Attraction, and the Z Parameter" (SFI preprint 94–04–025, 1994), 193.

32 Since I wrote this essay in 1997, this position has been newly championed by Stephen Wolfram, *A New Kind of Science* (Champaign, Ill.: Wolfram Media, 2002).

33 Here and elsewhere in this essay, at their request, I omit specific information on people interviewed.

34 Stevan Harnad, "Levels of Functional Equivalence in Reverse Bioengineering," *Artificial Life* 1(3) (1994): 297–298.

35 Bruce MacLennan, "Synthetic Ethology: An Approach to the Study of Communication," in Langton et al., *Artificial Life II*, 638.

36 Brian Cantwell Smith, *On the Origin of Objects* (Cambridge: MIT Press, 1996).

37 Ray, "Evolutionary Approach," 184.

38 Ibid., 183.

39 Ibid., 182–183.

40 Edwards, *Closed World*, 13.

41 MacLennan, "Synthetic Ethology," 637.

42 N. Katherine Hayles, "The Closure of Artificial Worlds: How Nature Became Virtual," paper presented at "Vital Signs: Cultural Perspectives on Coding Life and Vitalizing Code," Stanford University, 2–4 June 1994, 4.

43 Langton, lecture at "Artificial Life IV," Massachusetts Institute of Technology, 6 July 1994.

44 Ray quoted in Kevin Kelly, *Out of Control: The Rise of Neo-Biological Civilization* (Redwood City, Calif.: Addison-Wesley, 1994), 297.

45 Christopher G. Langton, "SimLife from Maxis: Playing with Virtual Nature," *Bulletin of the Santa Fe Institute* 7 (1991): 4.

46 Carol Delaney, "The Meaning of Paternity and the Virgin Birth Debate," *Man* 21 (1996): 494–513.

47 N. Katherine Hayles, "Narratives of Evolution and the Evolution of Narratives," in J. L. Casti and A. Karlqvist (eds.), *Cooperation and Conflict in General Evolutionary Processes* (New York: John Wiley and Sons, 1994), 125.

48 Stanislaw Lem, *The Cyberiad*, trans. Michael Kandel (1967; San Diego: Harcourt Brace Jovanovich, 1985).

49 Ibid., 167.

50 Santa Fe Institute, *Simple Rules*.

51 Le Guin, "Do-It-Yourself Cosmology," 120.

52 Langton, "SimLife from Maxis," 6.

53 David Noble, *The Religion of Technology: The Divinity of Man and the Spirit of Invention* (New York: Knopf, 1997), 159–160.

54 Robert N. Bellah, *The Broken Covenant: American Civil Religion in Time of Trial*, 2d ed. (Chicago: University of Chicago Press, 1992).

55 Ziauddin Sardar, "alt.civilizations.faq: Cyberspace as the Darker Side of the West," in Sardar and J. R. Ravetz (eds.), *Cyberfutures: Culture and Politics on the Information Superhighway* (New York: New York University Press, 1996), 14–41.

56 D. Lestel, L. Bec, and J.-L. Lemoigne, "Visible Characteristics of Living Systems: Esthetics and Artificial Life," in J. L. Deneubourg, S. Goss, G. Nicolis, H. Bersini, and R. Dagonnier (eds.), *ECAL 93: Self Organisation and Life: From Simple Rules to Global Complexity* (photocopied proceedings from the Second European Conference on Artificial Life, Santa Fe Institute, 11–13 November 1993), 598.

57 Carolyn Merchant, *The Death of Nature: Women, Ecology, and the Scientific Revolution* (New York: Harper and Row, 1980).

58 Donna Haraway, "Situated Knowledges: The Science Question in Feminism and the Privilege of Partial Perspective," in Haraway, *Simians, Cyborgs, and Women: The Reinvention of Nature* (New York: Routledge, 1991), 188–189.

59 Christopher G. Langton, introduction to Langton et al., *Artificial Life II*, 7–8; also Walter Fontana and Leo Buss, "What Would be Conserved if 'The Tape Were Played Twice'?" in G. Cowan, D. Pines, and D. Meltzer (eds.), *Complexity: Metaphors, Models, and Reality* (Redwood City, Calif.: Addison-Wesley, 1994), 223–244. The notion of evolution as a "tape" is discussed in Stephen Jay Gould, *Wonderful Life: The Burgess Shale and the Nature of History* (New York: W. W. Norton, 1989). Artificial Life scientists hope to make Gould's metaphor do literal work, but do not demonstrate how "rewinding" simulated evolutionary history can "make up" in any straightforward way for the " 'lack' of reality" in computer models.

60 Demetri Terzopoulos, Xiaoyuan Tu, and Radek Grzeszczuk, "Artificial Fishes with Autonomous Locomotion, Perception, Behavior, and Learning in a Simulated Physical World," in R. Brooks and P. Maes (eds.), *Artificial Life IV* (Cambridge: MIT Press, 1994), 17–27.

61 Timothy Mitchell, *Colonising Egypt* (Berkeley: University of California Press, 1988).

62 George Lakoff, "Artificial Worlds" workshop, Sante Fe Institute, 13 November 1993.

63 Lars Risan, personal communication, July 1994; see also Risan, *Artificial Life: A Technoscience Leaving Modernity? An Anthropology of Subjects and Objects* (Ph.D. diss., University of Oslo, 1996).

64 Evelyn Fox Keller and Elisabeth A. Lloyd, introduction to Keller and Lloyd (eds.), *Keywords in Evolutionary Biology* (Cambridge: Harvard University Press, 1992), 1–2.

Claus Emmeche

There is nothing wrong with a good illusion as long as one does not claim it is reality.—H. H. Pattee, "Simulations, Realizations, and Theories of Life"

An explanation is always a reproduction.—H. R. Maturana and F. J. Varela, *Autopoiesis and Cognition*

By explaining things, we change them. They do not remain the same in our conceptions. Explanations should satisfy our quest for understanding. There are, of course, forms of understanding that do better without explanation, like jokes. In general, explanations are a good thing to pursue in science; they are called upon in courtrooms, and they are often required in the upbringing of children and in everyday life; we should, however, be careful about their use and about the idea that all explanations of any kind share a unique conjunctive set of properties.

Explaining life in biology

I will consider here the role of explanations in the quest to understand life as a coherent phenomenon, as pursued in traditional biology and within one of the "sciences of complexity": the interdisciplinary field of Artificial Life (Alife). Though the approach is conceptual, not historical or sociological, the motivation for this study is a set of very general assumptions about scientific activity, which should be subject to closer scrutiny within the history, sociology, and philosophy of science. These assumptions include the following. (1) From science, the assumption that modified forms of reductionism are called for when a science is faced with the problem of complexity. Thus it is often assumed that "growing explanations" or constructing emergent patterns as the outcome of computer simulations— presumably representing (even in a very abstract sense) patterns of biological processes—is a case of extending the traditional reductionist explanatory strategy to the complex phenomena of life and mind. (2) From philosophy, the assumption that a general notion (or definition, understanding, conception, coherent view) of life has not been established. (3) From Alife research, the belief that Alife deeply changes our concept of life or the scientific approach to studying complex living systems. (4) From biology, the opinion that any general definition or concept of life is irrelevant to research. (5) From sociology of science: the view that we can only discuss the eventual relevance of a concept or a notion if we can empirically assess the "work" that such a notion can "do" in practical research. (6) From a perspective inspired by Thomas Kuhn's theory of science, the assumption that we see, understand, and define the objects of science from a certain perspective or paradigm which includes metaphysical components, and furthermore, that no clear demarcation can be drawn between explicit and implicit components of a paradigm, and thus, no clear demarcation between explicit theoretical "definitions" of the concepts most important to a paradigm and the "work done" or the "understanding provided" by such concepts.

Far from addressing all questions about explanation, I will focus on a single aspect: the explanatory role of *ontodefinitions*. For the moment, these can be thought of as certain very broad categories—such as matter, life, mind, or society—which not only simply denote huge phenomenologies in a vague manner, but also refer to some categories of the modern world picture whose symbolic and cognitive content science has helped to deepen. Ontodefinitions, of which I shall consider only the concept of life, are basic for the paradigmatic character of scientific activity. They belong to the metaphysical component of a paradigm (the disciplinary matrix in Kuhn's sense). They are mixed explanations and definitions in the sense that they define what scientists are looking for (thus constituting what is relevant and what entities should be dealt with in experiments) and at the same time they provide some basic understanding of the very nature of these objects, a narrative, an explanatory story of some kind. This is all implicit in a given paradigm. Scientists do not explicitly consider ontodefinitions to be important or to have anything to do with the usual everyday experimental activity. Biologists are typically reluctant to define the general notion of life in explicit terms; nonetheless, such notions exist within distinct paradigms of contemporary biology. Here I will consider their role in the quest to understand life universally as an emergent phenomenon, as pursued in traditional biology and within Artificial Life.

Standard ideas in philosophy of science inform us that different explanatory strategies are used in different sciences (e.g., mechanical explanation in physics, functional explanation in addition in the life sciences, intentional explanations in the social sciences, etc.). Views differ about the derived or nonfundamental status of some of these forms. However, for such interdisciplinary fields as complex systems research, Artificial Intelligence, and Artificial Life, the received schemes do not give adequate pictures of the work actually going on and the kinds of explanatory approaches that are at stake. With its special research agenda of trying to study "life-as-it-could-be" (not simply "life-as-we-know-it" in its earthly instances), Alife focuses on the "natural" generation of complex objects, which (by their very development or evolution as represented by a computational model) cross the boundaries between traditional classes of objects as constituted by physics, biology, and psychology.[1] Examples include models of the origin of life from a purely chemical "soup" of elements, models of the generation of a multicellular organism from a single fertilized cell, and models of the creation of internal representations (boldly called "mental") of the outer world within an artificial agent such as a wall-following robot with sensors, motor, and a neural network. By these approaches, Alife research crosses the boundary between the area where only causal and mechanical explanations are used and the areas where functional and intentional explanations are allowed or even requisite; this is consistent with the field's own emphasis on emergent phenomena. Typically, one can identify two levels of interpretation of a given: a lower level of prespecified rules for local interactions (the generalized genotype, or gtype), and a global level exhibiting a structural or behavioral ptype (generalized phenotype) that emerges as a result of nonlinear interactions, and thus, one may even study the "important feedback mechanism between levels in such systems."[2]

From a scientific and epistemological point of view, the basic promise of Alife research is to use the new computational resources (or new robotic techniques, or new biotechnology) to construct completely new and bona fide forms of life (software, hardware, and wetware forms) to overcome what Carl Sagan has called the fundamental handicap of biologists, that they basically know only one particular instance of life, namely life on Earth, which thus constitutes the notion of "the same" on the biochemical and metabolic level, governed by the same mechanisms of inheritance, the same types of metabolic patterns, and so forth. By constructing quite new instances of life, Alife research promises to add to our zoo of life-form archetypes and thus to enable us to explain or understand what is universally true about life and what is simply contingent upon the particular way life evolved on Earth.

The intuition at work here is twofold: (1) What we can construct we are also able to explain. We can imagine an isomorphy between constructing and explaining. This is probably a very common sentiment in physical science. (2) Complex things in nature construct themselves as wholes via long processes of local interactions between simple entities; this emergence of wholes (or collective behavior of units), should be mimicked in our explanations. Instead of "top-down" reductive explanations, complexity research provides "bottom-up" explanations of emergent phenomena. Even though these explanations are still reductive (in the methodological sense that one can in principle show exactly what is going on from step to step in a simulation of, e.g., the evolution of new species), the complexity of the system makes prediction impossible, that is, computational shortcuts (to predict the future state) cannot be found for every case, and hence some simulations are computationally irreducible.

The intuitions behind the computational Alife research program show, first, that even though strict reduction as an explanatory strategy is admittedly not possible for complex living things, the phenomena are considered to be open for explanation of another kind; second, these explanations have an important *narrative* aspect. One could of course argue that any explanation has a narrative aspect, in that the story to be told must have, as it were, the logical and semiotic power to generate a believable representation of the story's subject matter. Scientific explanations are narratives produced within a rule-governed game of language. The game may generate a representation of a phenomenon by logical mechanisms, such as deduction, induction, or abduction, or it may employ more elaborate computational and hermeneutic procedures, such as constructing and interpreting computer models of emergent phenomena. Explanations are also constructive: they reproduce a system by producing another one.[3] Though this is true of any explanation, the extensive use of computer simulations and related visualizations has made this constructive aspect particularly salient. In Alife research the scientist literally animates the representation of life as a collective phenomenon generated by the rule-governed behavior of individual units.

In the year 2000, the term *biology* celebrated its second centennial. The modern idea of a unified science of living systems is about the same age. Biology before Artificial Life was (and still is) considered an autonomous science with its own methods, theories, and basic assumptions about its subject matter—living systems, life, organisms. Biologists have officially been reluctant to define life so clearly that it could be held up as a standard for judging the Alifer's claim that a simulation is a "real" living thing. Is this because previous attempts to formulate a universally valid notion of life have failed? In what follows I will argue that the standard view of the definition of life is flawed; that, in practice, contemporary biology uses two implicit definitions of life (here called *ontodefinitions*); and that both of them constitute life as an emergent phenomenon. Emergence as a candidate for explanation of life, however, requires a clear conceptual framework. This can be supplied by considering the observer's role in such a framework.

It is hardly necessary to do a systematic survey of practicing researchers to observe that most of them are extremely skeptical of attempts to clearly define living beings. They simply assert (with some justification) that a definition is of no use in solving the various experimental puzzles of normal research. On the theoretical side, from the point of view of biology as a general field of inquiry, it may be surprising that relatively few attempts have been made to reflect systematically and critically upon the nature of living systems. A remark made by Ernst Mayr is representative of the skeptical and empiricist attitude: "attempts have been made again and again to define 'life.' These endeavours are rather futile since it is now quite clear that there is no special substance, object, or force that can be identified with life."[4]

Nevertheless, Mayr, known as a main contributor to "the modern synthesis" in neo-Darwinian evolutionary theory, transcends the pure refusal to define life. Thus he immediately proceeds by acknowledging a kind of definition of life: "The process of living, however, can be defined. . . . Living organisms possess certain attributes that are not in the same manner found in inanimate objects. Different authors have stressed different characteristics, but I have been unable to find in the literature an adequate listing of such features."[5] Mayr then informs the reader that his own list is probably both incomplete and somewhat redundant, but it will "illustrate the kinds of characteristics by which living organisms differ from inanimate matter." The list has as its key words: complexity and organization, chemical uniqueness, quality, uniqueness and variability, possession of genetic program (with the notable remark that "nothing comparable to it exists in the inanimate world, except for manmade computers");[6] historical nature, natural selection, and indeterminacy. Thus, Mayr thinks that, on the one hand, definitions have to be essentialist and that attempts to define life by one single "essence" or crucial characteristic are futile, but that, on the other hand, it is possible to define life *as a process* by a qualitative and possibly redundant list of about eight properties.

The central assumptions in Mayr's approach are the following, which I will call the standard view of the definition of life (SVDL). It seems to constitute the received view, not only in modern evolutionary and theoretical biology but in most branches of biology:

1. Life as such cannot be defined, thus a clear definition is missing.
2. The question of defining life is not important for biology.
3. Living processes may be defined, however, or at least approximately

demarcated from inorganic processes through a list of characteristic properties (nonessentialism).

4. Difficulties in delimiting such a set of properties are recognized, but are not considered to be serious. Particular living beings may not hold all properties given, so the list may not be a list of necessary and sufficient properties; it may be more vague or redundant.

5. Even though life is a physical phenomenon, biology deals with systems of such a vast complexity that we cannot in practice hope to reduce it to physics. Among crucial properties of living processes are complexity, organization, self-reproduction, and metabolism.

The central claims are (1) and (2), and most biologists will usually not feel committed to elaborate further on the consequences of this stance. This may be due to a misconceived view of definitions as being by necessity Aristotelian, or essentialist in character. Biologists of today are emphatically nonessentialist and nonvitalist, and they shy away from any claim about the nature of life that seems to have the slightest stain of vitalistic ideas. This attitude may be justified if one faces attempts to substitute quasiscientific or pseudoholistic notions of life for scientific study. However, it is a sad consequence of the defeat of vitalism that many biologists still conceive the foundation of their field to be mechanistic, rooted in classical or quantum physics, rather than having its own autonomy as a science with a set of paradigmatic ideas, which are better described as *organicist*.[7] One can also say that the attitude of the SVDL is this: "Don't talk too much of definitions; enough has been said." The neo-Darwinian philosopher Michel Ruse provides a clear example of the SVDL: "efforts to find some distinctive substance characterizing life have proven as futile as they have been heroic. The one thing which is clear is that any analysis of life must accept and appreciate that there will be many borderline instances, like viruses. Inconvenient as this may be for the lexicographer, this is precisely what evolutionary theory would lead us to expect."[8]

Though the SVDL is standard for most biologists of various breeds, theoretical biologists attracted to Artificial Life, as well as computer scientists, physicists, and mathematicians within this field, share a more open attitude to the possibility of defining, redefining, or inventing broad new notions of biological life. In a sense, the very idea of studying life-as-it-could-be enforces an interest in more general conceptions of life. Within the neo-Darwinian tradition, John Maynard Smith's view is an exception to the SVDL. He argues for two criteria, namely (1) metabolism and (2) functions ("the parts of organisms have 'functions,' that is, the parts contribute to the survival of the whole").[9] Both must be integrated within the modern version of Darwinism so that life can be understood as something that possesses those properties necessary to ensure Darwinian evolution, that is, "entities with the properties of multiplication, variation, and heredity are alive, and entities lacking one or more of those properties are not."[10] This notion is nearly identical with the first of two candidates I will discuss later for a theoretically satisfying concept of life in theoretical biology.

We have noticed that biologists are skeptical about the use of an explicit concept of life. They do not appreciate previous attempts and think that to define life in an exact way is impossible. What I want to emphasize at this point, however, is that even though this SVDL may seem sound, it is only justified for those parts of a general notion (or ontodefinition) of life that are explicitly visible, which manifest themselves with a badge saying "I am the definition of life." But for any concept of life, this is just the tip of the iceberg. It is on this very superficial and quasi-Aristotelian level that most discussions of the concept have been conducted. Below the water surface of explicit manifestations, there is a huge realm of implicit rules of reasonable behaviour that are not reducible to explicit algorithms for open observable behavior but which are imbued with conceptually meaningful structures that take years to master.[11] When we consider science as a concrete set of activities, we find that explanations, theoretical ideas, idealizations, models, experiments, concrete observations, good exemplars of how to write down a biochemical reaction or set up a Southern blotting for detecting a DNA sequence, and so forth are woven tightly together. The very notion of biology's subject matter is extremely complex, but one can nevertheless prescind this notion from its comprehensive paradigmatic and practical background, and, as it were, lift it from the submarine level to explicate its intrinsic structure and thereby describe the ontodefinition of life within a given biological paradigm.

Thus, it should not surprise us that those espousing SVDL attitudes never clearly state the requirements of a possibly general notion of life. We could ask: What requirements should such a notion or compound idea meet? Are such requirements specific to biology or general for any scientific term? These questions are, of course, enormously complex, and in this essay I can only give a few hints. The idea here is to make explicit (tip of the iceberg) what is already known in the paradigmatic "background" sense of knowledge (the total iceberg). The characteristics of an ontodefinition of life can be stated as the following list of requirements, which can be justified on theoretical and pragmatic grounds: generality, coherence and nonvitalism, comprehensive elegance, and specificity.

1. An ontodefinition of life should be *general* so as to encompass all possible forms of life, not just the contingent products of Darwinian evolution on this planet. Life on other planets may not have its genetic material stored in DNA molecules, or it may not have a metabolism based on

proteins with enzymatic function, but it will probably have both a metabolism and genetic memory of some kind. It is very hard to imagine forms of life which do not have (or which are not parasitic on other forms with) a kind of genotype-phenotype duality. Generality must be ontological as well as epistemic: the idea of life covers all possible forms of life in the universe; it is not just a simple reflection of the particular disciplinary framework in which the notion is given.[12]

2. An ontodefinition of life should not involve notions that conflict with what we already know of living things and their inorganic components, that is, it should be, to some extent, *coherent* with the general understanding of living systems based on modern biological research, and based on this tradition it should be *nonvitalistic* in having no reference to supernatural directing forces, even though it does not have to entail an ontological reductionism.

3. An ontodefinition of life will have what we might call a conceptual organizing *elegance*, that is, it can organize the field of knowledge within biology and crystallize our experience with living systems into a clear structure, a kind of schematic representation that summarizes and gives further structure to the field. The role of an ontodefinition is not simply the role of definitions of more or less technical terms within specific subfields; rather, its role is to give biology's general object of study its own profile, to organize our cognitive models and theories of living systems in a coherent way (as a kind of rational root metaphor), and to distinguish the scientific study of life from other sorts of inquiry, such as investigations of phenomenological aspects of human existence in a society, the scientific study of the human psyche, or the study of physical matter. It holds true for multicellular organisms with an immune system, such as the vertebrates, as well as for single cells, and with it one can relate specific components of life at the subcellular level, such as ribosomes, to its general properties. It gives a comprehensive view of life compared with matter, mind, and society, and it enables us to comprehend the internal unity in the biologic diversity of life.

4. Though able to give an idea about any kind of system that has the ability to live, metabolize, replicate its own kind, or carry on other functions considered relevant to life, an ontodefinition should be *specific* enough to distinguish life from obviously nonliving systems. Of course, a moment's reflection may indicate a circularity in the demand for a definition to demarcate life from nonlife by appeal to what is taken as "obviously" living or nonliving. We have different intuitions, and our evidential knowledge may fail. How can we be sure that a crystal that can grow may not, to a certain extent, be alive? We could manufacture various concepts of life that meet different people's intuitions. However, as part of a scientific structure, an ontodefinition is not simply a deliberately built conceptual scheme

that we impose on the malleable world of experience. Such a view would in effect block the way of inquiry. The world is not infinitely malleable, and we should not dismiss the genuine and detailed knowledge of living systems that has been gained during this century. A more pragmatic view takes for granted that there are real things (and real differences between categories of things) whose characters are entirely independent of our opinions about them, and that we can come to know these realities through scientific methods. Our understanding of living systems is based on fallible but nevertheless scientific knowledge of the distinctive characteristics of cells and organisms.

Thus we can conjecture that an ontodefinition of life is general enough to deal with life as a universal phenomenon, not just earthly carbon life, and not just life-as-a-thermodynamic-system or life-as-genetic-processes; that it is coherent with current knowledge of biology and physics; that it makes no appeal to vitalistic forces; that it shows some conceptual elegance and a cognitive organizing capacity; and that it is specific enough to catch the basic primary characteristics of biological life, no more and no less.

Such a set of properties is not required for the definition of every scientific term; most terms represent many more specific kinds of objects or processes. For instance, in molecular biology, we may preliminarily and operationally define a specific polynucleotide involved in protein synthesis as the behavior of an isolated "factor" whose presence changes the translational properties of the ribosome. When explicating the concept of life, however, we wish to demarcate a very broad class of processes, a very general and organized mode of physical systems (different from culture, society, mind, or matter). The cognitive scientific function of this concept is of another kind, so it would be a mistake to demand the same level of operational or conceptual concreteness of such a notion.

Whether the individual scientist subscribes to an instrumentalist, a realist, or a pragmatist philosophy of science, he or she will usually recognize the importance of a clear and comprehensive understanding of biological systems' distinctive features, even if one does not believe that science by necessity reveals the final reality of life. I will discuss just two definitions of life that are scientifically cogent; life as the natural selection of replicators, and life as an autopoietic system.

Life as the natural selection of replicators

A significant aspect of the concept of life as the natural selection of replicators is its peculiar status within biology. The idea that life basically can be defined as the natural selection of self-copying entities is often ignored or not recognized as a definition at all, due to the SVDL's skepticism. Even though neglected, this notion lives an implicit life of its own within evolutionary biology. It is relatively easy to make explicit, and once this is done, it would likely be accepted by the majority of evolutionary biologists, who are used to thinking of life not on the level of the individual organism but as lineages of organisms connected by the processes of reproduction and selection.

This concept can be formulated from the generalization of statements about the kind of entities that undergo variational change, that is, that evolve by natural selection. Maynard Smith's 1986 contribution mentioned earlier can be generalized in this way. Life is a property of populations of entities that (1) self-reproduce, (2) inherit characteristics of their predecessors by a process of informational transfer of heritable characteristics (implying a genotype-phenotype distinction), (3) vary due to random mutations (in the genotype), and (4) have the propensity to leave offspring determined by how well the combination of properties (inherited as genotype and manifested as phenotype) meets the challenge of the environment's selective regime.

This formulation can be made even more abstract if we emphasize that by "genotype" and "phenotype" we do not necessarily refer to particular genes made of DNA or organisms made of cells but to any kind of "replicators" and "interactors." The term *replicator* comes originally from the zoologist Richard Dawkins, who believed evolution to proceed primarily by selection at the level of genes (replicators), which by the very process of replication preserved their structure in time. For Dawkins, life on Earth began with the appearance in the primordial soup of molecules that could replicate, that is, catalyze the production of their own kind by template copying processes. This led to the evolution of cells and multicellular organisms as "survival machines" for their replicating genes, the perpetual information sequences written in the nucleotides of DNA molecules. Thus Dawkins's defining characteristic of life amounts to the natural selection of more and more effective replicators.[13]

Dawkins embedded his view of gene selection in a reductionist metaphysics that identified only the replicating structures as the real entities in evolution, while organisms were like ephemeral, transient epiphenomena. In contrast, the philosopher David Hull sought to broaden the ontological framework of the theory of evolution by introducing for the entities at work in Darwinian evolution the general terms *replicators* (entities that pass on their structure directly by replication), *interactors* (entities that produce differential replication by interacting as cohesive wholes directly with their environments) and *lineage*. "A process is a selection process," in Hull's view, "because of the interplay between replication and interaction. The structure of replicators is differentially perpetuated because of the relative success of the interactors of which the replicators are part. In order to perform the functions they do, both replicators and interactors must be discrete individuals which come into existence and cease to exist. In this process they produce lineages which change indefinitely through time. Because lineages are integrated on the basis of descent, they are spatio-temporally localized and not classes of the sort that can function in laws of nature."[14]

Apart from philosophical discussion of the species concept, Hull's contribution has been much overlooked. This is unfortunate, since it conforms pretty well to the conceptual requirements (1) to (4) of an ontodefinition of life discussed earlier. It is easily conceivable that all life in the universe evolves by a kind of Darwinian selection of interactors whose properties are in part specified by stored information that can be replicated. No nonphysical forces are involved in this process, yet the very notion of natural selection and replication (the transfer of information sequences that specify biological activity of macromolecules) seems to be specific to biological entities; lineages of interactor-replicator entities are not the kinds of things described in physics textbooks. Hull's notion is simple, elegant, general, and crystallizes our ideas about the general mechanisms that create living systems from an evolutionary perspective.

Dawkins has been most successful as popularizer of his idea that life is a kind of naturally selected informational system of replicators. A problem with Dawkins's idea (in contrast to Hull's) is its bias toward a purely informational conception of life. We normally consider life to be both form and matter—something with both informational-organizational and material-physical aspects—but Dawkins's idea emphasizes the informational aspect, namely, the replicators as self-propagating patterns of information. This may mislead weak souls to a nearly Platonic concept of life defined simply as any instantiation of some specific set of abstract informational properties (such as an entity's ability to replicate its own informational description).

Such instantiations can be found in Artificial Life, where the computational and man-mediated representations of life processes are seen not just as simulations, but realizations.[15] This "strong artificial life" view implies that we can synthesize genuine examples of life, simply by having strong enough computational support and good enough (informational) criteria to decide whether our simulations are mere approximations or whether they realize enough properties to be considered alive. The strong thesis depends upon (1) a notion of life as a very general process that can be instantiated—for example, as systems of replicators—in various material media (which may be correct if we allow for the specific dependence of each of these forms of life on their specific medium), and (2) a claim (similar to functionalism in the philosophy of mind) of medium-independence of life, the idea that realized life is essentially the same in all media—characterized only by its informational aspects independent of the material "substratum" supporting these processes (which seems to be incorrect).[16]

Before we leave the first ontodefinition we should ask if it relates to the notion of life as an *emergent* phenomenon. It does not in itself explicate the notion of emergence in a standard sense (in which "a property of a complex system is said to be 'emergent' just in case, although it arises out of the properties and relations characterizing simpler constituents, it is neither predictable from, nor reducible to, these lower-level characteristics").[17] Describing life as the natural selection of replicators, however, seems to imply a notion of life as an emergent phenomenon in two respects. First, it presupposes genetic systems with "digital" codes for information transfer, that is, systems characterized by properties that supervene on but cannot be reduced to physical properties (the genetic code is a good example of a system of properties that supervene on its molecular or chemical base). Second, during adaptive Darwinian evolution, genuine new properties appear that cannot (even in principle) be predicted in advance (due to random mutations) or explained merely by physical or chemical theories. Thus, conceptualizing life as the natural selection of replicators implies that it is an emergent phenomenon.

Life as an autopoietic system

We have seen that the informational ontodefinition of life as the natural selection of replicators helps to explain the origin of the functional characteristics of organisms as interactors. Now along with function, Maynard Smith mentioned metabolism as a basic feature of life. The closed network of metabolic components within a cell is a point of departure for understanding the second candidate for a general conception of life: life as an autopoietic system, as defined by Humberto R. Maturana and Franscisco J. Varela. Autopoiesis literally means "self-production" or "self-creation," and is a term for the "self-defining," "circular" organization (organizationally closed but structurally, i.e., materially and energetically open) of a living system (such as a cell), consisting of a network of component metabolites that produce the very network and its own components as well as the network's boundary.

Life as autopoietic is a distinct notion from the first one in several respects. (1) It has been deliberately invented as part of a general, abstract theory of life, a theory that attempts not only to capture the biological phenomenon of life in its most general sense, but also to give a biologically founded epistemology whereby the distinctions the observer makes among living systems and any other units are reflected in the theory from its very beginning. This epistemological feature of the theory is interesting but transcends the scope of this essay. (2) It rejects the notion of genetic or biologic information as something intrinsic to the autopoietic system; rather, it sees information as being ascribed to the system from an observer's point of view. Any form of intrinsic teleology is also rejected. (3) Even if referential relations in the strict intentional sense are prohibited within the theoretical framework, it illuminates a self-referential aspect of life through the closed topology of internal relations, the idea that living systems can only be characterized with reference to themselves.[18]

The explication of life as autopoietic system is formulated within a mechanistic framework. The living cell is an *autopoietic machine*, which is "a machine that is organized (defined as a unity) as a network of processes of production, transformation and destruction of components that produces the components which: (i) through their interactions and transformations regenerate and realize the network of processes (relations) that produced them; and (ii) constitute it (the machine) as a concrete unity in the space in which they (the components) exist by specifying the topological domain of its realization as such a network."[19] Furthermore, "the biological phenomenology is the phenomenology of autopoietic systems in the physical space and a phenomenon is a biological phenomenon only to the extent that it depends in one way or another on the autopoiesis of one or more autopoietic unities." Autopoiesis is an all-or-nothing property; a system cannot be "more or less" autopoietic. Man-made machines are not autopoietic in that they do not by themselves generate their constituents. According to the theory (and its many, precise, interrelated, and unusual definitions of terms), such central biological phenomena as evolution, self-reproduction, and replication are phenomenologically secondary to the constitution of autopoietic units in the physical space.

Is an autopoietic system by implication an *emergent* one in relation to its physical constituents? Clearly, the theory of autopoiesis adds something to the physicalist intuition of causal closure in the physical domain, namely, organizational closure in the biological domain. One may portray the theory as an antiteleological strategy to account for the teleological nature of living systems—their seemingly purposefulness—by explicating in what sense such systems have (emergent) properties that are not reducible to properties of physical systems. In fact, the phenomenological framework in which the theory was formulated emphasizes precisely that though the origin of autopoietic organization may be explicable "with purely mechanistic notions. . . . Once the autopoietic organization is established it determines an independent phenomenological subdomain of the mechanistic phenomenology, the domain of the biological phenomena." Thus, "one phenomenological domain can generate unities that define a different phenomenological domain, but such a domain is specified by the properties of the new different unities, not the phenomenology that generates them."[20] Again, though the concept of emergence is not explicated, the general idea is implicitly in this ontodefinition's theoretical context.

Both ontodefinitions of life have all the required properties given earlier for the most general types of scientific objects that are at the same time the objects of ontology (generality, coherence, comprehensiveness, specificity). They belong, however, to two separate paradigms of biology. "Life as the natural selection of replicators" is rooted in neo-Darwinian evolutionary biology, which is nowadays perceived as being based on a thoroughly molecular-genetic description of heredity (in spite of the metaphorical complex of informational terms such as "the genetic code," "biological information," etc.). "Life as autopoietic systems" belongs to a separate, minor, but very important branch of theoretical biology with origins in systems theory, cybernetics, and neurobiology. Both ontodefinitions implicitly construe life as an intrinsically emergent phenomenon.

Implicitly defined general objects

Despite some incommensurability between the theoretical terms of the two paradigms and their different metaphysical commitments, partial communication between their respective advocates is indeed possible. It is natural to ask if it is feasible to integrate the two ontodefinitions into one even more general perspective, and in fact, important work has been done in this direction.[21] One should also point to a third, new, and promising paradigm of biology, biosemiotics, which gives us another way to comprehend life as based not on the organization of molecules, but on the communication of signs in nature.[22] We may even expect a major recasting of our view of life within the next decade from this sign-theoretical perspective. However, the discussion of the two more established conceptions suffices to support my central point. Many philosophers and biologists have tended to believe that it is a simple fact that all attempts to formulate a coherent notion of life have failed. Some would even add that this is completely insignificant for research, and that one should simply study concrete cells and organisms and give particular molecular, functional, and evolutionary explanations for these systems; any attempt to define life would result in overly general concepts. I have argued that this opinion can be contested for several reasons.

First, even though one embraces the standard view of the definition of life, one does not necessarily deny that "living processes" may be defined, demarcated, or characterized in a general way, for example, through a list of shared properties of living beings, even though such a list may be vague, incomplete, redundant, or may not constitute a set of universally necessary and sufficient conditions.

Second and more important, because of the manifold character of research, definitions of scientific concepts cannot be restrained to a single type of definition (e.g., operational, mathematical, ostensive, Aristotelian, ontological, or whatever). Talking about biological life is talking about a very general set of objects—the whole subject matter of the biosciences—and we should not too rigidly demand precise definitions, especially where the cognitive and theoretical function of an ontodefinition serves as a root metaphor for the whole field. In contrast to the standard view, we have nevertheless seen that general notions of life can be explicated with some theoretical precision.

Third, biology in the twentieth century has not only been empiricist and fact-oriented, but has given us rich conceptual tools to construct a coherent picture of at least some of the *universal* properties of living systems (cells, multicellular organisms, and systems of such organisms), viewed within an evolutionary frame as evolved, highly organized, adaptive systems with some autonomy and specific informational properties, that is, with properties that are *emergent* but no less material than chemical and physical properties. In this sense, organisms are genuine ontological units, and well defined as objects of biology.

Fourth, even though interpreting the results of Alife research remains controversial—with respect to distinguishing the contingent properties of life on Earth from the universal properties of life of any form, and more philosophically, deciding if some simulations may be considered realizations—this field represents a set of inspiring approaches and methods. They serve not only to synthesize vehicles, animats, new kinds of self-replicating molecules, and new virtual universes of complex informational forms, but also to produce a more general understanding of the principles of complexity-that-is-alive, and perhaps a general feeling for the laws of form that universally constrain the biological processes of living.

I mentioned the possibility of looking at life as a semiotic phenomenon, as a system of signs mediated by interpreting organisms. We should remember that a concept of life is itself a sign—a sign of the quest for simplicity, comprehension, and scientific understanding. Wittgenstein remarked that "every sign *by itself* seems dead. *What* gives it life?—In use it is *alive.* Is life breathed into it there?—Or is the *use* its life?"[23] Only as abstracted and isolated from the practice of biology will a certain concept of life fail to give us insight. As dead metaphors they have been handed over to philosophical analysis. However, good definitions of life do exist and enjoy a life of their own. The meaning of concepts is in their use and an ontodefinition of life can be used as a condensed paradigmatic expression for a whole view of living beings.

In this sense, the two ontodefinitions also implicitly explain what kind of physical systems the living ones are. Whereas definitions have often been considered as explications of particular concepts required prior to an explanation, the ontodefinitions come with a built-in understanding of life. Both are in this sense paradigmatic: they provide the biologist with a way to "see" life and explain particular instances.

It may appear to some as a *contradictio in adjecto* to speak of living phenomena (as specified in the two ontodefinitions) as "implicitly well-defined general objects" of biology. Let me stress that by "well defined" I do not suggest that, for example, the problem of borderline cases will disappear (since cases of conceptual vagueness as Peirce noted may reflect the existence of vague boundaries in nature); they are only well defined relative to the criteria for adequacy given above. This does not mean that such objects cannot be more clearly understood if we untangle some of their implicit properties. One such is emergence.

Emergence as explanatory strategy: the observer reappears

After a long period of oblivion, the notion of emergence was revitalized at the end of the twentieth century by the sciences of complexity, focusing on the complex emerging properties of life and mind. Emergence is no longer perceived as something mysterious, in conflict with a scientific worldview. We saw that in Artificial Life one of the basic intuitions was that we can computationally imitate emergent processes of construction observed in nature as the creation of new wholes on higher levels of organization. Thus, instead of top-down reductive explanations of constituent structure, one can as a complementary approach search for bottom-up explanations of emergents. What we can construct, we should be able to explain, because the constructions seen in nature are completely material, and the corresponding computational constructions must be, on the basic level, completely algorithmic. This is the idea.[24]

What form does such an explanation have? It is not "hypothetical-deductive" or predictive; it is often acknowledged that the complexity of (the model of) the system makes prediction impossible. This is due to the computationally irreducible nature of the mathematical model and the fact that answering questions computationally about future states of a model that exhibits deterministic chaos requires definite (arbitrarily high precision) initial conditions, which cannot be provided by measurement of parameters of the real system. Further, in Artificial Life the focus is on systems with nonlinear interactions. In contrast to linear systems, nonlinear systems must be treated as wholes in which the behavior of the whole is "more than" (or better: "different from") the behavior of the parts. As Langton put it, "*Behaviors themselves* can constitute the fundamental parts of nonlinear systems—*virtual parts*, which depend on nonlinear interactions between physical parts for their very existence. . . . It is the virtual parts of living systems that Alife is after: the fundamental atoms and molecules of behavior."[25]

A "growing explanation" in Alife is thus seen as growing from the bottom up, that is, *from the simple rules of behavior of simple agents* (e.g., simulated molecules in a virtual soup) for which we can reasonably justify the algorithms that correspond to the agents' real behavior, *to the complex collective pattern*, often surprising, that emerges during the simulation, and that represents a higher-order form of life-as-it-could-be. An example would be the emergence of primitive cells with membranes and a connected metabolism.

The Alife form of bottom-up explanation may more appropriately be called "interpretational emergent explanations" (or maliciously, "jumping to the conclusion") because the model involves two levels of interpretation and *the observer* doing the interpreting is crucial for establishing the emergent phenomena. This is often ignored: the as-if-emergent higher-level patterns of behavior in the model (often represented visually on the computer screen as a 2-D virtual world) based on low-level computation of interactional primitives (the rules representing local interactions) are often seen as simply realizing emergent behavior, where one tends to forget that these patterns are not real in any trivial sense. If there is nothing intrinsically biological in the emergent phenomena of the model, the emergence may simply be in the eye of the beholder.[26]

To determine more precisely whether something is emergent, Nils Baas has proposed a more formal framework (based on the theory of categories). His proposal is interesting in the present context because it explicitly recognizes that an observer is required at any level to establish an emerging property. Baas considers his idea as a step toward a general theory of hierarchies, complexity, emergence, and evolution. These four interrelated phenomena (the "hyperstructure" of the theory) are always found not only in biological systems, but also in Alife's computational systems, and in dynamical systems. Whenever we encounter life, it must be hierarchically organized; hierarchies have had the time to evolve from simple to complex structures, where complexity is here used in the algorithmic sense of needing a long "programme" for the specification of the system or a long route for computational development. Life cannot do with just one distinction between the macro- and the microlevel; hierarchies make complexity manageable through several levels of organization. Evolution by natural selection is the process which gives rise to new levels. By evolution, the environment, as it were, acts as an observer that "sees" or "acts upon" higher level properties, thereby establishing recurrent forms of interactions within and between the different levels.

According to Baas, for something new to be created we need some dynamics, or better, some interaction, between the entities.[27] But to register that this new thing has come into existence, we need mechanisms to observe the entities. So emergent properties must be observable, but they appear because of the system of interactions among the lower level objects, not because of observation. Baas does not specify the nature of the observing subject or the observational mechanism because he intends only general and formal requirements for emergence. The process of emergence of properties on several levels may be considered as a result of a series of abstract construction processes, similar to mathematical constructions. Given a set S_1 of first-order structures, one can, by some kind of observational mechanism $Obs_1(S_1)$ obtain or "measure" the properties of the structures at this level. The structures in S_1 can then be subjected to a family of interactions, Int, using the properties registered under observation (this could be a dynamic physical process). Hence, one gets a new kind of structure, $S_2 = R(S_1, Obs_1(S_1), Int)$, where R stands for the result of the construction process. The interactions may be caused by the structures themselves or imposed by external factors. Obs is related to the creation of new categories in the systems. S_2 is a second-order structure, a new unity whose properties may now be observed by another observational mechanism Obs_2, which may also observe the first-order structures of which it consists.

Baas defines P as an *emergent* property of S_2 if and only if P belongs to the set $Obs_2(S_2)$, and P does not belong to the set $Obs_2(S_1)$—which may be interpreted as saying that the whole is more than the sum of the parts. The idea of emergence as a function of interaction and observation is indicated in figure 1. Baas distinguishes between different types of emergence: (a) *deducible/computable* emergence, which means that there is a deductional or computational process D such that P can be determined by D and $Obs_1(S_1)$; and (b) *observational emergence*, which is the more profound type, characterized by the condition that if P is an emergent property, it cannot be deduced as in (a). Type (a) emergence clearly indicates that the defining characteristic of an emergent property P—that it belongs to $Obs_2(S_2)$ but not $Obs_2(S_1)$— does not entail that it could not be determined by $Obs_1(S_1)$ in an explanation using D. This is close to Kincaid's idea that the irreducibility of a higher level theory does not entail that lower level theories, with respect to some questions, cannot *explain* higher level phenomena.

Baas exemplifies deducible emergence by nonlinear dynamical systems in which simple systems interact to produce new and complex behavior:

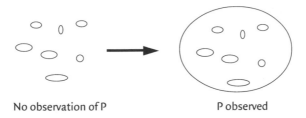

No observation of P   P observed

1. A depiction of Baas's conception of emergence as a function of observation and interaction.

phase transitions, broken symmetries, and many types of engineering constructions where coupled components interact in known ways so that we can calculate the new compositional properties of the system. Chaotic dynamical systems are considered to be borderline. A genuine example of observational emergence in a formal system, according to Baas, is found in Gödel's theorem: in some formal systems there are statements which are true, though this cannot be deduced. Here observation is the truth function (ascribing the truth value "true" to a string of symbols which is not deducible from the axioms and rules of inference of the formal system itself). Starting from first-order calculus—which is complete in the sense that every true statement can be deduced from the axioms—one can add further axioms to cover the theory of arithmetic, and this "adding" is a kind of "added interactions" among well-formed expressions which creates emergent properties. The property of membership in the Mandelbrot set may also be observationally emergent. Furthermore, the semantic noncompositionality of a language (for instance natural language) would imply that the meaning of sentences in such a language is observationally emergent.

Thus Baas concludes that even in formal, abstract systems (including models of lifelike processes), profound kinds of emergence may occur. Baas's contribution shows us how scientific explanation as a strictly deductive argument can be reinterpreted and explanations can be seen in a more dynamic and context-dependent setting, eventually themselves being emergent structures, "emergent explanations," and that this intuitive idea can be made precise and formally explicated. Questioning traditional notions of explanation may lead to a more general view of what constitutes scientific understanding of complicated phenomena. By paying attention to the general framework for description of higher order structures which includes the mechanisms of observation, and which eventually allows for self-generation in such systems of new observational frames, new observers may also emerge.[28]

It is a sensible intuition that the *autonomy* of biology in relation to physical science is grounded in the *observational emergence* of specific properties of biosystems, such as the self-reproduction of living cells. One might even understand the concept of life-as-an-autopoietic-system as the observational emergence in the physical space of systems that realize their own self-production, boundary, and self-observation (through the boundary's distinctive or selective property, based on molecular recognition reactions by membrane-bound proteins). Thus the system could be "cognitive" in this primitive sense. If this is valid, an autopoietic system is autonomous because it realizes observational emergence of itself as an observer.

It should be clear that the notion of emergence, as specified by the formal framework, does not in itself suffice for an ontodefinition of life. It is rather the other way around: once such a complex notion of a living system has been obtained, the notion will imply that such a system is emergent.

A more general inference from this discussion is that the increased interest in emergence in the sciences of complexity necessitates a deeper understanding of the nature of the modeling relation and the role of the observer in specifying the properties modeled and interpreting the resulting constructions.

Postmodern biology?

If biology uses concepts of life as a general phenomenon, as I have argued, does this vindicate the unity of knowledge principle? Or can we see in Alife signs of replacing the unitary view with a postmodern, pluralistic "principle of juxtaposition" in which one gives up the search for universality and generality? Complex dynamical systems and Alife research invite different interpretations which, for want of better terms, may be called (1) deconstructive, (2) postmodern, and (3) universalist.

1. According to *a deconstructive interpretation*, Alife can be seen as deconstructing our present rational conception of life as a unitary phenomenon, constituted by a single universal set of "generic" properties. Alife research reveals that our concept of life is not so singular as we wish to think, and that no simple set of fundamental criteria can decide the status of our models and constructs when these are already embedded in specific preconceptions of what constitutes the aliveness of natural and artificial creatures. This deconstructive move helps one to realize that at least the following different "conceptual models" of life as a biological phenomenon exist: an older idea about the organism as living animal; the modern idea of the cell as the simplest living thing; life as an abstract phenomenon; and life in the cybernetic sense as a machine process that can be made by natural selection or by an engineer.

We normally think that life in a biological sense has something to do with good old fashioned organisms, either single cells, or made up of cells, having a metabolism, constituted by specific macromolecules, and so forth. Now, if we wanted to construct life artificially by abstracting its logical form and subsequently realizing this form in various media, it is most likely that we would get different things depending on our point of departure. From the concept of good old animals, we get *neocybernetic life:* animats, robots, and other lifelike devices, that is, machines that are living in the sense of seeming to move autonomously, with sensors and effectors and an internal structure that coordinates input information and output behavior. What, then, do we get from the more fundamental concept of a cell taken from molecular biology? Attempts to realize its content artificially would lead to a material copy of a cell. But if an artificial material cell, to be alive, has to be made up of the same kind of biochemical compounds as a natural cell, it will probably be just another instance of the same kind. The point in making it will disappear; it will be merely a replica. We might, however, eventually make a formal version of a cell, simulated in a cellular automata model, realizing the formal properties of the physical cell in an abstract cellular-automata space. This is the idea of strong artificial life: any lifelike phenomenon can be realized in other media because life is form, not constituent materials; it is *an abstract phenomenon;* an informational structure emerging from lower-level local interactions. But what should be apparent here is that this notion functions as a separate, intuitive, conceptual model of life!

So life seems to be a multiplicity of phenomena if we list these conceptual models of life, or language game–specific constructs, some of which are associated with the Alife research program:

—GOFBO, good old-fashioned biological organisms, or life as characterized by the properties known partly from the common sense of everyday life, partly from the fields of biology, physiology, genetics, and so on. Most often, this is life conceived of as animals—thus GOFBA: good old-fashioned biological animals.

—MOMACE, modern macromolecular-based cells as characterized by molecular biology: You know it when you have it in your test tube.

—ABLI, abstract life, life as a space-time pattern that "realizes" some formal properties of biosystems either within a biochemical medium or in a symbolic formal space.

—ROLI, robotic life, that is, neocybernetic life, animats, nanorobotic life, and so on.

—CYBERlife, the idea of creating lifelike structures in a virtual reality to which we can relate through hypermedia.

—Other nonscientific definitions and conceptions of life.

From the point of view of modern biology—as the study of general principles of life—this is shocking and farfetched. But Alife helps us to see that the idea that the fundamental principles of life are universal may be a presupposition, a metaphysical prejudice with a questionable basis. Traditional biology has been haunted by conceptual dualisms and metaphysical contradictions:[29] the dualisms between structure and process, form and function, part and whole, inheritance and environment, contingency and necessity, holism and reductionism, vitalism and mechanism, and concept and metaphor. The construction of Alife may help to dissolve these dualisms or combine them in more fruitful ways, not so as to approximate the real nature of living beings but to catalyze new research, including reflections upon research. In this perspective Alife leads us to a new way of "reading" the science of biology. It deconstructs good old-fashioned biological life. The very opposition between living and dead nature, the organismic and the inorganic domains—which have constituted the science of biology since its definition in the beginning of the nineteenth century—may be reconstructed in a new framework drawing on insights gained from disciplines outside Artificial Life, for example, the thermodynamic study of self-organization and dissipative structures, the mathematical study of complexity and randomness, Artificial Intelligence, robotics, neural networks, and so on. It is possible but not assured that this will lead to the discovery of common principles for living as well as nonliving phenomena.

2. According to a similar interpretation of Alife, *as postmodern science*, this field of research drastically reveals the process of derealization, the distancing of ourselves from the real and material that is a part of the postmodern condition.[30] Several developments within the arts can be interpreted as representing this movement away from reality as object toward immaterial forms as something to be produced or, rather, simulated. Art has long transcended the classical idea of naturalistic depiction, where the picture was simply a faithful copy of the real (art as mimesis). Today's postmodern art and architecture also transcend the modern idea of the creating artistic subject who in a sovereign fashion generates originals by natural creativity (art as *poiesis*). Instead, postmodern art becomes a simulation where copies enter into a combination of significations that are not actually new but which represent small games that can be transmitted onward in a timeless infinity of circulating signs. The picture becomes a simulacrum, a picture without precedent object, model, or ideal; a picture that meets no requirement for truth, individuality, or utopian hope. Simulacrum connotes that which has broken with any reference to a primary reality and which is alienated from any original meaning or material substance. As image, the simulacrum is neither copy nor original. It has put

behind it any precedent, understood as representation of something *other* than itself.

It is characteristic for those simulations of life produced by Alifers and cellular-automata specialists that these simulations do not derive from any natural domain, let alone an experimental *physical* system. The reference to substance is lost. Of course, even an idealized mathematical model in traditional physics allows for an antirealist interpretation; such "a model is a work of fiction," as Nancy Cartwright states.[31] But for strong artificial life, this fictional character is even stronger because the models help to produce a fictive reality, where the distinction between copy and original, model and reality, is meaningless. The model no longer seeks to legitimate itself via any requirement for truth, accuracy, or (for the decent antirealist) empirical adequacy. It creates a simulacrum, its own universe, where the criteria for computational sophistication replace truth or adequacy and only have meaning within the artificial reality itself.

In this scenario, the relation to natural objects no longer constitutes a point of departure for efforts to achieve a cohesive understanding of a nature independent of us. Science loses its mimetic function. Instead, the capacity to simulate emergent phenomena entails an increased production of new, immaterial realities whose complexity is apparently not surpassed by nature herself. The exploration of this universe certainly lies in an extension of the traditional empirical natural sciences. At the same time, however, it has freed itself from the constraints of a given material substance with a limited set of natural laws. Artificial life and artificial physics make it possible for the researcher to be the cocreator of those "natural laws" or rules of the game that he or she wishes to investigate. The researcher creates a virtual nature, which is in itself an interesting complex object for investigation.

From this perspective, Artificial Life signifies the emergence of a new set of postmodern sciences, postmodern because they have renounced or strongly downgraded the challenge of providing us with a truthful image of one real world and instead have taken on the mission of exploring the possibilities and impossibilities of virtual worlds. It is a case of *modal* sciences, passing freely between the modalities of necessity and possibility. Science becomes "the art of the possible" because the interesting question is no longer how the world *is*, but how it could *be*, and how we most effectively can create other universes, given this and that set of computational resources.

3. According to *a universalist interpretation*, the attempt to achieve general understanding in science cannot be dispensed with. Alife has been viewed by its founders as a contribution to reforming and universalizing theoretical biology so as to explain life of any kind, form, or medium, and to discov-

ering the general principles that govern evolution, adaptability, growth, development, behavior, and learning. In this interpretation, the analysis of the implicit and very general ontodefinitions of life would be seen as supporting the claim that biology as a science of general processes of life should profit from interdisciplinarity and from the search for universal principles of organization. Alife is simply a tool in this process, just as mathematics and computer simulation are tools in physics and chemistry. The origin of order in the universe and the emergence of biological organization on Earth and on other planets should be understood in a single (causal, historical, process-oriented) frame. The emergence of special principles of organization (e.g., a genetic code, and thus, of biological information) may grant biology conceptual autonomy and may grant organisms a special ontology and mode of being—but the evolution of the universe, life, and mind should ultimately be explained in a grand narrative provided by science. Discovery of new laws of self-organization and evolution may eventually reform our picture of the cosmos in a more "organic" direction, in which our perception of the world may be reenchanted. But we should not give up the search for a single unified world picture in science.

A general concept of life is highly relevant in some contexts, as within general evolutionary biology, protobiology (research on the origin of life), artificial life, extraterrestrial life, philosophy of biology, and bioethics. A general concept of biological life may still vary within these different contexts. But we should not in general dismiss generality. In many metascientific fields, such as sociology of scientific knowledge and science studies, much emphasis is placed on the study of concrete scientific practices. This is pertinent, but knowledge of the concrete must involve the universal as well as the specific. We might be tempted to perceive "the disorder of things"[32] as signifying the impossibility of general knowledge, but we should not resign ourselves to accepting the disorder of thought. Unity of science in the positivistic sense is neither possible nor desirable, but postmodern abandonment of the search for general principles and concepts will not do. Order as well as disorder is inherent in mind and nature. This last stance does not have to be in conflict with a sober critique of any form of scientism because the search for universality in science does not commit us to universalism in politics, religion, or ethics. Furthermore, universality in science (as a value, or as a sign of modernity) is not the same as universality in politics or in culture.

Notes

1   C. Emmeche, Introduction, *The Garden in the Machine: The Emerging Science of Artificial Life* (Princeton, N.J.: Princeton University Press, 1994); and M. A. Boden (ed.), *The Philosophy of Artificial Life* (Oxford: Oxford University Press, 1996).

2 C. G. Langton, "Artificial Life," in C. G. Langton (ed.), *Artificial Life*, Santa Fe Institute Studies in the Sciences of Complexity, vol. 6 (Redwood City, Calif.: Addison-Wesley, 1989), 31.

3 H. R. Maturana and F. J. Varela, *Autopoiesis and Cognition: The Realization of the Living*, Boston Studies in the Philosophy of Science, vol. 42 (Dordrecht: Reidel, 1980), 55.

4 E. Mayr, *The Growth of Biological Thought: Diversity, Evolution, and Inheritance* (Cambridge, Mass.: The Belknap Press, 1982), 53.

5 Ibid., 53.

6 Ibid., 55.

7 C. Emmeche, "Autopoietic Systems, Replicators, and the Search for a Meaningful Biologic Definition of Life," *Ultimate Reality and Meaning* 20 (1997): 249; also S. F. Gilbert and S. Sarkar, "Embracing Complexity: Organicism for the 21st Century," *Developmental Dynamics* 219 (2000): 1–9.

8 M. Ruse, "Life," in T. Honderich (ed.), *The Oxford Companion to Philosophy* (Oxford: Oxford University Press, 1995), 487.

9 John Maynard Smith, *The Problems of Biology* (Oxford: Oxford University Press, 1986), 1–8; compare I. Kant, *Critique of Judgement*, trans. J. H. Bernard (New York: Hafner Publishing Company, 1951), 222 (§66).

10 Smith, *Problems of Biology*, 7.

11 See C. Taylor, "To Follow a Rule," in M. Hjort (ed.), *Rules and Conventions—Literature, Philosophy, Social Theory* (Baltimore, Md.: Johns Hopkins University Press, 1992), 168–185.

12 Carl Sagan, "Life," in *The Encyclopaedia Britannica*, 15th ed., *Macropaedia*, vol. 10 (London: William Benton, 1973), 893–911.

13 R. Dawkins, *The Selfish Gene* (New York: Oxford University Press, 1976).

14 David L. Hull, "Units of Evolution: A Metaphysical Essay," in U. J. Jensen and R. Harré (eds.), *The Philosophy of Evolution* (New York: St. Martin's Press, 1981), 41.

15 As critically discussed by H. H. Pattee, "Simulations, Realizations, and Theories of Life," in Langton (ed.), *Artificial Life*, 63–77; and by C. Emmeche, "Life as an Abstract Phenomenon: Is Artificial Life Possible?," in F. J. Varela and P. Bourgine (eds.), *Toward a Practice of Autonomous Systems: Proceedings of the First European Conference on Artificial Life* (Cambridge, Mass.: MIT Press, 1992), 466–474.

16 See Emmeche, "Life as an Abstract Phenomenon"; and Emmeche, *Garden in the Machine*.

17 J. Kim, "Emergent Properties," in Honderich (ed.), *Oxford Companion to Philosophy*, 224; see also A. Beckermann, H. Flohr, and J. Kim (eds.), *Emergence or Reduction? Essays on the Prospects of Nonreductive Physicalism* (Berlin: Walter de Gruyter, 1992).

18 Maturana and Varela, *Autopoiesis and Cognition*, xiii.

19 Ibid., 135; see also 78–84, 97.

20 Ibid., 116.

21 See, e.g., A. Moreno, A. Etxeberria, and J. Umerez, "Universality without Matter?," in R. Brooks and P. Maes (eds.), *Artificial Life IV* (Redwood City, Calif.: Addison-Wesley, 1995), 406–410.

22 J. Hoffmeyer, *Signs of Meaning in the Universe* (Bloomington: Indiana University Press, 1996); also T. A. Sebeok, *Global Semiotics* (Bloomington: Indiana University Press, 2001); K. Kull (ed.), *Jakob von Uexküll: A Paradigm for Biology and Semiotics* (Berlin: Mouton de Gruyter, 2001; reprinted in *Semiotica* 134 (1–4): 1–828); see also www.nbi.dk/7Eemmeche/p.biosem.html.

23 L. Wittgenstein, *Philosophical Investigations* (1953; Oxford: Basil Blackwell, 1972), §432.

24 The asserted isomorphy between causal processes in nature and algorithmic, syntactic processes in Alife (and other dynamical) models faces fundamental problems as indi-

cated by Robert Rosen, "Effective Processes and Natural Law," in R. Herken (ed.), *The Universal Turing Machine: A Half-Century Survey* (Oxford: Oxford University Press, 1988), 523–537.

25 Langton, "Artificial Life," 41.

26 Pattee, "Simulations"; also Emmeche, *The Garden in the Machine*; G. Kampis, *Self-Modifying Systems in Biology and Cognitive Science* (New York: Pergamon Press, 1991).

27 Nils A. Baas, "Emergence, Hierarchies, and Hyperstructures," in C. G. Langton (ed.), *Artificial Life III*, Santa Fe Studies in the Sciences of Complexity, Proceedings Volume 17 (Redwood City, Calif. Addison-Wesley, 1994), 515–537. See also N. A. Baas and C. Emmeche, "On Emergence and Explanation," *Intellectica* 2 (1997): 67–83.

28 N. A. Baas, "A Framework for Higher Order Cognition and Consciousness," in S. Hameroff, A. Kaszniak, and A. Scott (eds.), *Towards a Science of Consciousness* (Cambridge: MIT Press, 1996), 633–648.

29 See J. H. Woodger, *Biological Principles: A Critical Study* (London: Kegan Paul, 1929); and S. Oyama, *The Ontogeny of Information* (Cambridge, England: Cambridge University Press, 1985).

30 I have discussed this interpretation at more length in Emmeche, *The Garden in the Machine*, chapter 7, from which the material in this section derives.

31 N. Cartwright, *How the Laws of Physics Lie* (Oxford: Oxford University Press, 1983), 153. Cartwright likewise relates the antirealist interpretation to the usual notion of simulacrum (from *The Oxford English Dictionary*) as "something having merely the form or appearance of a certain thing, without possessing its substance or proper qualities."

32 J. Dupré, *The Disorder of Things: Metaphysical Foundations of the Disunity of Science* (Cambridge: Harvard University Press, 1993).

Afterword

The essays in this volume begin to suggest the range of subjects from which the theme of a historic turn in explanatory goals in the sciences needs to draw its content. Intentionally, the papers have focused largely on areas outside the traditionally reductionist strongholds of elementary particle physics and molecular biology, for the movement—to the degree that it is a movement—has been as diverse and nonunified as complex phenomena themselves. We have wanted also to draw out the constantly contested character of antireductionist claims, whether in fuzzy logic, cybernetics, immunology, or Artificial Life. But it has become increasingly apparent in the twenty-first century that the claims of emergence are hammering at the inner gates of reduction in both physics and biology. In drawing together some of the strands of this volume, it may help to look briefly at how the debate is shaping up in these two central areas of contemporary science.

The challenge to the model of science promoted by elementary particle physics is not new; it has only taken on a new confidence. Way back in 1972 Philip Anderson (Nobel Laureate in 1977) famously declared the physics of condensed matter independent from the reductive explanations demanded of any "fundamental" theory by elementary particle people, which he classed as mere "arrogance." "The more the elementary particle physicists tell us about the nature of the fundamental laws, the less relevance they seem to have to the very real problems of the rest of science, much less to those of society." Anderson's jump here from physics to society signals the sweeping nature of the epistemological shift he advocated, for he was concerned with the irreducibility of many-body systems quite generally. "At each level of complexity entirely new properties appear, and the understanding of the new behaviors requires research which I think is as fundamental in its nature as any other."[1] Here and in later papers he discussed the emergent properties arising from many-body complexity: rigidity, superfluidity, superconductivity, and ferromagnetism, extending even to speculations on the requirements for life and consciousness.

The world of science that Anderson envisaged was not a world of pure forms; it was the everyday world of impure substances with hybrid properties. Semiconductors, for example, are neither conductors nor nonconductors but conceptual and material hybrids made of doped silicon. In such boundary regions he found the beauty and excitement of an open future for science, a future that he and other members of the Santa Fe Institute hoped to realize through cross-disciplinary work. "We believe the growth points of science lie primarily in the gaps between the sciences."[2] Thus, as several of the essays in this volume have emphasized, the investigators would have to live on the borderlands of the disciplines, as hybrids, like the objects they studied.[3]

Since Anderson's early manifesto, the physics of complexity has blossomed in many directions, with ever more insistent calls for recognition of its fundamental role. A good example comes from a conference on the past, present, and future of physics held in 1996 to celebrate the 250th anniversary of Princeton University. Among the analysts of complexity was the solid-state theorist James Langer, soon to be president of the American Physical Society. Speaking on "Nonequilibrium Physics and the Origins of Complexity in Nature," he expressed his conviction (much like Anderson, who was one of the organizers) "that the physics of complex, every-day phenomena will be an extraordinarily rich and multi-faceted part of the science of the next century"; that it will involve "close collaboration with many kinds of scientists and engineers"; and that "we must place the understanding of life, and human life in particular, at the top of the list of our most compelling intellectual challenges." Langer's optimism about reaching a "deep new understanding of a wide range of every-day phenomena" did not rest on new unifying principles but on the development of new "tools": experimental, computational, and conceptual. He called on his fellow physicists "to modify our innate urge to speculate about unifying principles at very early stages of our research projects" and instead to pay much closer attention to the diversity of specific conditions and mechanisms. Conscious that his enthusiasm for contingency and diversity deviated from "the traditions of twentieth-century physics, which insist on grand unifications and underlying simplicity," Langer nevertheless insisted that his view did not imply pessimism about discovering new basic principles and unifying concepts. Rather, any such principles and concepts would have to be able to accommodate the specificity requisite to conditions of instability, where small quantitative differences make large qualitative differences and complexity shows its characteristic irreducibility, but also shows the growth of relatively stable morphologies. Exemplary for

Langer were the growth of snowflakes and dendritic growth in alloys and earthquakes.[4]

The views of solid-state physicists like Anderson and Langer have gradually been acquiring explanatory strength and credibility over the last thirty years. They seem now to have reached a critical point in the competition with the elementary particle people for possession of fundamentality. Two other solid state theorists marked the opening of the twenty-first century as a new era. "The central task of theoretical physics in our time," asserted Robert Laughlin (1998 Nobelist) and David Pines in the 4 January 2000 issue of *Proceedings of the National Academy of Sciences*, "is no longer to write down the ultimate equations but rather to catalog and understand emergent behavior in its many guises, including potentially life itself. We call this physics of the next century the study of complex adaptive matter. For better or worse we are now witnessing a transition from the science of the past, so intimately linked to reductionism, to the study of complex adaptive matter . . . with its hope of providing a jumping-off point for new discoveries, new concepts, and new wisdom."[5]

Laughlin, Pines, and others are now challenging the elementary particle physicists on their own ground, suggesting that the so-called fundamental laws and elementary particles themselves may be emergent phenomena, emerging perhaps from the chaotic soup of virtual particles thought to constitute the vacuum of space. This line of questioning leads to the cosmological theory of the Big Bang and to whether it was an explosion like other explosions. If so, it would have been highly unstable and unpredictable. "It could well turn out that the Big Bang is the ultimate emergent phenomenon, for it is impossible to miss the similarity between the large-scale structure recently discovered in the density of galaxies and the structure of Styrofoam, popcorn, or puffed cereals." Rather than a Theory of Everything based on a small set of underlying physical laws, they envisage "a hierarchy of Theories of Things, each emerging from its parent and evolving into its children as the energy scale is lowered" (to the level of everyday objects and organisms).[6]

This is a state of "things" deeply troubling to the former arbiters of good science, for it challenges not only their epistemology but their professional identities and values (much as in Galison's account, in this volume, of a controversy in string theory). As David Gross, a leading elementary particle theorist, said in a *New York Times* interview, "I strongly believe that the fundamental laws of nature are not emergent phenomena." He holds to his beliefs passionately: "Bob Laughlin and I have violent arguments about this."[7] It is impossible to judge which side will emerge victorious from the epistemological battle but it is apparent that a new contender for scientific explanation is taking up a great deal of space in contemporary physics. It falls squarely within the broad spectrum of explanatory strategies that we have called Growing Explanations. They extend to biology.

When we recall, following Laughlin and Pines, that the problem of emergence arises because the "fundamental law" of ordinary matter (the Schrödinger equation of nonrelativistic quantum mechanics) is not accurately solvable for systems consisting of more than about ten particles, then biological reduction is hard to imagine, or in their more polemical language: "Predicting protein functionality or the behavior of the human brain from these equations is patently absurd."[8] Thus it is no surprise that molecular biologists have also been learning the limitations of reduction, in this case of reducing phenotypes to genotypes by appeal to the existence of genes taken to be elementary units. The alternative appeal to complexity, however, had not become prominent, at least not publicly, until the great event of February 2001, when the two competitors in mapping the human genome simultaneously announced their draft sequences. The big surprise was the small number of genes on the map. The expected number of more than 100,000 came down to about 30,000, or only about 10,000 more than a roundworm and 300 more than a mouse. Suddenly it became very difficult to believe that the difference between humans and lower animals depended essentially on the number of genes. Instead, as the leaders of the two projects—J. Craig Venter of Celera Genomics and Francis S. Collins of the International Human Genome Sequencing Consortium—announced at press conferences, talk shows, and interviews, the difference lies in the complexity of gene-protein networks of interaction.[9]

In a sense, there is nothing very new here. As noted in the introduction to this volume, it has been known for some time that genes cannot be regarded as simple units that code uniquely for a single protein, that the active coding bits of DNA are spread over a considerable distance and can be spliced together in alternative ways to produce a variety of different proteins, that the proteins themselves can be modified by rearrangement and by adding new parts, and that the interactions between genes through a wide variety of regulatory functions of the noncoding parts of a DNA strand are enormously sensitive. So complexity, emergence, and contingency were already in play before the mapping result became known. But the reduced number of genes has brought the full panoply of gene-protein scenarios into the spotlight. Other findings have intensified their interest. The proportion of coding regions in the genome, previously only 5 to 8 percent, has been reduced to about 1 percent. And the clumpy character of the distribution of genes on and between chromosomes has come to look like a rugged mountain range rather than a moderately hilly landscape. These and other qualitatively new features of the genome as a whole have made it look ever more like the "complex adaptive matter" of the solid-state physicists. Indeed, terms like *dynamic, plastic,* and *fluid* commonly describe the highly adaptive character of the genome.

All of this emphasis on the dynamic and adaptive nature of both physical and biological matter raises an issue of a broader sort. Many physicists and biologists, like others discussed in this volume, are turning into historians.[10] In treating their objects of investigation as emergent systems, which might well have developed otherwise if conditions had been slightly different, they are treating them as historically evolved objects, subject to contingencies analogous to those familiar to historians. This historicity goes beyond the biologists' traditional evolutionary perspective, which has allowed them to treat the genomes of organisms as essentially static structures. As complex adaptive matter, genomes, like snowflakes and superconductors, demand explanations of how they come into existence, what maintains their relative stability, and how they can change, all within a framework in which no underlying laws of the parts can explain the emergent dynamics of the whole. This is not to say, however, that explanations cannot be had. They can be grown.

Notes

1 P. W. Anderson, "More Is Different: Broken Symmetry and the Nature of the Hierarchical Structure of Science," *Science* 177 (1972): 393–396; reprinted in Anderson, *A Career in Theoretical Physics* (Singapore: World Scientific Publishing, 1994), 1–4.

2 P. W. Anderson, "Theoretical Paradigms for the Sciences of Complexity," Nishina Memorial Lecture, Dept. of Physics, Keio University, Japan, May 1989; reprinted in Anderson, *Career in Theoretical Physics*, 585.

3 For another example and further discussion of the views of Anderson and Langer, see M. Norton Wise and David Brock, "The Culture of Quantum Chaos," *Studies in the History and Philosophy of Modern Physics* 29 (1998): 369–389.

4 J. S. Langer, "Nonequilibrium Physics and the Origins of Complexity in Nature," in V. A. Fitch, D. R. Marlow, and M. A. E. Margit (eds.), *Critical Problems in Physics: Proceedings of a Conference Celebrating the 250th Anniversary of Princeton University, Princeton, New Jersey, October 31, November 1, November 2, 1996* (Princeton: Princeton University Press, 1997), 11–27.

5 R. B. Laughlin and David Pines, "The Theory of Everything," *Proceedings of the National Academy of Sciences* 97 (2000): 30.

6 Ibid., 29–30. Although the new physics would describe a diversity of things, each equally fundamental and obeying its own laws, it might still be said to be unified and reducible through the hierarchy of laws. A considerably more radical version would do away with the hierarchy and the notion of building up (or down) in energy. Instead, there would be only different sorts of things, with no lower or higher. The philosopher Nancy Cartwright calls that vision of science a "patchwork of laws"; see "Fundamentalism versus the Patchwork of Laws," in *The Dappled World: A Study of the Boundaries of Science* (Cambridge: Cambridge University Press, 1999), 23–34.

7 George Johnson, "New Contenders for a Theory of Everything," *New York Times*, 4 December 2001, F-1.

8 Laughlin and Pines, "Theory of Everything," 28. See also the discussion of biological

emergence in R. B. Laughlin, David Pines et. al., "The Middle Way," *Proceedings of the National Academy of Sciences* 97 (2000): 32–37.

9   The full reports, along with other interpretive articles, appeared in the 15 February 2001 issue of *Nature* for the Consortium, and the 16 February 2001 issue of *Science* for Celera. Very accessible discussions appeared in the *New York Times*, 13 February 2001, sec. F.

10 See Amy Dahan Dalmedico, "Chaos, Disorder, and Mixing: A New Fin-de-Siècle Image of Science?" in this volume, 83–84.

Contributors

David Aubin is maître de conférences in the history of mathematics at the Université Pierre-et-Marie-Curie (Paris 6). His chapter in this volume derives from A Cultural History of Catastrophes and Chaos: Around the Institut des Hautes Études Scientifiques, France (Ph.D. diss., Princeton University, 1998). His recent research concerns the history of the observatory sciences in nineteenth-century France.

Amy Dahan Dalmedico is director of research at the Centre National de la Recherche Scientifique (CNRS, France) and joint-director at the Centre Alexandre Koyré. Her works include Mathématisations. Augustin-Louis Cauchy et l'Ecole française (Paris : A. Blanchard; Argenteuil: Editions du Choix, 1992) and (coedited with Umberto Bottazzini) Changing Images in Mathematics (London: Routledge, 2001). Currently she is focusing on applied mathematics, modeling, and dynamical systems in the twentieth century.

Richard Doyle is an associate professor of rhetoric and science studies in the English department at Pennsylvania State University, focusing on the futuristic literatures of science and science fiction. He is the author of On Beyond Living: Rhetorical Transformations of the Life Sciences (Stanford: Stanford University Press, 1997) and has completed a manuscript titled Wetwares! Experiments in Post-Vital Living.

Claus Emmeche is an associate professor and head of the Center for the Philosophy of Nature and Science Studies, hosted by the Niels Bohr Institute, University of Copenhagen. A theoretical biologist, he authored The Garden in the Machine: The Emerging Science of Artificial Life (Princeton, N.J.: Princeton University Press, 1994). He teaches philosophy of nature for science students; his research is part of the Copenhagen school of biosemiotics.

Peter Galison is the Mallinckrodt Professor of the History of Science and of Physics at Harvard University. His other work includes How Experiments End

(Chicago: University of Chicago Press, 1987); *Image and Logic: A Material Culture of Microphysics* (Chicago: University of Chicago Press, 1997); and *Einstein's Clocks, Poincaré's Maps* (New York: W. W. Norton, 2003).

Stefan Helmreich, an assistant professor in the anthropology program at the Massachusetts Institute of Technology, studies relations among science, technology, and culture. His book *Silicon Second Nature: Culturing Artificial Life in a Digital World*, 2d ed. (Berkeley: University of California Press, 2000), examines the cultural dimensions of Artificial Life. His current project concerns the remaking of marine biology in the age of genomics and biotechnology.

Ann Johnson is an assistant professor in the history department at Fordham University, Bronx, New York. Trained in theater, engineering, and history, her interest focuses on the culture and epistemology of modern engineering, as in her "Engineering Culture and the Production of Knowledge: An Intellectual History of Anti-lock Braking Systems" (Ph.D. diss., Princeton University, 2000).

Evelyn Fox Keller is a professor of the history and philosophy of science in the Program in Science, Technology, and Society at the Massachusetts Institute of Technology. Her recent works include *Refiguring Life: Metaphors of Twentieth Century Biology* (New York: Columbia University Press, 1995); *The Century of the Gene* (Boston: Harvard University Press, 2000); and *Making Sense of Life: Explaining Biological Development with Models, Metaphors, and Machines* (Boston: Harvard University Press, 2002).

Ilana Löwy is a senior research fellow (directrice de recherche) at the Institut National de la Santé et de la Recherche Médicale (INSERM), Paris. She has published extensively in the history of medicine and biomedical sciences. Her most recent book is *Virus, moustiques et modernité: La fievre jaune au Brésil entre science et politique* (Paris: Editions des Archives Contemporaines, 2001).

Claude Rosental is an associate professor and research fellow at SHADYC, a research center of the CNRS and the École des Hautes Études en Sciences Sociales in France. His work in historical sociology concerns modes of representation and proof in logic. He is the author of *La trame de l'evidence: Sociologie de la démonstration en logique* (Paris: Presses Universitaires de France, 2003).

Alfred I. Tauber is director of the Center for Philosophy and History of Science at Boston University. He is author of *Immune Self: Theory or Metaphor?* (Cambridge: Cambridge University Press, 1994); with Leon Chernyak, *Metchnikoff and the Origins of Immunology* (Oxford: Oxford University Press,

1991); and with Scott Podolsky, *The Generation of Diversity: Clonal Selection Theory and the Rise of Molecular Immunology* (Boston: Harvard University Press, 1997).

M. Norton Wise is a professor of history at the University of California, Los Angeles, where he focuses on science in culture in the nineteenth and twentieth centuries. Previous works include, with Crosbie Smith, *Energy and Empire: A Biographical Study of Lord Kelvin* (Cambridge: Cambridge University Press, 1989), and, as editor, *The Values of Precision* (Princeton, N.J.: Princeton University Press, 1995). He is currently completing a book with the working title *Bourgeois Berlin and Laboratory Science*.

Index

abduction, 16, 256–60, 264–71

abstraction materialized, 163–64

actors: human and nonhuman, 161

aerospace industry: speculative research in, 136–37

AIDS (Acquired Immuno Deficiency Syndrome), 15, 240–41; as autoimmune disease ornot, 226–28, 231, 234, 238; and AZT therapy, 231, 236; and hepatitis B, 223; and measurement of viral load, 231–38; and PCR, 231–41; and pharmaceutical industry, 236–38; as superantigen disease, 228; and technological drive, 231–40; as viral infection vs. complexnetwork, 223–41; and Working Group on AIDS, 225–26

air du temps: of complexity 67–68, 78; key elements of, 68

Air Force (U.S.): and academic research, 135–36; aircraft vs. missiles in, 136; outsourcing vs. arsenal research in, 135, 155n14

airframes: coevolution of, 139; complexity of, 9, 133, 139, 149; efficiency considerations in, 140; modeling of, 133, 141–45, 149–51, 156n30, 157n34; and use of computer, 10, 140, 156n27

algebraic geometry, 25, 41–43; and enumerative geometry, 41–44; and Hodge conjecture, 42

Ali, Ahmed, 27–28

Alife (artificial life), 3, 16–19, 251 56, 257–63; as life-as-it-could-be, 195, 258, 265, 277, 306, 317; as real life, 281–87, 295–

96, 304, 311–12, 315, 321; as simulacrum, 16–18, 251, 256, 258, 322–23, 326n31. *See also* resources of Alife: cultural; rhetorical; technical and scientific

Anderson, P. W., 28, 61n8, 327–28, 331nn1-2

artificial life. *See* Alife

Ashby, Ross, 191–92

attractor, 75; of Lorenz, 72–73; vs. strange attractor, 123; of Thom , 8, 96, 107, 109–16

Aubin, David, 7–8, 90n8, 94n55, 124n6, 124n9, 124n11, 127n47

autocatalytic system, 253–54

autoimmunity: and AIDS, 226–28, 234, 238; as pathological, 14–15, 201, 207–8; as systemic, 14, 210–11

automaton, 67, 91n19, 181, 192, 264; cellular, 283–4, 321

autonomy: in Alife, 267–70, 296, 321; in autopoietic systems, 18, 267–70; of biology, 306, 319–20; in explanation, 97; of mathematics, 88; of organism or living system, 253–54, 273n25, 274n48, 274n52, 306, 315

autopoiesis, 13, 15; and autonomy, 18, 267–70; and cybernetics, 192

autopoieticnetwork or sytem: in Alife, 17–18, 267–69, 309, 312–14, 320; closed or bounded character of, 312–13

Baas, N. A., 18, 317–19, 326n27–28

Baltimore, David, 227, 232, 242n14, 243n36

Barbieri, Marcello, 254–55

basin of attraction, 77, 112, 115

Bernard, Claude, 181, 185

Birkhoff, George D., 95

Bloor, David, 176n18, 177n26

Boeing: aircraft and missiles in, 136, 155n16; computers at, 10, 142–46, 157n47; and Clough, 9, 138–42

Boolean network: as Alife, 16, 253–54, 268, 283

borderland (also crossroads, trading zone), 5, 6, 24–25, 27, 50–51, 55, 59, 60n1. *See also* hybridity

bottom-up explanation, 10, 19, 279, 284, 289, 304, 316–18, 321

Bourbaki, Nicolas, 7, 59, 73, 85–8, 93n54, 99–103, 115, 119, 121

Brian, Eric, 175n12, 176n15

Brule, James, 161

Burnet, F. M., 15, 206–8, 218n18, 219n22

Burroughs, W. S., 251, 255–56, 265

butterfly effect, 69, 72, 78

Calabi, Eugenio, 32

Calabi-Yau manifolds, 32–51, 59; and Betti numbers, 37; and Euler numbers, 33, 38–40; and generations of particles, 33–39, 42; and Hodge conjecture, 42–43; and Hodge diamond, 37–38, 42; and Hodge numbers, 37–39, 42; and interests of physicists vs. mathematicians, 40–51, 59; mirror symmetry of, 38–46

Candelas, Philip, 32, 36–43, 45–49, 61n15, 62n27, 62n36

Cartan, Henri, 99, 103

Cartwright, Nancy, 20n2, 323, 326n31, 331n6

catastrophe theory, 7–8, 95–123; as beyond experiment, 99, 113, 122–23; and biology, 96, 98, 107–15; and chaos, 96–97, 107, 112, 123; chreods in, 109–10, 113, 115–118; and elementary catastrophes, 103–6, 108, 113, 117, 120, 123; and embryology, 96–97, 107–9; and genericity, 103–4, 123; idealism of, 120–21; as language or method, 99, 107, 122; and linguistics, 107, 115–19; as qualitative, 98, 107, 120–22; and structural stability, 98, 103–5, 108–9, 119–23; and structuralism, 96–97, 115–19

causality: circular, 71, 187–88; linear, 69; me-

chanical, 185; necessary vs. sufficient in disease, 241; purposive, 183, 187; structural, 118; as viral infection vs. complex network in AIDS, 223–41; whole-making, 185

cellular automaton, 283–84, 321

chaos, 67–70; and biological and social science, 76; and catastrophe theory, 96–97, 107, 112, 123; control of, 76; and critical phenomena, 79–81; deterministic, 6, 71, 74, 316; experimental character of, 73; and immunology, 217; and life, 188, 196; and order and disorder, 6, 69–73; quantitative study of, 75–76; road to, 76, 84; and simplicity, 72; and social science, 77–78; universality of, 75; of virtual vs actual event, 262

chreods, 109–10, 113, 115–18

Church-Turing thesis, 17, 284

Clemens, Herbert, 42, 45, 48–49, 62n31

clonal selection theory, 15, 206–8, 224

Clough, R. W.: at Berkeley, 141–42, 146–47; at Boeing, 138–42; as pioneer in MIAC, 138. *See also* FEA (finite element analysis); matrix methods

COGP (Candelas, Ossa, Green, Parkes), 43–44, 46–48, 62n36

cold war: reordering of engineering during 9, 133–35, 154

Collins, Harry, 175n12, 176n117

complexity: of airframes, 139, 143; in Alife, 17, 255, 264, 275–97, 301–17; in biology, 13, 187, 189–90; of complex adaptive matter, 329–30; in finite element analysis, 9–10, 149–50; in gene-protein networks, 330; in immunology, 14–16, 206, 208–10, 216–17, 223–31, 238–39; irreducibility of, 68, 328; of life, 12–14, 187, 189–96; in physics, 327–29; in sciences of disorder and mixing, 67–89; and simplicity, 7, 10, 72, 78–84, 96, 119; in technology, 8–11. *See also* chaos; emergence

computer: in Alife, 16–17, 251–71; in aeronautical engineering, 139; analogies in immunology, 208–9, 212; at Berkeley, 146–48; at Boeing, 10, 142–46; and bottom-up computation, 10; connectionist and parallel processing with, 13, 189, 193; and finite element analysis, 9–10, 139–54; IBM, 10, 145–47; and impact on

genotype (gtype), 8, 13, 16, 109, 254–55, 259, 264, 303, 308, 310

Georgi, Howard, 28–30

Gepner, Doron, 34–40, 61n21

Ginsparg, Paul, 29, 61n10

Ginzburg, Carlo, 85, 93n45

Glashow, Sheldon, 28–29, 55, 61n10–11

god: Alife researchers as, 288–97

Goody, Jack, 163, 176n14, 177n25

Green, Paul, 43. *See also* COGP

Greene, Brian, 39–40, 43, 45–46, 49, 61n21, 62n26, 62n41, 63n52, 63n54

Griffiths, Philip, 42, 62n33

Gross, David, 25, 32, 60n1, 61n15

Grossman, Z., 213–14, 220n39, 220n41

growing explanations, 1–4, 10, 15–19, 222, 239–41, 251, 264, 297, 301, 319, 330–31

Guattari, Félix, 262, 271n7, 272n25

GUTs (grand unified theories), 29–30. *See also* string theory

Haraway, Donna, 267–70, 274n41, 295, 300n58

Harnad, Stevan, 285–86, 290, 299n34

Harris, Joseph, 49, 62nn33–34, 63nn50–51

Hayles, N. K., 92n28, 279, 289, 298n16, 299n47

hegemony in mathematical sciences, 67

Helmreich, Stefan, 16–18, 298n17

Heraclitus, 120–21, 129nn96–99

Hilbert, David, 73, 86–87, 99, 125

historicity: in contemporary science, 7, 68, 83–85, 331; of dynamical systems, 77; as property of organisms, 305

history: alternative and nonlinear, 181–82; of civilization, 167–68; contest over histories, 59; as demonstration, 168; as factual, 161–63; of ideas, 160–61, 163; of immunological self, 201–14, 224, 239

HIV (Human Immunodeficiency Virus), 15, 223–41

Ho, David, 232–36, 243n35, 244n43, 245n59

holism, 10, 71, 127n55, 186–93, 205, 216, 306, 318

Hopf, Heinz, 100–101, 125n23

Hull, David, 310, 325n14

Hurd, Cuthbert, 145

hybridity: of machine and organism, 256; in

physics, 6, 328; in string theory, 4–6, 25, 50–60

IBM: early computers of, 10, 145–47, 157nn43–46

identity: boundaries of, 265–66; as core of immunology, 14, 202–8, 217; as dynamic process, 206; uncertainty of, 4–5

ideotypic network theory, 15, 209–11, 239

images: of AIDS, 233, 238–41; of science and mathematics, 67–68, 78–80, 85–89

immunological models: clonal selection, 15, 206–8, 224; complex network, 15, 224–28, 238–39; contextual, 212–17; ideotypic network, 15, 209–11; linguistic, 210–12; mechanist, 215–17; neural or cognitive, 15, 208–14; self and nonself, 14–16, 201–17, 224, 227–28; war against viral load, 14–16, 223, 233–37

immunology, 14–15, 201–6; as decentered, 206; historiography of, 203; instruments and measurements in, 15, 225, 229–31; and lymphocytes (T4 or CD4+), 207–14, 222–41; practical use in AIDS, 228–31; at Santa Fe Institute, 206; as science of self and nonself, 201, 206; terrain in, 224; and war metaphors, 223, 233–40

interactor, 310–12

interpreter of Alife, 276. *See also* narrator of Alife; observer in Alife

intuition: vs. abstract formalism, 99–101; biological, 115; of construction, 303–4, 316; as feminine, 281; geometrical and topological, 99–103; of life, 278, 296, 308; mathematical, 52; physical, 28, 53; vs. proof, 7; mathematical and physical, 59

inversion of sciences, 1–3, 68, 78–83, 96–97, 327–31

irreversibility, 74

Jacob, François, 181–82, 194, 196n2, 252, 272n3

Jaffe, Arthur, 53–60, 63n62, 63n70

Jerne, N. K., 15, 209–11, 217, 219nn26–28, 224, 239, 242n8, 246n74

Johnson, Ann, 9–10

Judeo-Christian cosmology, 288–97

Matzinger, Polly, 213, 219n25, 219n29, 220nn41–42
Mayr, Ernst, 305, 325n4
mechanism, 12, 182–87, 215–17
mediation: as conference, 169; and counter-mediation, 11, 169, 173; defined, 164; as demo, 170–73; as denunciation, 169; as graphic display, 167; as historical narrative, 11, 167–68; as publication, 168
Metchnikoff, Elie, 15, 203–5, 217n1, 218n12, 239
MIAC (military-industrial-academic complex), 9, 134–37, 153, 155n9, 155n10, 155n19
mind: analogy to immune system, 208–17; and self-organization, 188–89, 191, 195
mirror symmetry, 23, 25, 36–51; for physicists vs. mathematicians, 40–51; of size and shape, 39, 41, 44, 62n38. See also Calabi-Yau manifolds; string theory
mixing: sciences of, 67–89. See also chaos
modeling airframes. See airframes
modeling practices in catastrophe theory, 96–98, 106–7, 115, 123, 124n11
models, immunological. See immunological models
modernism vs. postmodernism: in immunology, 14–16, 202, 215
molecular biology, 8, 11–13, 107, 113–14, 186–88, 190–94, 251–53, 309, 327, 330
Monod, Jacques, 113–14, 128n69, 252
morphogenesis in catastrophe theory, 7–8, 95–98, 107–15
morphology, 2; in catastrophe theory, 7–8, 95–123; creation and destruction of, 98; dynamical, 96–98, 118–19, 122; and ends, 110; of life as network, 270; stability and instability of, 98, 108, 110; and structuralism, 116–19; and topology, 4–8, 19, 95–123
Morrison, David, 34, 46–50, 62n45, 63n66
Morse theory, 103–4
mundane world. See everyday world

narration of Alife. See resources of Alife, rhetorical
narrative: in physical science, 58, 84, 89
narrator of Alife, 260–63, 273–74n32. See also interpreter of Alife; observer in Alife

naturalization, 6, 11, 166, 176n22
networks: of actors, 86; of Alife, 252–6; autocatalytic, 253–55; autopoietic, 267–69, 312; Boolean, 16, 253–54; heterogeneous, 161–63, 174; ideotypic, 15, 209–11; of immune system, 224–28; of self-reference, 18; of universe, 266
nonlinear dynamics, 6–8, 13, 70–76
nonlinearity: in Alife, 17, 260, 277, 282, 284, 291, 303, 316–18; in biology, 194; and butterfly effect, 69–73; in engineering sciences, 76; in history, 182; in immunology, 206, 208, 216; of interactions, 2, 6; in morphogenesis, 110; in structural engineering, 9, 148, 156n27; of the world, 68, 70, 75

observer in Alife, 17, 19, 305, 312–13, 316–20. See also interpreter of Alife; narrator of Alife
ontodefinition, 301–24; concept of, 302; requirements of, 307–9. See also life, definition of
order and disorder, 6, 69–75
organicism, 185, 188, 191
organism, 11, 13, 181–96; in Alife, 251–71, 275–78, 289, 294; definition of, 182–83; with form and ends, 109–10; and immune self-identity, 201–17; understanding by growing, 2; as a whole, 108–9. See also life; living system
organization: as emergent property, 3; hierarchical, 317; of organisms, 181–96, 305–6, 313

Pantaleo, Giuseppe, 234, 244n37, 244n38, 244n39, 244n40, 244n45, 244n48
paradigm, 67, 95, 301–2, 306, 316
paradox: barber's, 165–67, 169; logical, 161
Parkes, Linda, 43. See also COGP
Pask, Gordon, 192, 197n32, 197n34
PCR (polymerase chain reaction), 231–41
Peirce, C. S., 16, 256–58, 262, 272nn13–14, 316
phenotype (ptype), 8, 13, 16, 109, 254–55, 259–64, 273n32, 303
Pines, David, 329, 331n5, 331n8
Planck length (scale), 31, 50
Plesser, M. R., 39–40, 43, 45, 62n26, 62n41

classes, 84–85; in economics, 77; in epigenetics, 113; and individuality, 85; of life, 303–4, 307–9, 311–12, 315, 320–24; of mathematics, 88, 121; of scale, 79–80; of unfoldings, 105
universe. *See* world or universe

Vafa, Cumrun, 36, 39–40, 46, 61nn21–22
Varela, Francisco, 13; and Alife, 17–18, 258, 267–69, 274n40, 301, 312, 325n3; and immunology, 15, 192, 198n35, 206, 208, 218n14, 219n32
viral load in AIDS, 231–41
visualization: in Alife, 295–96; in chaos, 72; of epigenetic landscape, 109–13
vitalism and nonvitalism, 187–8, 194, 204, 306–9
vitality. *See* liveliness
von Foerster, Heinz, 191, 193, 197n31, 198n40
von Neumann, John, 87, 121, 264, 268, 283

Waddington, C. H., 8, 96, 108–113, 117, 124n13, 127nn57–59
Warwick, Andrew, 176n16, 178n36
Whitney, Hassler, 103–4, 106, 126nn36–37

Wilson, E. W., 147–48, 152–54, 157nn59–60
Wilson, Kenneth, 79–80, 92n34
Witten, Edward, 29, 32, 46, 50–52, 59, 61n15, 63nn55–60
Wittgenstein, Ludwig, 166, 176n22, 177n25, 315, 325n23
Woodger, J. H., 190, 197n25
world or universe: in Alife, 17, 275–97, 323; as cellular automaton, 283; closed or bounded, 279–81, 285–87, 296; as computation, 284; definition of, 279, 282–83; as frontier, 276, 293–95; as green world, 287–88; as microworld, 279–81; as nature, 287–88; as real world, 281–87, 295–96, 315. *See also* second nature
World War II: and aircraft industry, 135–37; and computer, 12, 186–88; and mathematics, 87; and military-industrial complex, 135–38; and simulations at Los Alamos, 70, 279, 282

Yau, Shing-Tung, 32, 40, 46, 49, 62n36, 63n52

Zadeh, Lotfi, 160–62, 175nn3–4
Ziegler, John, 227, 242n17